T3-BEE-847

579.8 Tay

97378

Taylor.
Marine algae of the northeastern coast of North America.

Date Due

ILL			
2/28/79			

The Library
Nazareth College of Rochester, N. Y.

MARINE ALGAE OF THE NORTHEASTERN COAST OF NORTH AMERICA

MARINE ALGAE
OF
THE NORTHEASTERN COAST
OF NORTH AMERICA

BY

WILLIAM RANDOLPH TAYLOR

ILLUSTRATED BY

CHIN-CHIH JAO

REVISED EDITION

ANN ARBOR

THE UNIVERSITY OF MICHIGAN PRESS

Copyright © by The University of Michigan 1937, 1957
All rights reserved
Second revised edition 1957; Second printing 1962 with corrections
Fourth printing 1969
Library of Congress Catalog Card No. 57-7103
Published in the United States of America by
The University of Michigan Press and simultaneously
in Don Mills, Canada, by Longmans Canada Limited
Manufactured in the United States of America
97378

579.8
T24

In appreciation
of the patience and encouragement
shown by my wife
Jean Grant Taylor
through more than twenty-five summers during which the studies for this book have limited our horizon and laid first claim on my time and attention

PREFACE

MORE than seventy years have elapsed since W. G. Farlow published the first comprehensive catalogue of the marine algae of any section of the American coasts, and nearly a century since W. H. Harvey completed his *Nereis Boreali-Americana*. Since knowledge of the algae shared in the great advance made early in this century, Harvey's work has become of only historical importance, and even Farlow's handbook of New England algae presents but a fraction of the species known in that well-studied area, and those inadequately. It was very evident that need existed for a descriptive catalogue of the marine algae of the northeast coast which would bring together all scattered records, revise the descriptions and the nomenclature, and adequately illustrate the diverse types represented. A start upon this considerable task was made in the summer of 1933, when it became possible to secure the services of Dr. Chin-Chih Jao to prepare the illustrations, upon which he worked at intervals during the two following years.

In the preface to the first edition the writer explained in detail his indebtedness for the use of unpublished materials of the late F. S. Collins and his reliance on the major works referring to American algae. These, the works of Harvey, Farlow, Collins, Setchell, and others, remain of fundamental value. Nothing has supplanted them, but many studies of algal morphology and life history have appeared which now very greatly supplement what was available fifteen years ago. In fact, the publications of Kylin, Fritsch, and others now make this information so accessible that the writer is enabled in this edition to concentrate on characters directly aiding description and identification of the algal species. Except where American studies were involved, citations of such literature have generally been removed. To have incorporated all the new data would have increased the size of this book unduly. The information is reflected in changes in classification and nomenclature, but here the author has tried to be cautious, since most of the discoveries, particularly in regard to alternations of dissimilar generations, remain unconfirmed. Doubtless most of them will prove well founded, but one does not yet know how far some of these observations apply to related species, and they affect the organization of a manual very much.

For many other helpful services, particularly in the loan of material and the inspection of critical specimens, the writer is greatly indebted to the late Dr. M. A. Howe, and since his death Dr. D. P. Rogers, curator of cryptogams at the New York Botanical Garden, has been most generous. The curators of the Farlow Herbarium of Harvard University, the late Drs. D. H. Linder, and W. L. White, and their successor, Dr. I. Mackenzie Lamb, have most cordially permitted the examination of the fundamental collections there assembled. The writer's associates in botany at the Marine Biological Laboratory at Woods Hole have been of frequent and significant help, particularly Dean Ivey F. Lewis, to whom he is indebted for many years of encouragement and advice, and Dr. Hannah T. Croasdale, who for several summers expended great effort in the securing of necessary material for study. Most recently Dr. John L. Blum has helped with information respecting Vaucheria, and Mr. Robert T. Wilce has contributed numerous records of algae from northernmost Labrador and Quebec, improving very much our knowledge of the southern ranges of arctic species. Grateful mention must be made of the generous services of Dr. Chin-Chih Jao,[1] who in energetic collection of material, study and establishment of new records, and beautifully critical illustration has contributed importantly to this account. Miss Llewellya W. Hillis has handled most of the clerical work through the long process of revision of this text, and her competent help is most gratefully acknowledged. The author is also indebted to the Faculty Research Fund of the University of Michigan for appropriations for technical assistance and to the National Science Foundation for an appropriation for the manufacturing cost of the book.

[1] The illustrations were drawn by Dr. Jao, with the exception of Plates 26, 49, 50, 53, and 60, and portions of Plates 5, 7, 11, 35, 40, 43, 44, 45, and 48. Plates 29 (in part), 32, 33, and 34 are republished from the *Bull. Torrey Bot. Club,* with the consent of the editor, Professor M. A. Chrysler.

CONTENTS

INTRODUCTION

AREA SURVEYED

THIS catalogue has a wide range, including records from Virginia to the far north in the eastern American arctic, and westward in the arctic to include Hudson Bay. This does not involve as diverse floristic areas as one might suppose. Where the tropical American types drop out north of North Carolina is unknown. The very few algae recorded from Virginia are quite undistinctive, belonging to the temperate flora of southern New England. The cold-water flora of northern New England and eastern Canada merges with and is displaced at high latitudes by the relatively few types of the arctic, and the western species hardly reach as far east as Hudson Bay. This great area can, therefore, be covered by inclusion with the New England flora of a few additional, mostly arctic, species. The plan has been adopted of citing distribution by such political units or geographical entities as seemed most clear. However, fairly continuous distribution has been cited inclusively to save space, usually disregarding absence of records from New Hampshire's short coast line.

GEOGRAPHICAL DISTRIBUTION

Harvey a century ago and Farlow seventy-five years ago were aware of the general character and distribution of the marine algae within our area. They recognized that Cape Cod constitutes a point of division between a northern flora with subarctic tendencies and a southern flora adapted to warmer waters, but they did rather overemphasize this. In details, the distribution of the algae permits of a somewhat closer analysis than they attempted. Although the flora of the extremes of our range is too little known to afford material for a full comparison with that of Europe, the general features are clear even in these parts. Since our own station data are insufficient safely to establish floral districts of our own, with full lists of component plants, a basis for analysis was sought in Europe, and the paper by Børgesen and Jónsson (1905) proved very helpful, although the ranges of some species seem to differ materially in America from those recorded there. Without attempting to involve all species common to the two coasts, and introducing species important in American rather than in European waters, we come to a segregation of types to be discussed in the following paragraphs.

We may first consider the arctic flora (Taylor 1954), a term which is to be understood in a somewhat restricted sense. Few of the component species reach New England; they range from northern Newfoundland and Labrador northward and include:

Omphalophyllum ulvaceum
Laminaria groenlandica
Laminaria nigripes
Laminaria solidungula
Cruoria arctica
Turnerella pennyi
Dilsea integra
Pantoneura baerii
Phyllophora interrupta
Polysiphonia arctica

The flora which dominates the Maritime Provinces and northern New England to the latitude of Provincetown is a less restricted, more widely ranging group of plants, many of which are subarctic in their preferences. Just as in Europe some of these extend but to the Faeroës and others south to England and France, so on our coast some range but to northern New England, and others, although common northward, range to the Long Island Sound area. The species grouped together to illustrate the characteristic floras in America are not associated quite as in Europe.

The first grouping for consideration is one of plants of the Maritime Provinces and northern New England which range northward, even into the arctic, are to be thought of as essentially subarctic, and do not appear as an established element in the flora south of Cape Cod. This group is as follows:

Monostroma fuscum blyttii
Ectocarpus tomentosoides
Ralfsia fungiformis
Chaetopteris plumosa
Agarum cribrosum
Alaria pylaii
Laminaria longicruris
Laminaria saccharina
Saccorhiza dermatodea
Fucus edentatus
Fucus filiformis
Porphyra miniata
Euthora cristata
Rhodophyllis dichotoma
Gigartina stellata
Halosaccion ramentaceum
Antithamnion boreale
Ptilota serrata
Membranoptera denticulata
Odonthalia dentata

The plants characteristic of the Maritime Provinces and northern New England which range into southern New England are more adaptable to warmer waters than those in the list preceding. Ranging westward into Long Island Sound (a few going beyond), these represent an important northern element in the flora of that area but are more typical of the waters to the north of Cape Cod. The group is as follows:

Codiolum gregarium
Chaetomorpha atrovirens
Cladophora rupestris
Urospora penicilliformis
Spongomorpha arcta
Spongomorpha lanosa
Ectocarpus fasciculatus
Ralfsia verrucosa
Chordaria flagelliformis
Elachistea fucicola
Punctaria plantaginea
Dictyosiphon foeniculaceus
Chorda filum
Laminaria digitata
Fucus evanescens
Rhodochorton purpureum
Polyides caprinus
Phyllophora brodiaei
Phyllophora membranifolia
Antithamnion pylaisaei

There are numerous forms which, although in part unimportant or doubtfully even present in the extreme portions of our range, do otherwise extend nearly through its full extent, and likewise range widely in Europe. These are not distinctive of any district, and some reach into the tropics as a northern element. Of them there may be mentioned:

Enteromorpha intestinalis
Enteromorpha linza
Enteromorpha prolifera
Ulva lactuca
Chaetomorpha melagonium
Cladophora flexuosa
Rhizoclonium riparium
Ectocarpus confervoides
Ectocarpus siliculosus
Pylaiella littoralis
Scytosiphon lomentaria
Petalonia fascia
Ascophyllum nodosum
Fucus vesiculosus
Bangia fuscopurpurea
Porphyra umbilicalis
Ahnfeltia plicata
Ceramium rubrum
Rhodomela confervoides

The flora from southern New England to New Jersey is boreal rather than tropical in its affinities. There is a great change in the algal flora between New Jersey and North Carolina. The coastline of Delaware and Virginia is so sandy and inhospitable to marine algae that few are recorded from this region, and as yet no transition zone has been defined. There seem to be few species of which we may confidently write that their center of abundance is south of Cape Cod, and yet that they show a consistent range north of that boundary. We may, however, tentatively associate for such a group:

Enteromorpha clathrata
Enteromorpha plumosa
Percursaria percursa
Cladophora gracilis
Giffordia granulosa
Ectocarpus tomentosus
Eudesme virescens
Desmotrichum undulatum
Punctaria latifolia
Fucus spiralis
Ceramium diaphanum
Plumaria elegans

Plants distinctive of southern New England which range southward are more easily determined, and a list would include:

Chaetomorpha aerea
Eudesme zosterae
Stilophora rhizodes
Goniotrichum alsidii
Fosliella lejolisii
Agardhiella tenera
Gymnogongrus griffithsiae
Champia parvula
Lomentaria baileyana
Antithamnion cruciatum
Ceramium fastigiatum
Griffithsia globulifera
Seirospora griffithsiana
Spermothamnion turneri
Spyridia filamentosa
Grinnellia americana
Dasya pedicellata
Chondria baileyana
Chondria tenuissima
Polysiphonia denudata

The few species common both in the West Indies and in southern New England are exceedingly wide-ranging forms. They are certainly not distinctively tropical, though their center of range is below our southern limit. A small group may be designated which is to be considered primarily tropical, but which often appears in southern New England and occasionally in isolated warm stations farther to the north. Our list would include:

Bryopsis plumosa
Giffordia mitchellae
Sphacelaria furcigera
Sargassum filipendula
Gelidium crinale
Fosliella farinosa
Hypnea musciformis
Gracilaria foliifera
Gracilaria verrucosa
Caloglossa leprieurii
Bostrychia rivularis
Chondria dasyphylla
Chondria sedifolia
Polysiphonia subtilissima

The number of species whose range is limited to a narrow portion of our territory is rather small. Few of these are to be termed endemic, for most are, rather, introductions from abroad, or rare and sporadically occurring species whose limits cannot yet be set. The records of their occurrence are of various sorts. Several new species that were catalogued in the early days, especially in Ectocarpus, cannot be confirmed today. There are also a number of records of familiar European species, such as *Fucus ceranoides* and *Plocamium coccineum,* which do not seem to be supported by any convincing evidence, whereas others, like *Fucus serratus,* have long been known to inhabit limited areas. A few species of European affiliation have been found but once or twice in America; the records are confirmable by authentic specimens, as *Platoma bairdii* and *Tilopteris mertensii.* Yet other exotic species have appeared on our shores, some repeatedly at intervals of a few years, disappearing meanwhile, as *Brachytrichia quoyii* and *Derbesia vaucheriaeformis,* or become potentially permanent components, as *Prasiola stipitata, Striaria attenuata, Dumontia incrassata, Asparagopsis hamifera,* and *Trailliella intricata.* These examples have been selected from the flora of Massachusetts only because there alone have observations been sufficiently continuous to enable one properly to relate these elements to the permanent flora. The fluctuation from year to year even in the "permanent" constituents is sufficient to confuse short-term observations very much.

These lists must not be judged too critically. Although the writer is confident that they are substantially correct in the light of what we know today, there are many sources of apparent or real error possible. In the first place, they deal with the commoner species only, and local areas may be characterized by other and less common species. However, only by so restricting the lists could he eliminate gross errors due to incorrect identification in published accounts. Again, it is well known that there are local colonies of warm-water species in the Maritime Provinces and cold-water colonies in deep water off Marthas Vineyard and in Long Island Sound, the stations not fitting into the general distribution of the species concerned and causing some species to be assigned unduly wide ranges. The lists do, however, give a general idea of the distinctive elements of the flora for each portion of our coast.

ALGAL HABITATS

A discussion of the algal habitats and associations or, in effect, of the ecology of our territory is an impossibly large task for the purpose of this volume. Furthermore, the data are available for very few stations. The only considerable study of recent years is that of Johnson and Skutch (1928) on the algae of one portion of the Maine coast. Our knowledge of the ecology of American marine algae is based on the general familiarity of collectors with their field rather than on planned studies, though we may obtain data from the papers of Johnson and York (1912), Davis (1913b), Hoyt (1920), Lewis (1924), Taylor (1928a), Bell and MacFarlane (1933b), Prat (1933), Le Gallo (1947), the Stephensons (1954), and the phycological writings of Collins. A brief sketch of the characters of the vegetation in different areas of our territory alone can be attempted here.

In the southern part of our range and as far north as northern New Jersey the shore line is essentially sandy, with salt marshes behind the beaches or about the estuaries. Stone jetties or structures of pilings alone and artificially give firm support for larger algae. As a result, the flora is generally of such species as can grow on shells, mud, or Zostera, since little attaches on the sand or on Spartina. The flora here is relatively poor, and one may mention, as of the marsh ditches and pools, Myxophyceae, Bostrychia, *Gracilaria verrucosa, Ulva lactuca latissima,* Enteromorphae, and *Ascophyllum nodosum,* and on Zostera in the open water of these marshes and sheltered bays Fosliella, small Ceramia, Callithamnia, *Polysiphonia harveyi,* and Dasya. Practically none of the higher algae grow on the sand itself, but where reefs of shells occur we may expect Enteromorphae, Ceramia, Polysiphoniae, and other small forms. Occasional artificial structures erected along the coast and driftwood partly embedded in sand give anchorage for Myxophyceae, Enteromorpha, Pylaiella, Ectocarpus, Bangia, Fucus, and a few other types, some of which would not otherwise exist on this shore line.

Although there is still a great deal of sandy shore, the submerged or emergent rocky reefs and stretches of boulders give a much better chance for the development of a rich algal flora in Long Island Sound. Relatively massive types become important, and Ascophyllum- and Fucus-covered rocks, or Laminaria beds in deeper water, are no novelty in this area. The variety of algae increases immensely. On the rocks a group of Myxophyceae, particularly Calothrix, appear about and above high-tide line. With them occur several minute Chlorophyceae and (at least locally) Prasiola. Below this there is likely to be a zone of Lyngbya, Bangia, Porphyra, or Enteromorpha, depending on season and place, with fuci or rockweeds in scattered clumps, which become more closely grown a bit lower, and overhang crusts of Ralfsia, Hildenbrandia, tufts of Chondrus, and *Ulva lactuca rigida* at low tide. Between the tide lines, where the rockweeds do not occupy the available space, we find in early summer Petalonia, Scytosiphon, and Nemalion, replaced later by Polysiphoniae and *Ceramium rubrum.* In somewhat deeper water the stones bear *Desmarestia aculeata, Chorda filum, Laminaria agardhii,* Chondriae, etc., rather than rockweeds. Warm, quiet bays have a flora with a distinctive cast, such as *Enteromorpha clathrata, Ulva lactuca, Cladophora expansa,* Stilophora, Hypnea, Gracilaria, Spyridia, Grinnellia, and *Polysiphonia harveyi.* Tide pools in this district are not so important as farther north, but when they occur they show a population of Ralfsia, Hildenbrandia, *Lithothamnium lenormandi,* Corallina, Chondrus, and often less ubiquitous things, such as Bryopsis, Dumontia, and Gloiosiphonia. From deeper water may be dredged Arthrocladia, Phyllophora, Rhodymenia, Phycodrys, *Plumaria elegans,* and *Polysiphonia nigra,* some of which grow nearer the surface farther north.

Cape Cod presents another long sand barrier, but there are sufficient bouldery areas to support much more of a flora than in New Jersey. The large offshore islands (Marthas Vineyard and Nantucket) show the same algal floras as does a similar terrain on the mainland, but with some deep-water stations for northern types. The boreal type of flora dominates both sides of the Cape to the latitude of Provincetown where, as opposite on the mainland, more northern types begin to be important.

Northward the subarctic flora gradually displaces the boreal types, which under favorable local conditions do reappear, however, at surprisingly isolated stations. Rocky reefs and cliffs, with attendant tide pools, become a more distinctive feature of the shore; long bouldery stretches also occur, but sandy beaches and rivers are less extensive and so likewise are the associated salt marshes. In general, the tidal range becomes greater, particularly about the Bay of Fundy. Near high-tide line great rock areas are covered with marine lichen as well as with Calothrix and, a little lower, with Codiolum. Between the tide marks the rockweeds (Fucaceae) develop particularly heavy growths if the bottom is bouldery, or Zostera forms wide meadows if it is muddy, and these broad tidal flats in parts of the Bay of Fundy are very striking, though poor in the variety of their algal population. The prominent tide pools on the open parts of this coast include in their vegetation those types in similar situations to the south and also *Ralfsia fungiformis, Fucus filiformis,* Petrocelis, and Gigartina, with other plants. The rockweed vegetation shows *Fucus edentatus* as an important item, with *F. spiralis* and *F. vesiculosus* to the south, but Monostromae and Spongomorphae are far more abundant. Below the low-tide line the bulk of the growth of kelps (Laminariaceae) is very much increased, and also the variety of genera presented. About the bases of these plants occur certain kinds not found in this situation to the south, as Euthora, Ptilota, and Membranoptera. Further, Rhodymenia becomes more common, and several lithothamnia absent from the southern flora occur in deeper tide pools or offshore. The delicate Rhodophyceae are much less in evidence.

Too little has been published on the algal flora of the northern section of our range to justify much comment. The St. Lawrence River offers an interesting type of habitat, since it carries a marine flora for a considerable distance from the sea. The plants are poor in variety and small in stature, but much more of a biological feature than in the lower reaches of the Chesapeake or Delaware rivers, for which the rockier shore line is in part responsible. Here, as in all northern areas, the action of drifting ice along the shore each winter in removing the algae from the exposed rocks is a factor in controlling the aspect of the flora. The Hudson Bay area is little known, but from the three or four papers which describe its flora we may judge that with a very slight tide we have a relatively poor flora of subarctic and arctic types, the fuci being particularly stunted.

The flora eastward along Hudson Strait, in Ungava Bay, through Gray Strait and, thanks to the influence of the cold south-tending Labrador current, along the Labrador coast, is even more arctic in character and far more productive. Southern Ungava Bay, with a great tidal range, shows vast Vaucheria-dominated mud flats broken by boulders and rocky reefs on which there may be good growths of fucoids, with Hildenbrandtia, Rhodochorton, and Rhodomela as an understory. In deep water huge kelps and luxuriant plants of Kallymenia and Phycodrys occur, the red algae only in late summer. Toward the east and north there is even more variety, and a conspicuous undergrowth of coralline algae occurs.

The algal flora at Killinek I. on the extreme northern tip of Labrador is representative of the most arctic environment well studied in eastern Canada. The uppermost algal zone, of such Myxophyceae as Calothrix, is persistent. The other algae down to extreme low tide are more or less seasonal, reaching maximum growth late in the open period. Just below the blue-green algae Enteromorpha dominates, maturing soon after the ice foot has disappeared. Chaetomorpha and Spongomorpha can be found in sheltered nooks of the intertidal area quite early but do not reach full development until later. Other green algae, as Ulothrix, Monostroma, and *Enteromorpha groenlandica,* flourish through the open season, but chiefly about the lowest tide levels. The brown and red algae are even more responsive to the severe environment of this bleak coast. Scouring by ice floes keeps the rocks in the intertidal zone clear of major vegetation, except in narrow fjords and until late in the season. After the ice break-up Sphacelaria, Chaetopteris, and Ralfsia are evident in crevices and pools, encroaching on the Enteromorpha zone, probably as perennials. Soon Pylaiella comes to dominate the ice-scoured rocks, also invading the Enteromorpha zone, later to be associated with many brown algae common to northern New England, in a dwarfed but crowded intertidal vegetation. Although sporelings of fuci may be numerous and regenerating plants may early emerge from crevices, these larger algae do not mature in this zone of abrasion. For the most part larger fucoids and kelps, requiring a longer growth period, mature only in the sublittoral, but there they dominate. The exposed rocks show little intertidal growth of red algae, except the late-season Halosaccion and Rhodomela, but in the tide pools a considerable variety of species occurs in dwarfed forms.

At Killinek I. *Fucus filiformis* is a dominant high tide-pool plant. *F. evanescens* is the dominant fucoid in the sublittoral, with Halosaccion, Rhodomela, and Rhodymenia intermingled; *F. vesiculosus* occurs only in the littoral of deep bays and fjords. The kelps are sublittoral and are luxuriant, of great size, especially in the fast tidal currents about the islets of the area. Here they reach lengths of 10–15 meters, exceeded only in the richer colonies of Ungava Bay. Along the east coast of Labrador these kelps and the red algae are usually in even deeper water than about Killinek I. In deep fjords near Hebron, with muddy bottoms and scattered rocks, *Phyllophora interrupta,* Kallymenia, Saccorhiza, and Agarum may be found, and these (excepting Kallymenia) have been collected washed ashore at Killinek I., and so probably occur there also. All in all, the floras of northeastern Labrador beyond Nain, and of eastern Ungava Bay, differ chiefly in response to local physiography, and approximately the same group of species characterizes the marine vegetation.

COLLECTION AND PRESERVATION

The simple equipment needed for the shore collection of marine algae is quite different from that required for other plants. First in importance is a pail, flexible plastic, wooden or fiber preferred, rather small, of 2–3 gallons capacity. To this should be added a few pint jars and smaller widemouthed bottles. As a substitute one may prefer a waterproof bag carried by a strap over the shoulder. Here the algae are kept wet rather than afloat, as in a pail of water. Some tool for scraping or breaking rocks is next in order; a suitable instrument can be made from a heavy file or a broken car spring by sharpening one end at a forge. Beyond these, the clothing of the collector is the only consideration. In cold regions stout hip boots are invaluable, but they are so inconvenient that they are dispensed with when possible. Heavy canvas shoes which come above the ankle are best, as a protection from barnacle cuts. The soles should be of thick rubber and deeply corrugated. Shoes with tough felt soles a half-inch thick or soles of coiled rope, though very hard to obtain, are better, being more efficient in preventing slipping. Denim trousers and shirt, again for protection from cuts and sun, are better than the customary bathing suit. In any case, the collector must plan to go where the algae grow, and this usually involves a wetting.

Knowledge of the character and the period of the tides is the next requisite. It is possible to do some collecting at any time, but certain tide stages favor certain types of work. No serious student collects material from the dried or decaying mass of seaweed lying above high-tide line. Living and healthy material alone is of any use. However, one should not be too scornful of what fate brings in the way of material drifting ashore. Algae from deep water cannot be obtained otherwise, except by dredging. Much just criticism has been leveled at records based on drift material, but if one uses discretion in this regard specimens of the utmost value can be had. One should choose a rising tide for this collecting, since then the algae will be borne in closest to the water's edge. One should also inspect the chosen coast to see in what sheltered spots the algae are accumulating, since wind and current tend to localize the richest harvests. Though an onshore gale brings much fine material, it is easier to see what specimens are worth collecting after the roughness of the waves has subsided. One may then often stand quietly at waist depth and select undamaged specimens from the great variety of plants floating by.

This type of collecting is admirable within its limitations, but only the locating of plants actually attached, growing in place, will acquaint the student with their growth habits and ecological relations and give sure data on their distribution. In the intertidal zone and a little below, this can be done afoot. Deeper waters can be explored only with underwater swimming or diving gear, or a dredge, but the waters of most of our area are too cold to favor the use of those independent diving outfits which contain their own air supply. With respect to the intertidal zone a few precautions must be observed. Work here can best be done as the tide reaches its lowest point, so that a few words of advice may be offered to those unacquainted with the sea. On the northeast coast the tide reaches high and low points of approximately equal levels twice each day, and on each succeeding day these extremes occur about fifty minutes later. Extreme or spring tides come twice monthly when the forces of moon and sun are in conjunction at new and full moon, and alternate with minimal or neap tides when these forces are in opposition near first and last quarters of the moon. Exact details for selected stations are available in annual tables published under government auspices, and from these the data for intermediate points may be calculated. The newspapers of most coastal towns publish daily the time of local low water. However, one should sharply observe local conditions, for they may greatly influence the success of a collecting expedition. For instance, a strong onshore wind causes the water to stand higher than it would otherwise for that stage of the tide. Near the narrow openings of coves or bays the strength of the current may be as important to safe collecting as is the tide level to access to the plants, and so one must ascertain the time and the duration of slack water, which does not necessarily occur when the tide is lowest. In the southern states, where the tidal amplitude is slight, there is little chance of being caught by it far from dry land, but in some northern stations, such as Passamaquoddy Bay and the Bay of Fundy, where the amplitude may reach twenty and thirty or even fifty feet (Sverdrup *et al.* 1942, p. 562) the phycologist should make sure that he will not be caught far out over a slippery boulder tidal flat, or at the foot of a long, possibly unscalable cliff, by the rapidly flooding waters, for in such case he may have to sacrifice his pails of material to attain safety—if he can.

In practice one goes out over the tidal flats or down over the rocks as the water falls, keeping at the receding water's edge, examining tide pools, rock crevices, overhanging rocks and cliffs, and looking under the tangle of rockweeds for small species. Crustose types may be chipped or scraped off and, like the more delicate species, placed in bottles of sea water. Coarser types may be placed loose in the pail with a little sea water. It is more important to keep the water cool than to have the specimens floating about in a large quantity of it; also, the pail is easier to carry so. Bits of strong paper bearing collection data written with a very soft black pencil should be placed with the specimens. The containers should not be crowded, and duplication of coarse species should be avoided, for algae decay quickly if unduly crowded or warmed. A few genera, such as Desmarestia, are extremely susceptible and should be kept separate if possible. If necessary, several pails may be used, kept in the shade, and the water changed at intervals. Most kelps and rockweeds can be stowed in a sack if kept moist and in the shade, and these need not crowd the collecting pails. Cliffs dropping into deep water and jetties and wharves which cannot be approached afoot may be inspected from a rowboat, with due respect for the severity of the waves. With discretion this kind of collecting is simple and safe in New England. As the tide rises to make unprofitable the search for shoal-water types, one may well turn to collecting in the drift.

The algae which grow below reach at low spring tides are difficult to study in position. They form a large proportion of the driftweed washed ashore, and in this state alone are ordinarily available. From a boat, if the water is very clear, some little of their habit can be seen, especially through a glass-bottomed bucket, and it is possible to secure specimens to a depth of about six feet or so if a heavy-tined right-angled digging fork (such as clams are dug with) is used on a handle about ten feet long. However, such a tool is very difficult to operate because of the resistance of the water. The extreme clarity of water which aids collecting in many Florida and West Indies stations is not a feature of the coast dealt with here. The turbidity of the water and, even more, the currents, tides, and cold render impracticable the use of a diving helmet, with which in tropical seas one may obtain so much more just an idea of the relation of the underwater vegetation to its surroundings. Even in Florida, where such an outfit is excellent for inspection, the writer has found by experience that it is poor for collecting, since one's movements are greatly retarded and small specimens are hard to handle.

On the whole, in deeper water a dredge must eventually be resorted to. Intelligently directed dredging is not easily accomplished, but when well conducted a dredging trip can be an exceedingly informative method of collection and study. A great deal depends on suitable equipment and the co-operation of the assisting boatmen. A box-type dredge, strongly built of angle iron, with the longer margins of the rectangular opening developed as two-foot cutting edges, will serve well on a three-fourths-inch hempen line, and with it one may dredge from a small launch to a depth of ten meters if sufficient weight is used to keep the dredge on the bottom. Deeper dredging should hardly be attempted without hoisting machinery, such as is found on fishing or lobstering boats. Because the turbid waters of northern coasts restrict the depths at which the algae grow, really deep-water dredging for these plants is hardly pertinent to the problems of our coast; somewhat larger dredges on a steel cable handled from special hoisting drums are best, and in tropical waters they bring up immensely valuable cargoes.

The type of bottom involved determines largely the bulk of the plant haul. In our area little is retrieved from sandy or muddy bottom. If scattered stones are too large, or admixed with boulders or reefs of rock, the dredge will probably catch sooner or later and, unless the boatman can extricate it, will be lost. The speed of the boat must be reduced to the point where the dredge will stay on the bottom most of the time, and then, with experience, a hand on the line will tell if it is drawing over cobbles or among dangerous rocks. There is no question, however, but that this type of collecting is a bit of a gamble—the loss of a dredge pitted against the chance of some rare specimens. Dredging for plants over rocks is more exciting than dredging up mud for worms! To go far afield with but one dredge is to invite its early loss, to a disastrous end of the day's endeavor, as the writer has bitterly experienced.

A few minutes' haul is best in shallow water. Better to catch a few specimens, save them in good condition, and try again, rather than haul a long time, packing sand or rocks on the plants to their destruction, and causing a heavy lift and, perhaps, a torn net. Also, each unnecessary minute below is an unjustified risk of getting the dredge jammed under or between rocks and lost. In deeper water, such as 200 meters, the time required to get on the bottom, the difficulty of keeping there, and the slow uphaul to prevent damaging material may require up to two hours. At such depths two or three times as much cable should be overside as the depth by sounding at the dredging station; in shoal waters somewhat less line is suitable, particularly in rocky situations, where the least should be used which will enable the dredge to keep in proper position. This will aid in preventing fouling among boulders; if it does get caught, it is usually most readily freed with the line taut and the boat nearly above the dredge. Then, strong and sharp pulls at various angles will eventually find one at which the dredge will become free—or break the line. Immediately the dredge comes aboard it should be emptied into a tub of cool sea water for sorting and sent below for another catch. All the while the collector's boat will lie and wallow in the troughs of the waves to the displeasure of unseaworthy passengers. To lean over and sort a tub of algae and moribund animals while the gear slats back and forth is a good test of anyone's resistance. However, it is of some comfort that dredging is not practicable in really rough water. A representative series of specimens should be segregated in sea water in pails and bottles, labeled, and put aside in the shade. Records should be made of the character of the bottom and of the depth of the water at the start and finish of the haul, the state of the tide if in very shallow water, and the location. Separate pails for each haul are necessary unless hauls are repeated over the same bottom with substantially the same type of catch.

The material collected will eventually reach the collector's base of operations, and the problem of disposal then appears. In general, the methods of preservation are two: in liquid or by drying. In the first a suitable supply of bottles, tins, or casks is requisite; in the second, herbarium equipment. Marine algae offer a difficult problem of liquid preservation because the fluids usually affect the cell walls rather drastically. For the most part, coarse species can be preserved for a time in neutral 4 per cent formaldehyde in sea water, if they are kept away from the light. More delicate species do better if alcohol amounting to 30 per cent is added to the weak formaldehyde, but then they shrink somewhat. All solutions containing acid should be avoided because they soften the intercellular substances, which allows the plants to fall apart. For this reason formalin solutions should have borax added until the solutions give a pink reaction when tested with phenolphthalein. Simple alcoholic preservation is best except for the risk of shrinkage. When practicable the material should first be placed in weak alcohol, perhaps 20 per cent, and then after a few hours in progressively stronger mixtures until a safe grade is reached. In weak alcohol the specimens fall apart; in strong, they shrink. The optimum for any given species is somewhere between 50 per cent and 85 per cent. If the material is to be stored indefinitely, 5 per cent of glycerine may be added to 70 per cent alcohol to prevent complete drying by accident. Specimens with a calcareous matrix preserved in formaldehyde may be altered by solution of the lime if the preservative becomes acid, so that an excess of borax is to be advised. Material destined for critical morphological or cytological study should be prepared by methods to be found in books on microscopical technique (as Taylor 1937a, p. 226). On expeditions far afield methods adopted will have to be more wholesale in type than is desirable after trips based on a permanent laboratory. Relatively few specimens can then be segregated in bottles. The home-canning devices whereby a lid is crimped onto a can body precisely as is done for preserved vegetables and fruits is excellently adapted to preserving algae on trips far afield (Taylor 1950, p. 42). The cans are far cheaper and lighter than glass jars and may be shipped great distances without protective packing. Algae in alcohol or formaldehyde may remain five years or more in these cans without mishap, provided they are fully filled. The writer has received in excellent condition shipments from as far as Brazil, the South Falkland Islands, and Chile, where each individual group of specimens from a common station was loosely wrapped, with a label, in a bag of cheesecloth, tied, and the whole packed with many others in five-gallon tins of 4 per cent formaldehyde, which were soldered shut, crated, and shipped north to be opened, sorted, and shifted to glass or paper many months later. Disintegration in formaldehyde does not soon become serious unless the material is exposed to strong light or unless the plants are very delicate. Heavy shells, stones, and lithothamnia should not be packed in the same containers with soft specimens and should be well wrapped to prevent rubbing or breakage. Kelps and rockweeds may be packed down in casks with abundant coarse salt, and if not left in this state too long they remain in good condition. Experiments have shown that if the brine first formed is drained and the algae relayered with fresh salt many species, and even many rather delicate ones, may be preserved in this way for several years. This method works excellently in the tropics and has been applied practically on arctic expeditions. Some marine algae may also be shipped alive, loosely wrapped in damp papers, during the winter of the northern states and Canada, to arrive by express, after a thousand-mile journey, in active reproductive condition. In summer they may be successfully shipped alive packed in ice like fish and shellfish and forwarded by express, particularly by air.

On the whole, except for special purposes, marine algae are best preserved for taxonomic study as dried specimens. They are as permanent in that state as vascular plants, for, though some are more brittle and so liable to damage in handling, they are not subject to attacks of dermestid or herbarium beetles. If kept in the dark they retain their color for a long time, but they fade in the light. Some seem able for an indefinite period to reassume their natural form when soaked in water; others give more trouble. Equipment for drying algae may be very simple. The massive coralline algae and smaller lithothamnia on stones and shells are simply momentarily rinsed in fresh water and dried in a shady place. When labeled they may be wrapped in newspapers and packed away for shipment. The kelps and the coarser rockweeds may be roughly dried on the ground and, when nearly dry, rolled or folded up for more convenient packing. If dried until crisp in the sun they should be gently removed to the safety of indoor tables, where during the night they ordinarily absorb enough moisture to permit them to be folded away. The phycologist should therefore make every effort to secure full-grown specimens, for a big adult plant is much more valuable and informative for taxonomic purposes than a young one or a few fragments cut from a large one—too often with the size not specified on the label! On our coast the problems presented by Nereocystis and Durvillaea do not arise! It is possible to soak up large specimens again, to study them, measure them approximately, and remount them for the herbarium. If after soaking they are transferred to a mixture containing about 30 per cent glycerine, 30 per cent alcohol, 10 per cent carbolic acid, and 30 per cent water until thoroughly penetrated, they may be dried, and thereafter retain their flexibility for a long time, particularly if stored in tin boxes, but should not be mounted as herbarium specimens.

For smaller species some method of support is requisite. Paper is the first essential, preferably good, medium-weight herbarium paper with a substantial rag content. Standard sheets (11.5 by 16.5 inches, or 29.5 by 42 cm.), halves, and quarters make a good selection. The larger looseleaf notebook drawing-paper sheets serve well, though limiting the mounting of the larger species to small portions. A number of pieces of unbleached, unsized muslin of the full-sheet size will be needed, some cheap thin-waxed paper, and herbarium blotters (or, failing these, folded newspapers), a pan big enough in which to immerse the mounting paper, and a few metal sheets to fit the pan (zinc preferred); a large camel's-hair brush, large-bulb pipettes, and dissecting needles complete the equipment. It is possible to mount the specimens in a shoreside tide pool, with a bit of slate or glass as a support for the paper, but it is much easier to work indoors. It is hard to mount on a ship unless she is tied up, because even a slight roll will send the water and the specimens across the pan to the utter destruction of characteristic and orderly arrangement. Although there is no excuse for careless mounting under ordinary circumstances, the traveler in out-of-the-way places should mount specimens whenever and however possible. Freshwater species should be mounted in fresh water, and marine species when alive in sea water, but salted or otherwise preserved marine specimens may be mounted in fresh water. In practice, enough water is placed in the pan to cover the support and the paper, but not deeply. The chosen specimen is washed free of dirt, laid on the paper, and roughly arranged. If it forms too thick a mat, branches should be removed from inconspicuous places, with care not to obscure the characteristic habit, but with the aim of so reducing the bulk that the individual branches will show clearly when the plant is dried. Next comes a final arrangement aided by brush, needles, or water from the pipette, the support being gradually lifted and stood aside on end for a few minutes to drain off the water. After draining, the sheets of paper with their plants should be laid face up on blotters and covered with either muslin or waxed paper. Which is chosen depends on the character of the specimen. Cloth should be used on all slippery and on most delicate things. Waxed paper is better on wiry specimens. Other layers of blotter, paper, specimen, and cloth are superimposed until the harvest is disposed of, and a moderate weight is then placed on the pile. After a couple of hours the blotters should be replaced by a dry set, and this repeated twice daily until the specimens are dry. When dry, the cloth or the waxed paper should be stripped very cautiously from the specimens and the mounts kept in folders under slight pressure until incorporated in the permanent herbarium. The specimens will stick to the paper more or less completely by virtue of their own adhesiveness. Wiry or non-lubricous specimens may often be made to stick by adding glue to the water in the mounting pans. If the cloths and the waxed papers are stored flat and free from wrinkles, they may be used indefinitely. The blotters may also be used repeatedly, but after they become much worn they should be used for the first change only.

These general directions must be amplified by special methods for certain types and conditions. If one is working on a considerable task of collection and mounting, it is important to use corrugated cardboard ventilators alternately between pairs of blotters. If this is done after one change of blotters, the pack of material may be placed above a heater, and the air circulating through the corrugations will dry the specimens in a few hours. The best type of corrugated board is faced on one side. Pieces are cut 12 by 34 inches, with the corrugations crosswise; when folded back to back, with the corrugations exposed, these are most effective. Various devices for heating are familiar to field workers. A tube of fireproofed canvas to tie around the suspended press is a good portable tool, one or two kerosene lanterns being placed on the ground at the bottom of the tube. The writer prefers to use galvanized sheet-iron boxes about 30 inches high and 15 by 20 inches at the top, open above and below. A stiff frame or wire grid should be fastened below the top of the box to support the loaded press, and a suitable baffle above the heaters if needed to spread the warmth evenly. If kerosene lamps or electric heaters are placed in the bottom of such a box (with allowance for free entrance of the air) and the press with specimens on the top frame (any unoccupied area being covered with a metal sheet), the rising hot air will dry the whole in 12 to, at most, 36 hours.

Certain special precautions pay in dealing with some of the algal types. Kelps and rockweeds do not stick well to paper or, if they do, shrink so much as to cause curling and breaking under pressure. These are best dried between two cloths in newspaper folders. Since kelps, if much folded, may decay before they desiccate, it is best to air-dry them partially before folding them into packets to complete the process between cloths in a press. After drying with heat, it is well not to pack bulky algae under pressure immediately; they should be left exposed to the air in a shady place for a short time to become more flexible and less brittle. However, they should never be left overnight except under pressure. It is very important for future critical study that selected portions of all delicate types be mounted on mica to be filed with the specimens. Each sheet should be split in two immediately before use and the sample mounted on the freshly exposed faces. Collection data may be scratched on the mica, and the specimens dried under cloth as if they were on paper. Only plants which adhere well should be mounted in this way.

When one is preparing the specimens for insertion in the herbarium, those which have been dried between cloths should be mounted on standard herbarium paper, but not until the climate of the repository is reached, for if mounted at the seashore they will shrink and crack at an inland station. If glue is used it should be as thick as possible; waterproof cements are generally preferable, for watery adhesives cause swelling and curling. Rubber cement and adhesive cellophane tapes should never be used, or any tapes with a flexible plastic adhesive. Good cloth tape with ordinary gum adhesive is quite safe. Mounts on small paper sheets should be trimmed and attached to the standard sheets; very small or fragile mounts and mica slips should be placed in envelopes, with the label on the flap. Small pebbles and shells with algae attached likewise go into envelopes on standard herbarium sheets, but bulky corallines or large stones must be stored, with their labels, in boxes. Data for the labels should include the exact locality, habitat of the plant (such as type of rock or other support, exposure to the air and light, and depth), date, collector's name, field number, and the name of the plant if it is known.

For identification it is frequently necessary to study microscopic sections of dried specimens. Although these may, if sufficiently tough, be sectioned after softening in water, it is more generally satisfactory to hold a bit of the dry specimen on a microscope slide and thinly shave the end of it with a sharp razor blade or a scalpel of good steel, collecting the sections in a nearby drop of water. Thin membranes such as those of Monostroma are best thus sectioned together with the supporting paper, bits 2–3 mm. square being cut for the purpose from desirable parts of the specimen and especially where the membrane is folded in overlapping layers. Often well-teased-out or gently crushed preparations will suffice, and sectioning will prove unnecessary. Most observations for taxonomic purposes can be effected without special staining, but it is sometimes helpful to stain with dilute aqueous solutions of iodine (Chlorophyceae), with methylene blue (especially for Phaeophyceae) or with Congo Red followed by a little weak caustic potash (Rhodophycean reproductive organs).

HISTORICAL SURVEY

The history of the study of American algae previous to 1847 is almost exclusively to be sought in occasional references in European scientific journals. Only one study specifically dealing with our territory stands out as a work of major importance, that of De la Pylaie on the Newfoundland algae. The real awakening of interest developed with the visit of Professor W. H. Harvey, of Dublin, on a lecture tour in 1849–1850. He took advantage of the opportunity to work at stations from Florida to Nova Scotia and, assembling material from the collections of several other investigators, produced the three volumes of *Nereis Boreali-Americana*. These are fundamental to our knowledge of the algae of both west and east coasts, even though but a very small fraction of the flora is represented.

With the incentive of these volumes more persons gathered material, and some even published short lists. The names associated with this period include Curtiss, Ashmead, Hall, and Hooper in the Florida area; Ashmead in New Jersey; Bailey, Durant, and Hooper in the vicinity of New York; Olney in Rhode Island; and Kemp in Canada. These men were all more or less directly encouraged and advised by Harvey. Toward the latter part of the century a new group appears, including as still active some of those who collected for Harvey and in addition: Martindale and Morse in New Jersey; Eaton and Holden in Connecticut; Hay, Kemp, and MacKay in Canada; and the two whose works are most influential, Farlow and Collins, with activities centered in Massachusetts. This period was one characterized by the exploratory cataloguing of our flora.

With the close of the century, papers dealing with features of structure and development began to appear, but these are not the primary material of this account. For nearly the first two decades of the twentieth century phycology in northeastern America was dominated by F. S. Collins, whose active work resulted in an algal flora of Bermuda, a complete survey of the North American green algae (exclusive of desmids and charophytes), and a series of shorter papers leading up to a survey of the New England marine algae, which he never finished. He was also chief partner in the preparation and issue of the *Phycotheca Boreali-Americana,* largest exsiccata of algae ever published. He took the leading place, vacated by W. G. Farlow when the latter's interests became centered in mycology, working always in close co-operation with Farlow and relying on the important collections of specimens and books which he had assembled at Harvard University. Collins' labors carried phycology into its present phase. On his death his herbarium, which had become of great consequence, was purchased, together with his manuscripts, for the New York Botanical Garden, which M. A. Howe had developed as an algal research center of the first importance.

Though the herbarium of Frank S. Collins is probably the largest of eastern American algae which has ever been privately assembled, the earlier phycologists made collections important in their time. The writer endeavored to ascertain by correspondence where these were deposited, but was not always successful; he is greatly indebted, however, to numerous persons who have generously contributed information. Many botanists who gathered specimens did not themselves publish upon them, and it seems hardly practicable at the present time to record these herbaria. The chief collection of Samuel Ashmead was not definitely located, but specimens known to lie in the collections of the Academy of Natural Sciences of Philadelphia may represent it. It is not likely that Jacob W. Bailey had any large collection of marine algae, but specimens of his gathering occur in the herbaria of Brown University and the New York Botanical Garden. Mrs. Floretta A. Curtiss is more noted for her collections of Florida seaweeds than in connection with the algae of the New England coast; her specimens were deposited by her son in the United States National Herbarium. The algal collections of C. F. Durant were made up into volumes; sets exist at the Brooklyn Botanical Garden, Farlow Herbarium, and a few other institutions. Those of D. C. Eaton are in the herbarium of Yale University, but duplicates have been distributed. The algae of G. U. Hay were destroyed by the great explosion in Halifax harbor in 1917. The algae of Isaac Holden are chiefly in the possession of the New York Botanical Garden and the Farlow Herbarium. The phycological collection of Franklin W. Hooper is in the herbarium of the Brooklyn Botanical Garden, in several large volumes. It is not certain that A. B. Klugh kept any considerable algal collection, and if he did it may have been destroyed in a fire at the St. Andrews Biological Station. Some specimens deriving from A. F. Kemp are in the Farlow Herbarium. The collections of A. H. MacKay at Dalhousie University met with the same fate as those of Hay. The New Jersey collections of Isaac C. Martindale are at the Philadelphia College of Pharmacy and Science, as are some of the algae of S. R. Morse, although much of the latter's material is reported to be in the custody of Rutgers University. Brown University Herbarium houses the collection of S. T. Olney, and the New York Botanical Garden that of Nicolas Pike.

These collections are of relatively little taxonomic importance as compared with those of William H. Harvey, which lie at Trinity College, Dublin. Harvey was an energetic correspondent, and in addition to his own material his herbarium probably contains material from the hands of many of the American students of his day. Specimens with his authentication or of his collecting exist in considerable numbers in the collections of the Farlow Herbarium and the New York Botanical Garden. In fact, these two, the chief algal collections in the eastern part of the country, which are due to the exertions and the collections of William G. Farlow and Marshall A. Howe, include numbers of specimens from the various phycologists whose main deposits were made elsewhere. It has not been possible for the writer to complete this census of algal materials or personally to inspect many of the collections. Most of the older ones are sadly deficient in data and accuracy of determination. Since the ordinary ranges of our algae are commonly established by more modern records, a thorough revision of the identifications is not urgent, but it would doubtless eliminate some of the more improbable extensions.

PURPOSES AND LIMITATIONS

The first edition of this work offered revised descriptions of the northeastern American algae and attempted to incorporate all pertinent information on their structure and life histories. Most of this was supported by citations of appropriate source literature. The great increase in knowledge of these algae has rendered it impossible to combine features of a taxonomic manual with those of a more general nature within practicable compass. Since good morphological books are now available, the less necessary data of these sorts together with the supporting citations have been eliminated. Many errors recognized since the appearance of the first edition have been corrected, but doubtless others remain. Where he could do so the writer has adopted nomenclatural changes proposed by other authors, but in some cases he has felt it best to delay incorporating the changes until all that may be involved is more clearly understood. The illustrations have been designed to afford a satisfactory idea of the aspect of all the commoner genera and to elucidate the detailed structural features used to distinguish the genera and the species. The bibliography was selected to give substantially all catalogues and lists of algae of the area covered and scattered reports of algae of the district in American journals.

This book is planned to relieve the algal student from considering scattered sources of information in the older literature in making routine determinations. It is not itself a final authority, for only the original specimens and descriptions underlying the various species names can so qualify. When a student finds agreement between a specimen and a description here included, the most that he can assume is that his plant conforms with a species concept adopted by the present writer. In questionable instances the type specimen must be the final authority, if it exists. If it does not, and if the original description is inadequate (as is all too frequently the case), the opinions of those who have studied the plant in question may serve as a guide. For the better-known species original citations seem superfluous, for the interested student will be able to reach them, together with descriptions and supplementary notes, through De Toni's *Sylloge Algarum.* This is a district manual, not a critical taxonomic monograph, and it does not carry the literature citations appropriate to a monograph. The references in the present book have a somewhat different purpose. First of all, there are given citations to the three important collections containing American algae which have been issued as uniform sets or exsiccatae: Hauck and Richter's *Phykotheka Universalis,* Farlow, Anderson, and Eaton's *Algae Exsiccatae Americae Borealis,* and Collins, Holden, and Setchell's *Phycotheca Boreali-Americana.* These permit reference directly to critically determined plants well distributed in the larger herbaria of the world. Then follows a series of references to books and papers designed to introduce the student to researches which have been done on American material of the species concerned and to the synonymy as used in the more recent American publications. There is little major literature available to relate our flora to that of the mainland to the south. The present writer some years since prepared a volume on the marine algae of Florida (1928a). Northward from that state the only area well studied below our range is that about Beaufort, North Carolina, reported by Hoyt (1920). It shows a considerable proportion of forms common to our flora. The practical southern limit to the reliability of the present manual is New Jersey, for beyond that state we have no useful published lists. To the north the dominant constituents of the flora of the Maritime Provinces of Canada are well known; beyond this point, the records are fragmentary for any one district, but for the subarctic and arctic areas the general characteristics of the flora are understood, and the writer has recently (1954) reviewed the pertinent literature.

SYSTEMATIC LIST

THE following tabulation preserves the advantages of some features of a check list. It presents at a glance the several species, genera, and families, etc., alphabetized to a considerable degree, yet with a systematic grouping, though not a natural order. It is hoped that this list will be a convenience for reference in ways in which neither the index of species nor the pages of full description can serve. Where a species is represented in our flora by some variety or form other than the typical one, that fact will appear in the Descriptive Catalogue which follows.

CHLOROPHYCEAE

TETRASPORALES

Palmellaceae

Gloeocystis
- scopulorum Hansg.
- zostericola (Farlow) Collins

Urococcus
- foslieanus Hansg.

Chlorangiaceae

Prasinocladus
- lubricus Kuck.

CHLOROCOCCALES

Chlorococcaceae

Chlorococcum
- endozoicum Collins

Codiolum
- gregarium A. Br.
- petrocelidis Kuck.
- pusillum (Lyngb.) Kjellm.

Endosphaeraceae

Chlorochytrium
- inclusum Kjellm.
- moorei Gard.
- schmitzii Rosenv.

Oöcystaceae

Palmellococcus
- marinus Collins

ULOTRICHALES

Ulotrichaceae

Stichococcus
- marinus (Wille) Hazen

Ulothrix
- flacca (Dillw.) Thuret
- implexa Kütz
- laetevirens (Kütz.) Collins

Chaetophoraceae

Acrochaete
- repens Pringsh.

Bulbocoleon
- piliferum Pringsh.

Ectochaete
- taylori Thivy

Entocladia
- flustrae (Reinke) Batt.
- perforans Huber
- testarum Kylin
- viridis Reinke
- wittrockii Wille

Ochlochaete
- ferox Huber
- lentiformis Huber

Phaeophila
- dendroides (Crouan) Batt.
- engleri Reinke

Pilinia
endophytica Collins
lunatiae Collins
minor Hansg.
morsei Collins
reinschii (Wille) Collins
rimosa Kütz

Pringsheimiella
scutata (Reinke) O. C. Schm. et Petr.

Protoderma
marinum Reinke

Pseudendoclonium
submarinum Wille

Tellamia
contorta Batt.

CHAETOPELTIDACEAE

Diplochaete
solitaria Collins

GOMONTIACEAE

Gomontia
polyrhiza (Lagerh.) Born. et Flah.

ULVACEAE

Capsosiphon
fulvescens (C. Ag.) Setch. et Gard.

Enteromorpha
clathrata (Roth) J. Ag.
compressa (L.) Grev.
cruciata Collins
erecta (Lyngb.) J. Ag.
groenlandica (J. Ag.) Setch. et Gard.
intestinalis (L.) Link
linza (L.) J. Ag.
marginata J. Ag.
micrococca Kütz.
minima Näg.
plumosa Kütz.
prolifera (Müll.) J. Ag.
torta (Mert.) Reinb.

Monostroma
fuscum (Post. et Rupr.) Wittr.
grevillei (Thuret) Wittr.
leptodermum Kjellm.
oxyspermum (Kütz.) Doty
pulchrum Farlow

Percursaria
percursa (C. Ag.) J. Ag.

Ulva
lactuca L.

PRASIOLALES

PRASIOLACEAE

Prasiola
stipitata Suhr

CLADOPHORALES

CLADOPHORACEAE

Chaetomorpha
aerea (Dillw.) Kütz.
atrovirens Taylor
cannabina (Aresch.) Kjellm.
linum (Müll.) Kütz.
melagonium (Web. et Mohr) Kütz.

Cladophora
albida (Huds.) Kütz.
crystallina (Roth) Kütz.
expansa (Mert.) Kütz.
flavescens (Roth) Kütz.
flexuosa (Dillw.) Harv.
glaucescens (Harv.) Harv.
gracilis (Griff. ex Harv.) Kütz.
hutchinsiae (Dillw.) Kütz.
laetevirens (Dillw.) Harv.
magdalenae Harv.
refracta (Roth) Kütz.
rudolphiana (C. Ag.) Harv.
rupestris (L.) Kütz.

Rhizoclonium
erectum Collins
kerneri Stockm.
riparium (Roth) Harv.
tortuosum Kütz.

Spongomorpha
arcta (Dillw.) Kütz.
hystrix Strömf.
lanosa (Roth) Kütz.
spinescens Kütz.

Urospora
collabens (C. Agardh) Holmes et Batt.
penicilliformis (Roth) Aresch.
wormskjoldii (Mert.) Rosenv.

SIPHONALES

CHAETOSIPHONACEAE
Blastophysa
rhizopus Reinke

DERBESIACEAE
Derbesia
vaucheriaeformis (Harv.) J. Ag.

BRYOPSIDACEAE
Bryopsis
hypnoides Lamour
plumosa (Huds.) C. Ag.

PHYLLOSIPHONACEAE
Ostreobium
quekettii Born. et Flah.

XANTHOPHYCEAE

HETEROSIPHONALES

VAUCHERIACEAE
Vaucheria
arcassonensis Dang.
compacta (Collins) Collins
coronata Nordst.
intermedia Nordst.
litorea C. Ag.
minuta Blum et Conover
piloboloides Thuret
sphaerospora Nordst.
thuretii Woronin

PHAEOPHYCEAE

ECTOCARPALES

ECTOCARPACEAE
Ectocarpus
confervoides (Roth) Le Jol.
dasycarpus Kuck.
fasciculatus Harv.
paradoxus Mont.
penicillatus (C. Ag.) Kjellm.
siliculosus (Dillw.) Lyngb.
subcorymbosus Farlow
tomentosoides Farlow
tomentosus (Huds.) Lyngb.

Giffordia
granulosa (J. E. Smith) Hamel
mitchellae (Harv.) Hamel
ovata (Kjellm.) Kylin
sandriana (Zanard.) Hamel
secunda (Kütz.) Batt.

Mikrosyphar
porphyrae Kuck.

Pylaiella
littoralis (L.) Kjellm.

Sorocarpus
micromorus (Bory) Silva

Streblonema
aecidioides (Rosenv.) Fosl.
chordariae (Farlow) De Toni
effusum Kylin
fasciculatum Thuret
oligosporum Strömf.
parasiticum (Sauv.) De Toni
sphaericum Derb. et Sol.

SPHACELARIALES

SPHACELARIACEAE
Chaetopteris
plumosa (Lyngb.) Kütz.

Cladostephus
verticillatus (Lightf.) Lyngb.

Halopteris
scoparia (L.) Sauv.

Sphacelaria
britannica Sauv.
cirrosa (Roth) C. Ag.
furcigera Kütz.
fusca (Huds.) C. Ag.
plumigera Holmes
racemosa Grev.
radicans (Dillw.) C. Ag.

TILOPTERIDALES

TILOPTERIDACEAE
Haplospora
globosa Kjellm.

Tilopteris
mertensii (Smith) Kütz.

CHORDARIALES

Myrionemataceae

Ascocyclus
- distromaticus Taylor
- orbicularis (J. Ag.) Magnus

Hecatonema
- maculans (Collins) Sauv.
- reptans (Crouan) Sauv.
- terminalis (Kütz.) Kylin

Microspongium
- gelatinosum Reinke

Myrionema
- balticum (Reinke) Fosl.
- corunnae Sauv.
- foecundum (Strömf.) Fosl.
- globosum (Reinke) Fosl.
- strangulans Grev.

Ralfsiaceae

Ralfsia
- borneti Kuck.
- clavata (Carm.) Crouan
- fungiformis (Gunn.) Setch. et Gard.
- pusilla (Strömf.) Batt.
- verrucosa (Aresch.) J. Ag.

Lithodermataceae

Lithoderma
- extensum (Crouan) Hamel

Sorapion
- kjellmanni (Wille) Rosenv.

Elachisteaceae

Elachistea
- chondri Aresch.
- fucicola (Vell.) Aresch.
- lubrica Rupr.

Giraudia
- sphacelarioides Derb. et Sol.

Halothrix
- lumbricalis (Kütz.) Reinke

Leptonema
- fasciculatum Reinke

Myriactula
- minor (Farl.) Taylor
- chordae (Aresch.) Levr.

Symphoricoccus
- stellaris (Aresch.) Kuck.

Chordariaceae

Chordaria
- flagelliformis (Müll.) C. Ag.

Eudesme
- virescens (Carm.) J. Ag.
- zosterae (Lyngb.) Kylin

Leathesia
- difformis (L.) Aresch.

Sphaerotrichia
- divaricata (C. Ag.) Kylin

Acrothricaceae

Acrothrix
- novae-angliae Taylor

Stilophoraceae

Stilophora
- rhizodes (Ehrh.) J. Ag.

DESMARESTIALES

Desmarestiaceae

Arthrocladia
- villosa (Huds.) Duby

Desmarestia
- aculeata (L.) Lamour.
- viridis (Müll.) Lamour.

PUNCTARIALES

Striariaceae

Isthmoplea
- sphaerophora (Carm.) Kjellm.

Stictyosiphon
- griffithsianus (Le Jol.) Holmes et Batt.
- subsimplex Holden
- tortilis (Rupr.) Reinke

Striaria
- attenuata (C. Ag.) Grev.

Punctariaceae

Asperococcus
- echinatus (Mert.) Grev.

Delamarea
- attenuata (Kjellm.) Rosenv.

Desmotrichum
- balticum Kütz.
- undulatum (J. Ag.) Reinke

Litosiphon
- filiformis (Reinke) Batt.

Myriotrichia
clavaeformis Harv.
densa Batt.
filiformis Harv.

Omphalophyllum
ulvaceum Rosenv.

Petalonia
fascia (Müll.) Kuntze

Phaeosaccion
collinsii Farlow

Punctaria
latifolia Grev.
plantaginea (Roth) Grev.

Rhadinocladia
cylindrica Schuh
farlowii Schuh

Scytosiphon
lomentaria (Lyngb.) J. Ag.

DICTYOSIPHONALES

DICTYOSIPHONACEAE

Dictyosiphon
chordariae Aresch.
eckmani Aresch.
foeniculaceus (Huds.) Grev.
macounii Farlow

LAMINARIALES

LAMINARIACEAE

Agarum
cribrosum (Mert.) Bory

Alaria
esculenta (L.) Grev.
grandifolia J. Ag.
membranacea J. Ag.
musaefolia (De la Pyl.) J. Ag.
pylaii (Bory) Grev.

Chorda
filum (L.) Lamour.
tomentosa Lyngb.

Laminaria
agardhii Kjellm.
cuneifolia J. Ag.
digitata (L.) Lamour.
groenlandica Rosenv.
intermedia Fosl.
longicruris De la Pyl.
nigripes J. Ag.
platymeris De la Pyl.
saccharina (L.) Lamour.
solidungula J. Ag.
stenophylla (Kütz.) J. Ag.

Saccorhiza
dermatodea (De la Pyl.) J. Ag.

FUCALES

FUCACEAE

Ascophyllum
mackaii (Turn.) Holmes et Batt.
nodosum (L.) Le Jol.

Fucus
edentatus De la Pyl.
evanescens C. Ag.
filiformis Gmelin
miclonensis De la Pyl.
serratus L.
spiralis L.
vesiculosus L.

SARGASSACEAE

Sargassum
filipendula C. Ag.
fluitans Børg.
hystrix J. Ag.
natans (L.) J. Meyen

RHODOPHYCEAE

BANGIALES

BANGIACEAE

Asterocytis
ramosa (Thwaites) Gobi

Bangia
ciliaris Carm.
fuscopurpurea (Dillw.) Lyngb.

Erythropeltis
discigera (Berth.) Schmitz.

Erythrotrichia
carnea (Dillw.) J. Ag.
rhizoidea Cleland

Goniotrichum
alsidii (Zanard.) Howe

Porphyra
leucosticta Thuret
miniata (Lyngb.) C. Ag.
umbilicalis (L.) J. Ag.

Porphyropsis
coccinea (J. Ag.) Rosenv.

NEMALIONALES

ACROCHAETIACEAE

Acrochaetium
alcyonidii Jao
amphiroae (Drew) Papenf.
attenuatum (Rosenv.) Hamel
dasyae Collins
daviesii (Dillw.) Näg.
emergens (Rosenv.) Weber-van Bosse
flexuosum Vickers
intermedium Jao
microfilum Jao
minimum Collins
radiatum Jao
sagraeanum (Mont.) Born.
thuretii (Born.) Collins et Herv.
zosterae Papenf.

Audouinella
efflorescens (J. Ag.) Papenf.
membranacea (Magnus) Papenf.

Colaconema
americana Jao

Conchocelis
rosea Batt.

Kylinia
alariae (Jóns.) Kylin
compacta (Jao) Papenf.
hallandica (Kylin) Kylin
moniliformis (Rosenv.) Kylin
secundata (Lyngb.) Papenf.
unifila (Jao) Papenf.
virgatula (Harv.) Papenf.

Rhodochorton
penicilliforme (Kjellm.) Rosenv.
purpureum (Lightf.) Rosenv.

HELMINTHOCLADIACEAE

Nemalion
multifidum (Web. et Mohr) J. Ag.

CHAETANGIACEAE

Scinaia
furcellata (Turn.) Bivona

BONNEMAISONIACEAE

Asparagopsis
hamifera (Hariot) Okam.

GELIDIALES

GELIDIACEAE

Gelidium
crinale (Turn.) Lamour.

CRYPTONEMIALES

DUMONTIACEAE

Dilsea
integra (Kjellm.) Rosenv.

Dumontia
incrassata (Müll.) Lamour.

RHIZOPHYLLIDACEAE

Polyides
caprinus (Gunn.) Papenf.

SQUAMARIACEAE

Cruoriopsis
ensis Jao
gracilis (Kuck.) Batt.
hyperborea Rosenv.

Hildenbrandia
prototypus Nardo

Peyssonnelia
rosenvingii Schmitz

Rhododermis
elegans Crouan
georgii (Batt.) Collins
parasitica Batt.

CORALLINACEAE

Corallina
officinalis L.

Fosliella
farinosa (Lamour.) Howe
lejolisii (Rosan.) Howe

Lithophyllum
corallinae (Crouan) Heydr.
macrocarpum (Rosan.) Fosl.
pustulatum (Lamour.) Fosl.

Lithothamnium
colliculosum Fosl.
foecundum Kjellm.
glaciale Kjellm.
laeve (Strömf.) Fosl.
lenormandi (Aresch.) Fosl.
norvegicum (Aresch.) Kjellm.

tophiforme Unger
ungeri Kjellm.
Melobesia
membranacea (Esper) Lamour.
Phymatolithon
compactum (Kjellm.) Fosl.
evanescens (Fosl.) Fosl.
laevigatum (Fosl.) Fosl.
polymorphum (L.) Fosl.
GLOIOSIPHONIACEAE
Gloiosiphonia
capillaris (Huds.) Carm.
KALLYMENIACEAE
Euthora
cristata (L.) J. Ag.
Kallymenia
reniformis (Turn.) J. Ag.
schmitzii De Toni
CHOREOCOLACACEAE
Ceratocolax
hartzii Rosenv.
Choreocolax
polysiphoniae Reinsch
Harveyella
mirabilis (Reinsch) Schmitz et Reinke

GIGARTINALES

CRUORIACEAE
Cruoria
arctica Schmitz
Petrocelis
middendorfii (Rupr.) Kjellm.
polygyna (Kjellm.) Schmitz
NEMASTOMATACEAE
Platoma
bairdii (Farl.) Kuck.
Turnerella
pennyi (Harv.) Schmitz
SOLIERIACEAE
Agardhiella
tenera (J. Ag.) Schmitz
RHODOPHYLLIDACEAE
Cystoclonium
purpureum (Huds.) Batt.
Rhodophyllis
dichotoma (Lepesch.) Gobi
FURCELLARIACEAE
Furcellaria
fastigiata (Huds.) Lamour.
HYPNEACEAE
Hypnea
musciformis (Wulf.) Lamour.
GRACILARIACEAE
Gracilaria
foliifera (Forssk.) Børg.
verrucosa (Huds.) Papenf.
PHYLLOPHORACEAE
Ahnfeltia
plicata (Huds.) Fries
Gymnogongrus
griffithsiae (Turn.) Mart.
norvegicus (Gunn.) J. Ag.
Phyllophora
brodiaei (Turn.) J. Ag.
interrupta (Grev.) J. Ag.
membranifolia (Good. et Woodw.) J. Ag.
traillii Holmes
GIGARTINACEAE
Chondrus
crispus Stackh.
Gigartina
stellata (Stackh.) Batt.

RHODYMENIALES

RHODYMENIACEAE
Halosaccion
ramentaceum (L.) J. Ag.
Rhodymenia
palmata (L.) Grev.
CHAMPIACEAE
Champia
parvula (C. Ag.) Harv.
Lomentaria
baileyana (Harv.) Farlow
orcadensis (Harv.) Collins

CERAMIALES

CERAMIACEAE
Antithamnion
americanum (Harv.) Farlow
boreale (Gobi) Kjellm.
cruciatum (C. Ag.) Näg.

floccosum (Müll.) Kleen
plumula (Ellis) Thuret
pylaisaei (Mont.) Kjellm.

Callithamnion
baileyi Harv.
byssoides Arnott
corymbosum (Smith) Lyngb.
roseum (Roth) Harv.
tetragonum C. Ag.

Ceramium
areschougii Kylin
circinatum (Kütz.) J. Ag.
deslongchampii Chauv.
diaphanum (Lightf.) Roth
elegans (Ducl.) C. Ag.
fastigiatum (Roth) Harv.
rubriforme Kylin
rubrum (Huds.) C. Ag.
strictum (Kütz.) Harv.

Griffithsia
globulifera Harv.
tenuis C. Ag.

Pleonosporium
borreri (Smith) Näg.

Plumaria
elegans (Bonnem.) Schmitz

Ptilota
plumosa (Huds.) C. Ag.
serrata Kütz.

Seirospora
griffithsiana Harv.

Spermothamnion
turneri (Mert.) Aresch.

Spyridia
filamentosa (Wulf.) Harv.

Trailliella
intricata (J. Ag.) Batt.

DELESSERIACEAE

Caloglossa
leprieurii (Mont.) J. Ag.

Grinnellia
americana (C. Ag.) Harv.

Membranoptera
alata (Huds.) Stackh.
denticulata (Mont.) Kylin

Pantoneura
angustissima (Turn.) Kylin
baerii (Post. et Rupr.) Kylin

Phycodrys
rubens (Huds.) Batt.

DASYACEAE

Dasya
pedicellata (C. Ag.) C. Ag.

RHODOMELACEAE

Bostrychia
rivularis Harv.

Chondria
baileyana (Mont.) Harv.
dasyphylla (Woodw.) C. Ag.
sedifolia Harv.
tenuissima (Good. et Woodw.) C. Ag.

Odonthalia
dentata (L.) Lyngb.
floccosa (Esp.) Falk.

Polysiphonia
arctica J. Ag.
denudata (Dillw.) Kütz.
elongata (Huds.) Harv.
fibrillosa Grev.
flexicaulis (Harv.) Collins
harveyi Bailey
lanosa (L.) Tandy
nigra (Huds.) Batt.
nigrescens (Huds.) Grev.
novae-angliae Taylor
subtilissima Mont.
urceolata (Lightf.) Grev.

Rhodomela
confervoides (Huds.) Silva
lycopodioides (L.) C. Ag.
virgata Kjellm.

From within our range 401 species have been authoritatively reported, and in addition 134 varieties and forms of these, of greater or less distinctiveness.

DESCRIPTIVE CATALOGUE

THE major systematic divisions of the algae are based chiefly on characters of reproduction and life history not readily ascertained without critical microscopic or even cultural studies. Consequently a key working through the greater groups and avoiding such seasonal or obscure features tends to loose distinctions and is not easy either to design or to use. A primarily artificial key leading direct to genera, such as Collins (1918c, p. 1) designed, has some advantages. In practice, the gross aspects of the common genera of marine algae are so readily learned that after a few days at the seashore the observer should be able to resort directly to the keys under the genera, which can be more accurately based on gross morphology.

KEY TO THE ORDERS OF THE MARINE ALGAE

1. Pigmentation of unmasked chlorophyll........................ 2
1. Chlorophyll masked by accessory pigments.................... 3
2. Food reserves of starch............... 4 CHLOROPHYCEAE, p. 36
2. Food reserves of oil.................. 9 XANTHOPHYCEAE, p. 96
3. Accessory pigments imparting a brown color; reproduction at some stage typically involving flagellate cells 10 PHAEOPHYCEAE, p. 101
3. Accessory pigments imparting a rose-red color (often fading or feebly developed); no reproductive stage with flagellate cells ....'................ 18 RHODOPHYCEAE, p. 200

CHLOROPHYCEAE

4. Plants of scattered or irregularly massed cells.................. 5
4. Plants generally forming filaments or membranes of vegetative cells 6
5. Cell divisions during the vegetative phase, which is generally colonial or gelatinous-invested............ TETRASPORALES, p. 36
5. Cell divisions absent from the vegetative phase, which is not invested with jelly..................... CHLOROCOCCALES, p. 39
6. Filaments or membranes multicellular......................... 7
6. Filaments remaining undivided by septa in the vegetative parts, though plurinucleate................. SIPHONALES, p. 91

7. Chromatophores stellate; reproduction by akinetes PRASIOLALES, p. 74
7. Chromatophores otherwise; reproduction various 8

8. Chromatophores usually large and simple; cells often with simple hairs .... ULOTRICHALES, p. 43
8. Chromatophores small and several in a cell; cells plurinucleate; simple hairs absent, filaments uniseriate CLADOPHORALES, p. 75

XANTHOPHYCEAE

9. Filamentous, plurinucleate; in our species with oögamous reproduction .... HETEROSIPHONALES, p. 96

PHAEOPHYCEAE

10. Filaments characteristically becoming polysiphonous............ 11
10. If filamentous, not polysiphonous, though often pluriseriate...... 12

11. Growth from apical cells.................. SPHACELARIALES, p. 117
11. Growth trichothallic or intercalary.......... TILOPTERIDALES, p. 124

12. Filamentous, filaments uniseriate, erect and free, or encrusting and consolidated .... ECTOCARPALES, p. 101
12. Massive, cylindrical, or phylloid, or slender but pluriseriate, or reduced types.. 13

13. Filamentous construction obvious throughout, but generally enveloped by a common jelly............. CHORDARIALES, p. 126
13. Filamentous structure lacking, obscured, or only shown in evanescent plant parts................................... 14

14. Sporophytes lacking free assimilators......................... 15
14. Sporophyte plants with evanescent free assimilatory filaments; gametophytes microscopic and oögamous. DESMARESTIALES, p. 152

15. Gametophytes and sporophytes, if known, both independent plants 16
15. Gametophyte reduced to cytological phases; sporophytes massive and branched........................... FUCALES, p. 188

16. Sporophyte slender or broad, usually macroscopic although often developing from a protonemal phase; gametophytes not oögamous 17
16. Sporophyte plants generally massive, not developing from a protonemal phase; gametophytes microscopic, filamentous, and oögamous .... LAMINARIALES, p. 174

17. Growth in the sporophyte from an apical cell; gametophyte microscopic, filamentous, isogamous.... Dictyosiphonales, p. 171
17. Growth otherwise; gametophyte like the sporophyte, or vegetatively preceded by a microscopic protonemal phase, which may partly or entirely assume the gametophytic function Punctariales, p. 155

Rhodophyceae

18. Intercellular connections absent, chromatophores stellate, cells uninucleate; carpogonia and spermatia derived directly from vegetative cells........................ Bangiales, p. 200
18. Intercellular connections usually present, cells often plurinucleate; carpogonia and, often, the spermatangia terminating special branch systems.................................... 19

19. Carpogonium giving rise directly to the gonimoblasts (carposporiferous filaments) 20
19. Carpogonium transferring the zygote nucleus to a special auxiliary cell, which in turn gives rise to the gonimoblasts..... 21

20. No special nutritive cells associated with the gonimoblasts; plants filamentous, slippery, or with hooked branch tips Nemalionales, p. 209
20. Vegetative nutritive cells associated with the gonimoblasts; plants small and wiry...................... Gelidiales, p. 230

21. Auxiliary cells established before fertilization; plants never filamentous in habit.. 22
21. Auxiliary cells not formed until after fertilization; plants filamentous or sturdier, sometimes delicately membranous, but not tough or massive...................... Ceramiales, p. 289

22. Auxiliary cells not directly connected to the carpogonium, receiving zygote nuclei by means of oöblast (sporogenous) filaments developed from the carpogonium.................. 23
22. Auxiliary cells developed from the cell supporting the contributing carpogonium; sometimes filiform, often phylloid or segmented .... Rhodymeniales, p. 282

23. Auxiliary cells constituted from specialized intercalary cells in vegetative filaments; crustose, slippery, or more often rather wiry or massive plants............. Gigartinales, p. 261
23. Auxiliary cells borne on specialized branches of the structural filaments, or forming part of each carpogenic branch; habit quite various, filiform, crustose-calcareous, slippery, wiry, or membranous .... Cryptonemiales, p. 232

CHLOROPHYCEAE

TETRASPORALES

Plants generally with a more or less prolonged quiescent or attached, often jelly-invested vegetative phase, readily passing into a flagellate, free-swimming state by simple emergence from the pectic investment; cells uninucleate, generally with a single chloroplast and pyrenoid; when motile, with two to occasionally four anterior flagella.

KEY TO FAMILIES

1. Plants in the attached phase in colonies enveloped in gelatinous layers, or else subsolitary................ PALMELLACEAE, p. 36
1. Plants in the attached phase in branched colonies CHLORANGIACEAE, p. 38

PALMELLACEAE

Plants in the attached phase subsolitary, or in groups of related cells successively enveloped in layers of gelatinizing membranes, or jelly general and not lamellose, the inner cell wall usually of cellulose; pseudocilia not present; chromatophore cup-shaped or radiating, with or without a pyrenoid; multiplication of groups by fragmentation; asexual reproduction by formation of zoöspores, aplanospores, or akinetes; sexual reproduction by formation of biflagellate isogametes.

KEY TO GENERA

1. Plants colonial, or gelatinous wall symmetrically thickened GLOEOCYSTIS, p. 36
1. Plants of scattered cells; wall asymmetrically thickened to form stalklike extensions..................... UROCOCCUS, p. 37

Gloeocystis Nägeli, 1849

Cells at first single, later united by a general gelatinous envelope, which often becomes lamellose; cells subspherical or laterally compressed, in structure showing cup-shaped chromatophores with one pyrenoid; vegetative multiplication by cell division in all planes, the cell walls persisting; biflagellate zoöspores formed directly from vegetative cells; akinetes also reported.

KEY TO SPECIES

1. Cells usually less than 6 μ diam....................... G. scopulorum
1. Cells usually over 10 μ diam.......................... G. zostericola

Gloeocystis scopulorum Hansgirg

Plants forming greenish-yellow gelatinous masses, of somewhat spherical cells united in colonies of two to eight, the investing gelatinous wall distinctly stratified; protoplasts 4–6 μ diam.

Maine; forming gelatinous masses in company with other minute algae near high-water mark.

REFERENCES: Phyc. Bor.-Am. 1579. Collins 1908c, p. 155, 1909b, p. 309.

Gloeocystis zostericola (Farlow) Collins

Plants forming brownish masses of subspherical or mutually compressed cells in groups of two to four forming a colony 40–100 μ diam.; gelatinous wall ample, lamellose; protoplasts when compressed 9–11 μ on the shorter axis to 19–26 μ on the longer axis.

Southern Massachusetts; forming brownish masses on Zostera.

REFERENCES: Algae Exsic. Am. Bor. 230; Phyc. Bor.-Am. 219 (as *G. chrysophthalma*). Farlow 1882, p. 68 (as *Gloeocapsa zostericola*); Collins 1900b, p. 44, 1909b, p. 309.

Urococcus (Hassall) Kützing, 1849

Plants of single cells or small colonies, the cells spherical or nearly spherical, with a lamellose wall often elongated into a stalklike extension on one side, and a cup-shaped chromatophore without pyrenoid; by gelatinization of the walls the cells form a jelly phase involving many protoplasts; zoöspores biflagellate.

Urococcus foslieanus Hansgirg

Plants of scattered cells, cells 8–18 μ diam., with a thick wall which increases the diameter to 15–25 μ; wall locally somewhat thickened and particularly lamellose, producing a stalklike extension; color green, at times becoming a dull orange.

Maine.

REFERENCE: Collins 1909b, p. 306.

CHLORANGIACEAE

Plants in the attached phase solitary or in branched colonies with a close gelatinous, often stalked, investment; cells with the anterior end downward, chloroplast cup-shaped at the posterior end of the cell, an eyespot, and often contractile vacuoles; cell division produces two to eight zoöspores which may be liberated or, retaining the gelatinous investment, may produce a branched colony; aplanospores formed in some genera; sexual reproduction by isogametes.

Prasinocladus Kuckuck, 1894

Plants branching, the living protoplasts mostly in the terminal segments of the branches; the gelatinous investment forming a close envelope, in which the protoplast moves upward toward the summit of its sheath at the time of division, the cells showing flagella at this time, remaining connected after division and forming new sheaths; the protoplast with a somewhat lobed cup-shaped chloroplast, a pyrenoid, an eyespot, and a non-contractile vacuole; asexual multiplication by the occasional release of a quadriflagellate zoöspore; sexual reproduction unknown.

Prasinocladus lubricus Kuckuck

Plants gregarious, branching, the filaments up to 250 μ long, the branching ditrichotomous; cells rounded oblong, 6–9 μ diam., 12–20 μ long.

Northern Massachusetts and Maine; forming a thin coating on pebbles, Spartina, etc., in salt marshes and sheltered pools.

REFERENCES: Phyc. Bor.-Am. 564 (as *Prasinocladus subsalsus*). Davis 1894b, p. 377 (as *Euglenopsis subsalsa*); Collins 1900b, p. 45, 1909b, p. 141 (as *Prasinocladus subsalsus*); Lambert 1930, p. 227; Smith 1950, p. 130.

CHLOROCOCCALES

Plants vegetating in a motionless phase, free, less often attached, never filamentous but sometimes coenobic or endophytic; cell wall usually thin or at least not gelatinous, often of elaborate and distinctive form; protoplast with one to several chloroplasts, pyrenoids, and nuclei; cell divisions absent in the vegetative phase; asexual reproduction by zoöspores, aplanospores, or other means; sexual reproduction by iso- or anisogametes.

KEY TO FAMILIES

1. Endophytes .... ENDOSPHAERACEAE, p. 41
1. Not endophytic 2
2. Cells usually laterally aggregated in large colonies CHLOROCOCCACEAE, p. 39
2. Cells solitary or united by enveloping walls of parent cells OÖCYSTACEAE, p. 43

CHLOROCOCCACEAE

Plants forming considerable aggregations, solitary, or endophytic; cells varying in shape from spherical to fusiform; wall thin, to thickened and irregularly produced into lamellose extensions; protoplast with a parietal netlike, a cup-shaped, or a radiating chromatophore, with pyrenoids, and one to plurinucleate; asexual reproduction by formation of zoöspores or aplanospores; sexual reproduction in a few genera by isogametes.

KEY TO GENERA

1. Minute, the round cells 25 μ diam. or less, our species endozoic CHLOROCOCCUM, p. 39
1. Larger, cells elongate but usually over 50 μ diam... CODIOLUM, p. 40

Chlorococcum Fries, 1825

Plants unicellular, solitary or massed; cells uninucleate, the chloroplast cup-shaped with one to several pyrenoids, wall variable in thickness; reproduction by formation of several biflagellate zoöspores, by aplanospores, or by gametes from little-modified cells.

Chlorococcum endozoicum Collins

Cells spherical, 10–25 μ in diam., thin-walled.

Maine; in the mantle of *Mytilus edulis;* summer. Of dubious affiliation.

REFERENCES: Phyc. Bor.-Am. 1323. Collins 1909b, p. 144.

Codiolum A. Braun, 1855

Plants unicellular, the cells endophytic or independent and forming wide aggregations; below elongate, above more or less expanded, the wall greatly thickened below to form a stalklike base and sometimes thickened over the apex in a lesser degree; cell with a peripheral netlike chromatophore containing many pyrenoids, and a single nucleus; reproduction by large quadriflagellate zoöspores; aplanospores and biflagellate gametes (?) have also been reported.

KEY TO SPECIES

1. Free, often forming a wide coating on rocks.................... 2
1. Endophytic .... C. petrocelidis

2. Upper part of the cell ovoid, much wider than the stalk.. C. gregarium
2. Narrowly clavate above, not much thicker than the stalk.. C. pusillum

Codiolum gregarium A. Braun

Cells growing crowded among other algae; clavate, ovoid above, where 65–100 μ diam., 250–500 μ long, the stalk below sharply delimited, 20–30 μ diam., 0.6–1.0 mm. long.

New York to Maine; growing on rocks and shells and generally forming the minor constitutent of a mixed assemblage of organisms, particularly Calothrix; summer.

REFERENCES: Phyc. Bor.-Am. 165. Farlow 1881, p. 58; Collins 1900b, p. 44, 1909b, p. 152.

From our range there has also been reported f. **intermedium** (Foslie) Collins (Collins 1909b, p. 152); this has a swollen, almost spherical upper portion to the cell, 55–110 μ diam., 150–300 μ long, and a short stout stalk, 25–40 μ diam., 170–250 μ long; it has been found in Maine with *Calothrix scopulorum.*

Codiolum pusillum (Lyngbye) Kjellman

Plants growing densely crowded together to form considerable strata; cells long, slender, gradually tapering from the obtuse top of the upper part of the cell to the stalk; width above 60–70 μ, length over-all about 2 (–5) mm., of which two-thirds constitutes the stalk.

Maine; growing on rocks at midtide level and forming wide, velvety, blackish-green patches; summer and fall.

REFERENCES: Phyc. Bor.-Am. 1126. Kjellman 1883, p. 318; Collins 1909b, p. 153.

The description given applies to the typical plant. Two additional forms are reported from our range. F. **americanum** Foslie (Phyc. Bor.-Am. 869; Foslie in Collins 1901b, p. 290; Collins 1909b, p. 153), with a stalk 5–10 times as long as the body of the cell, occurs in Massachusetts, forming an unmixed coating on rocks. F. **longipes** (Foslie) Collins (Phyc. Bor.-Am. 26 as *C. longipes;* Collins 1883, p. 55, 1900b, p. 44, 1909b, p. 153), sharply enlarging from the top of the stalk into the body of the cell, diam. above 100 μ, diam. of stalk 30–60 μ, total length to 1200 μ, appears on the Maine, New Brunswick, and Newfoundland coasts, forming wide coatings on rocks, when drying showing a spotted appearance due to local exposure of the stalks.

Codiolum petrocelidis Kuckuck

Plants scattered, endophytic, erect within the host tissues; upper part of the cell ovoid or obovoid, 20–30 μ diam., 65–90 μ long; stalk very slender, below terminating in a point.

Massachusetts to Maine; growing in the tissues of Petrocelis; summer.

REFERENCES: Collins 1900b, p. 44, 1909b, p. 152.

ENDOSPHAERACEAE

Plants generally unicellular, endophytes or parasites, scattered or at times gregarious, the form irregular; chloroplasts axial, radiating, pyrenoids present or absent; reproduction by biflagellate isogametes (or zoöspores?).

Chlorochytrium Cohn, 1872

Plants endophytic; cells rounded, with an irregularly thickened lamellated wall; chloroplast at first parietal, later irregular and radiating; vegetative cells frequently persisting in a starch-packed akinete-like condition; reproducing by bi- or quadriflagellate swarmers functional as gametes or possibly as zoöspores.

KEY TO SPECIES

1. Cells nearly spherical.. 2
1. Cells clavate with pointed base...................... C. schmitzii

2. Cells 16–26 μ diam., on Enteromorpha.................. C. moorei
2. Cells 80–100 μ diam., in Rhodophyceae................. C. inclusum

Chlorochytrium moorei Gardner

Vegetative cells spherical or nearly so, 16–26 μ diam., with a somewhat radiating chromatophore containing a single pyrenoid; developing into a sporangium containing motile quadriflagellate zoöids which are reported to be dimorphic, spherical and 6–7 μ diam., or pyriform and 2.5–3.5 μ diam., discharged through a pore in the cell wall.

Northern Massachusetts, Maine (?); partially embedded in the surface of Enteromorpha plants; early spring.

REFERENCES: Phyc. Bor.-Am. 565, 1121 (?) (as *Chlorocystis cohnii non* (Wright) Reinh.); Moore 1900, p. 100; Collins 1909b, p. 148, 1918a, p. 26 (all as *C. cohnii*); Gardner 1917, p. 382.

Chlorochytrium inclusum Kjellman

Vegetative cells globose or nearly so, or in confining host tissues irregular; 80–100 (–270) μ diam., chromatophore parietal, with many pyrenoids; wall becoming thickened; at maturity becoming sporangial and producing a conical projection to the host's surface for discharge of the spores.

Ellesmere Island; in tissues of Rhodymenia and Turnerella.

REFERENCES: Kjellman 1883, p. 320; Rosenvinge 1926, p. 15; Lund 1933, p. 16.

Chlorochytrium schmitzii Rosenvinge

Vegetative cells clavate to ovoid, with a rounded apex and a pointed base; diameter about 90 μ, length to 370 μ; chromatophore large, with two or more pyrenoids.

Maine; endophytic in crusts of Petrocelis.

REFERENCES: Rosenvinge 1893, p. 964; Collins 1900b, p. 43, 1909b, p. 147.

OÖCYSTACEAE

Cells solitary, or united by the enveloping persistent wall of the parent cell; wall simple or characteristically ornate; nuclei generally single, chloroplasts solitary or numerous, lenticular; reproduction by the internal formation of groups of daughter cells, each resembling the parent cell in miniature (autospores).

Palmellococcus Chodat, 1894

Plants unicellular, cells scattered or gregarious, spherical or usually ellipsoid, with one to several lenticular chloroplasts; pyrenoids present or absent; cells often congested with reddish oil; reproduction by autospores or by single aplanospores.

Palmellococcus marinus Collins

Vegetative cells nearly spherical, 10–40 μ diam., with a wall about 2 μ thick; color green to orange; autospores 8–64 in a sporangium, each forming a thick wall while still within the parent membrane, and the mass retaining the form of the parent cell for some time after the restraining wall has disappeared.

Maine; among various algae in warm marsh pools.

REFERENCES: Phyc. Bor.-Am. 1316. Collins 1907, p. 197, 1909b, p. 159.

ULOTRICHALES

Plants filamentous to foliaceous, simple or branched, without or with differentiation of attached base and free apex, in a few cases condensed toward a unicellular state or by irregular divisions, becoming somewhat parenchymatous; cells showing a distinct wall, usually a single parietal chromatophore, pyrenoids, and a single nucleus; asexual reproduction by zoöspores, aplanospores, or akinetes; sexual reproduction by isogametes, by anisogametes, or by oögamy.

KEY TO FAMILIES

1. Plants essentially filamentous 2
1. Plants of solitary cells, or cells loosely associated into gelatinous filaments .... CHAETOPELTIDACEAE, p. 58
1. Plants essentially membranous.................. ULVACEAE, p. 60

2. Distinctive sporangia present................ GOMONTIACEAE, p. 59
2. Sporangial cells often enlarged, but not distinctive in form........ 3

3. Filaments typically unbranched, uniseriate, without hairs ULOTRICHACEAE, p. 44
3. Filaments branched, crowded and disk-forming or free, often bearing simple hairs.................. CHAETOPHORACEAE, p. 47

ULOTRICHACEAE

Plants filamentous, filaments with or without an attaching base, cylindrical, typically unbranched, uniseriate; cells with thin to gelatinous walls, lateral platelike or bandlike chromatophores with or without pyrenoids, and a single nucleus; asexual reproduction by one to sixteen zoöspores, often of different sizes in the same species, by aplanospores or akinetes; sexual reproduction by iso- or anisogametes.

KEY TO GENERA

1. Chromatophores unilateral; one zoöspore formed in each sporangial cell .... STICHOCOCCUS, p. 44
1. Chromatophores nearly or quite encircling the protoplast; two or more zoöspores formed in each sporangial cell ULOTHRIX, p. 45

Stichococcus Nägeli, 1849

Filaments cylindrical, free, or cells sometimes solitary by disjunction of the filaments; cells uninucleate, with a platelike lateral chromatophore that extends no more than about halfway around the cell, and contains one pyrenoid; asexual reproduction by formation of zoöspores singly in each cell, or single aplanospores; iso- or anisogametes formed singly or two in a cell.

Stichococcus marinus (Wille) Hazen

Filaments dark green; cells cylindrical, 5–6 μ diam., 1–2 diameters long; chromatophore a roundish or oblong plate, the pyrenoid indistinct.

Connecticut, Rhode Island, and Maine; on banks and in salt-marsh creeks and estuaries; summer.

REFERENCES: Phyc. Bor.-Am. 615 (as *Ulothrix variabilis* f. *marina*), 1832. Collins 1900b, p. 45 (as *U. variabilis* f. *marina*), 1909b, p. 190; Hazen 1902, p. 161.

Ulothrix Kützing, 1833

Filaments at first attached by a sterile basal holdfast cell, later sometimes free, cylindrical, normally unbranched, with walls thin to subgelatinous; cells with one parietal bandlike chromatophore which more or less completely encircles the cell and shows one to several pyrenoids; quadriflagellate zoöspores formed two to many in each cell, discharging through a pore or forming aplanospores within the sporangium, or larger single aplanospores may be formed; sexual reproduction by the formation of eight to sixty-four biflagellate gametes.

KEY TO GENERA

1. Chromatophore occupying only the middle portion of the side wall of the cell; filaments usually 10–15 μ diam., cells 0.5–1.0 diameter long .... U. implexa, p. 46
1. Chromatophore occupying nearly or quite all the lateral face of the cell; cells rather broader and shorter..................... 2
2. Branches absent, the filaments not tending to become united; chromatophore occupying all the lateral face of the cell U. flacca, p. 45
2. Branches occasionally formed, the filaments in the lower parts of the tufts often conjoined laterally; chromatophore occupying nearly all the lateral face of the cell, thickened on the side which contains the pyrenoid................ U. laetevirens, p. 46

Ulothrix flacca (Dillwyn) Thuret Pl. 1, fig. 9

Filaments entangled, often in considerable skeins; color bright to dark green; filaments 10–25 μ diam., cells 0.25–0.75 times as long as broad; chromatophore covering the whole side wall of the cell, with 1–3 pyrenoids; sporangial cells swollen to 50 μ diam.

New Jersey to the lower St. Lawrence,[1] Hudson Bay and Strait, Davis Strait, and Baffin Island; common, growing on various objects between the tide marks, where it may form extensive expanses adjacent to the Calothrix zone; most abundant in winter and spring.

REFERENCES: Phyc. Bor.-Am. 17, 1123. Harvey 1858a, p. 90 (as *Hormotrichum boreale*); Farlow 1881, p. 45; Collins 1900b, p. 45, 1909b, p. 185; Hazen 1902, p. 155; Whelden 1947, p. 47.

Ulothrix laetevirens (Kützing) Collins

Filaments entangled or creeping, below 2–3 often laterally conjoined, or densely packed and the lower cells subparenchymatously united, the erect filaments tapering toward the base; not infrequently branched, the branches issuing at wide angles, more slender than the supporting filaments, of many cells 1–3 diameters long; filaments 10–25 μ diam., cells 0.25–0.75 diameter long, or rarely more; chromatophore occupying nearly all the lateral face of the cell, thickened on the side containing the pyrenoid; sporangial cells producing eight zoöspores; akinetes solitary.

Maine and the lower St. Lawrence, and Devon Island; on woodwork; summer.

REFERENCES: Phyc. Bor.-Am. 313. Collins 1909b, p. 186; Whelden 1947, p. 47.

Ulothrix implexa Kützing

Filaments tufted, the soft tufts 0.5–3.0 cm. long, adherent or free-floating; color pale yellowish green; filaments curled or spirally

[1] Throughout this volume references to Long Island and Staten Island apply to those islands which are part of New York State; the designation "Lower St. Lawrence" applies to a part of the river included in the Canadian Province of Quebec, where, also, the Gaspé is located. The designation Newfoundland applies to the island only, Labrador being referred to separately. Most old Labrador records cannot be localized readily, but where new records are available from northern Labrador, defined for our purposes as northward from Nain (approximately 56° 30″ N.L.), these are distinguished. No reference is made to records from Ile St. Pierre or Ile Miquelon if the same plants are recorded from nearby Newfoundland. For the sake of brevity, records from the coastline of northern Quebec east of Hudson Bay, including Ungava Bay and Gray Strait at Killinek Island, are all referred to Hudson Strait.

contorted and entangled, (6-) 10–15 (–18) μ diam., the cells 0.5–1.0 diameter long, or rarely somewhat longer; the chromatophore occupying only the middle part of the side wall of the cell.

Florida, New Jersey to New Hampshire, New Brunswick, and Devon Island; growing on woodwork and rocks near high-water mark; especially abundant in the spring.

REFERENCES: Phyc. Bor.-Am. 115. Collins 1900b, p. 45, 1909b, p. 185; Hazen 1902, p. 153; Whelden 1947, p. 47.

CHAETOPHORACEAE

Plants filamentous, erect from a basal holdfast, or decumbent, typically branched, generally uniseriate, but in some genera forming disks by lateral approximation of the filaments, or by subparenchymatous divisions; hairs frequently present, as terminal or lateral hyaline extensions of one or more cells; cells with thin to gelatinous walls, single nuclei and single parietal more or less dissected platelike or annular chromatophores generally with pyrenoids; asexual reproduction by formation of bi- or quadriflagellate zoöspores, often of different sizes in the same species, by aplanospores, or by akinetes; sexual reproduction iso- aniso- or oögamous.

KEY TO GENERA

1. Primary filaments creeping, producing distinct erect filaments PILINIA, p. 48
1. No distinctive erect filaments.............................. 2
2. Filaments more or less laterally united...................... 3
2. Filaments not laterally united.............................. 6
3. Forming an irregular loosely organized crust PSEUDENDOCLONIUM, p. 55
3. Forming a definite disk.................................... 4
4. Hairs more or less abundant.............. OCHLOCHAETE, p. 56
4. Hairs generally absent..................................... 5
5. Disk small, epiphytic; central cells enlarged to form sporangia PRINGSHEIMIELLA, p. 57
5. Disk growing into a wide thin coating on stones; filaments evident at the margin only, the central cells not enlarged PROTODERMA, p. 57

6. Hairs present; epiphytic, external or essentially interwoven in loose tissues of plants 7
6. Hairs absent; with plant or animal membranes 10
7. Hairs formed from ordinary vegetative cells 8
7. Hairs formed from small intercalary cells ..... BOLBOCOLEON, p. 52
8. Hairs from the ends of upright, usually lateral, vegetative cells ACROCHAETE, p. 51
8. Hairs from the sides of intercalary vegetative cells 9
9. Hairs straight .... ECTOCHAETE, p. 52
9. Hairs more or less spirally twisted .... PHAEOPHILA, p. 50
10. In the cell walls of various algae and the chitin of sertularians ENTOCLADIA, p. 53
10. In the outer coating on the shells of live mollusks .. TELLAMIA, p. 58
[10. In the calcareous shells of living or dead mollusks GOMONTIA, p. 59]

Pilinia Kützing, 1843

Plants of branched creeping filaments which bear simple or forked erect filaments ending in multicellular hairs; cells showing a chromatophore without pyrenoids; reproduction by numerous biflagellate zoöspores formed in terminal or lateral rounded to clavate sporangia.

KEY TO SPECIES

1. Endophytic .... P. endophytica, p. 50
1. Not endophytic 2
2. Erect filaments short and densely packed 3
2. Erect filaments longer 4
3. Dark green; filaments 8–12 μ diam., on live shells .. P. lunatiae, p. 48
3. Yellowish green; filaments 2–5 μ diam., on pebbles ... P. minor, p. 49
4. Forming a rather firm, spongy coating on woodwork, shells, or stones .... P. rimosa, p. 49
4. Forming a thin, soft coating 5
5. On woodwork; sporangia on the basal layer, rarely lateral on erect filaments .... P. morsei, p. 50
5. On shells, pebbles, etc.; sporangia on the erect filaments P. reinschii, p. 49

Pilinia lunatiae Collins

Plants creeping, deep green, the basal filaments crowded, soon subparenchymatously united, bearing densely branched, crowded erect filaments which increase in size toward the top; cells of the basal layer irregular, up to 15 μ diam., the erect filaments 5–6 (–10) cells long, 8–12 μ diam., the cells variable in size and shape; sporangia formed from the terminal cells of the erect filaments with little change.

Northern Massachusetts; in the living shells of Lunatia (and on stones?); spring and early summer.

REFERENCES: Phyc. Bor.-Am. 162 (as *Acroblaste reinschii non* (Wille) Collins). Farlow 1881, p. 57 (as *Acroblaste sp.*); Collins 1908a, p. 123, 1909b, p. 292.

Pilinia minor Hansgirg

Plants forming a thin membranous layer, yellow-green, the basal filaments densely crowded and hardly distinguishable, bearing sparingly branched erect filaments which are about 2 μ diam. at the base, to 7μ at the top; sporangia terminal, pyriform or irregular, 10–12 μ diam., 22–24 μ long.

Southern Massachusetts; on pebbles below a fresh-water spring, covered with sea water at high tide.

REFERENCES: Collins 1908a, p. 124, 1909b, p. 292.

Pilinia reinschii (Wille) Collins

Plants forming a soft coating, yellowish to olive, the basal filaments uniseriate, distinct throughout the plant and more or less disunited, bearing cylindrical or slightly constricted erect filaments to 500 μ long; cells of the basal filaments 5–8 μ diam., about two diameters long; sporangia terminal or apparently lateral on the erect filaments, ovoid, 16–20 μ diam., 20–25 μ long.

Southern Massachusetts; growing on stones and shells in quiet water.

REFERENCES: Phyc. Bor.-Am. 1685. Collins 1900b, p. 43 (as *A. reinschii*), 1908a, p. 125, 1909b, p. 293.

Pilinia rimosa Kützing

Plants forming a rather firm yet slippery, spongy olive-green coating; basal layer of somewhat irregular filaments, more or less contracted at the septa; erect filaments simple or branched, to

0.6–2.0 mm. tall, 5–10 (–19) μ diam., cells 1–2 diameters long, somewhat cask-shaped or cylindrical, with a lamellose membrane; sporangia 16–20 μ diam.

Maine; growing on wooden pilings, shells, and stones.

REFERENCES: Phyc. Bor.-Am. 971. Collins 1909b, p. 293.

Pilinia morsei Collins

Plants forming a soft coating, the basal creeping filaments irregularly contorted, more or less united, their cells rounded, often longitudinally divided and the basal layer by consequence subparenchymatous and of two or more layers, bearing erect cylindrical or slightly moniliform filaments to 2 mm. long; cells of the basal layer 8–15 μ diam., of the erect filaments 7–11 μ diam., 1–2 diameters long; sporangia borne on the basal layer, sessile or on few-celled pedicels, ovoid or pyriform.

New Jersey; growing on woodwork.

REFERENCES: Phyc. Bor.-Am. 2286. Collins 1908a, p. 126, 1909b, p. 293.

Pilinia endophytica Collins

Plants endophytic, creeping between the host cells, the basal filaments short, simple, or branched, 7–22 μ diam.; cells variable in form and size, often irregular, 1–5 diameters long, the light green chromatophore cup-shaped at the upper end of the cell, or sometimes filling it; sporangia terminal on the branchlets, spherical or ovoid, to 30 μ diam.

Connecticut and Maine; in the tissues of *Ralfsia borneti.*

REFERENCES: Phyc. Bor.-Am. 1838. Collins 1908c, p. 156, 1909b, p. 292.

Phaeophila Hauck, 1876

Plants epi- or endophytic, of branching uniseriate filaments the cells of which bear one to three long, unseptate hairs neither separated by a wall from the supporting cell nor swollen at the base; cells with lobed parietal chromatophores and several pyrenoids; reproduction by quadriflagellate zoöspores produced in sporangia which are little enlarged but often terminal on branchlets.

KEY TO SPECIES

1. In or on the surface of plant cell membranes; the branching habit irregular, often congested.................... P. dendroides
1. Near the surface of the shells of mollusks or the tubes of annelids .... P. engleri

Phaeophila dendroides (Crouan) Batters Pl. 2, figs. 4–5

Plants frequently and widely branched, epi- or endophytic, the cells 9–40 μ diam., 15–50 (–80) μ long, generally cylindrical, partly somewhat irregular, containing a parietal lobed and interrupted chromatophore with several pyrenoids; hairs 1 (–3) on each cell, the bases often somewhat spirally twisted; zoösporangia subcylindrical to irregularly swollen, intercalary or terminal on short branches, 16–40 μ diam., 30–85 μ long.

Florida, southern Massachusetts; on filaments of Callithamnion and Polysiphonia; summer.

REFERENCES: Collins 1918a, p. 73; Taylor 1928a, p. 58, 1937c, p. 51 (all as *P. floridearum*).

Phaeophila engleri Reinke

Plants freely branched, the branches near the colony center sometimes partly fused, peripherally free and spreading; cells cylindrical or a little lobed, 5–22 μ diam., 1–5 diameters long, or isodiametric and to 50 μ diam.; setae few or absent, straight or somewhat sinuous; sporangia usually intercalary, flask-shaped or irregular, 10–28 μ long, to one-half taller, with a narrow neck; zoöspores rarely replaced by aplanospores.

Southern Massachusetts; giving a green color to the shells of mollusks, Spirorbis tubes, etc., during the summer.

REFERENCE: Thivy 1943, p. 244.

Acrochaete Pringsheim, 1862

Plants creeping as epiphytes or endophytes, filamentous, irregularly branched, with short erect branches which may terminate in a delicate bristle; cell structure showing a platelike chromatophore with one to several pyrenoids; asexual reproduction by biflagellate zoöspores which are generally formed in terminal non-piliferous

97378

cells of the erect branches and are larger than the biflagellate isogametes, which are generally formed from lower cells of the plants.

Acrochaete repens Pringsheim

Filaments repent; branches short, straight; vegetative cells 7–9 μ diam., 2–6 diameters long, usually with several pyrenoids; sporangia elongate-ovoid, 8–12 μ diam., 20–40 μ long; hairs sometimes abundant, but sometimes very rare.

Southern Massachusetts; in the outer cortex of Chorda, Laminaria, and similar plants.

REFERENCES: Phyc. Bor.-Am. 1279. Collins 1906b, p. 124, 1909b, p. 282.

Bolbocoleon Pringsheim, 1862

Plants creeping as epiphytes or endophytes, filamentous, branched, but without erect axes; vegetative cells irregularly rounded, usually much swollen or even prolonged on the upper side; hair-bearing cells distinctively smaller, conical, bearing long hairs bulbous at the base; chromatophores in the vegetative cells irregular, parietal, perforate, with five to ten pyrenoids; reproduction by the formation of numerous biflagellate zoöids in any non-piliferous cell.

Bolbocoleon piliferum Pringsheim

Plants as in the generic description; vegetative cells 12–16 μ diam., 2–4 diameters long.

Long Island [1] to Maine, and Newfoundland; in the outer portions of soft marine algae; summer.

REFERENCES: Phyc. Bor.-Am. 1225. Farlow 1881, p. 57; Collins 1900b, p. 43, 1909b, p. 283; Hazen 1902, p. 227.

Ectochaete (Huber) Wille, 1909

Thallus microscopic, of endophytic ramifying filaments, branching irregularly alternate, cells uninucleate, often bearing a long seta which is not separated by a cross wall from the supporting cell; chloroplast parietal, incompletely encircling the protoplast, with

[1] See footnote on p. 46.

one or more pyrenoids; reproduction by zoöspores or by isogametes in sporangia or gametangia formed by enlargement of ordinary filament cells and development of necks through which the swarmers escape.

Ectochaete taylori Thivy

Thalli 0.3–0.65 mm. diam., formed in the jelly between the cortical host filaments; plants of interwoven, sometimes fused terete filaments of cells 8–18 μ diam., 1–5 diameters long; setae conspicuous, long and straight, usually 2.66 μ diam., to 0.8 mm. long; gametangia irregularly saccate, as wide as the vegetative cells and 1–3 diameters long, discharging through a conspicuous tubular neck.

Southern Massachusetts; growing in the jelly of Nemalion, Leathesia, and Sphaerotrichia.

REFERENCE: Thivy 1942, p. 97.

Entocladia Reinke, 1879

Plants forming small patches in the membranes of the supporting host, the filaments spreading, branched, in the center of the disk tending to become subparenchymatously congested; hairs and setae absent; cell structure showing a single, simple platelike chromatophore with one to several pyrenoids and a single nucleus; asexual reproduction by the formation of quadriflagellate zoöspores; sexual reproduction by formation of biflagellate isogametes.

KEY TO SPECIES

1. In plant cell walls 2
1. In animal tissues 4

2. In leaves of Zostera; cells 3–5 μ diam. .... E. perforans, p. 54
2. In marine algae; somewhat larger 3

3. Cells averaging 6 μ diam., irregular in form .... E. viridis, p. 54
3. Cells averaging 9 μ diam., generally cylindrical ... E. wittrockii, p. 54

4. In the chitin of sertularians; cells usually over 5 μ diam. E. flustrae, p. 55
4. In mollusk shells; cells except in the center of the colony usually less than 5 μ diam. .... E. testarum, p. 55

NAZARETH COLLEGE LIBRARY

Entocladia perforans (Huber) Levring

Plants endophytic, the filaments mostly 3–5 μ diam., but more or less irregular in form and the cells of varying length and width, the slender filaments among the epidermal cells of the host, the larger between the deeper tissues, where the cells reach 14 μ diam.; chromatophore encircling the cell, with a pyrenoid; reproduction by ovoid or subspherical quadriciliate zoöspores, eight formed in each of the larger sporangial cells.

Massachusetts to Maine; in dead or decaying leaves of Zostera, particularly in marsh pools; early autumn.

REFERENCES: Phyc. Bor.-Am. 1625. Collins 1909b, p. 279 (as *Endoderma perforans*).

Entocladia viridis Reinke Pl. 2, figs. 1–2

Plants endophytic, filamentous, branching arborescent, the filaments 3 to generally 6 μ diam.; cells variable in length and width, about 1–6 diameters long, sometimes cylindrical, oftener irregularly swollen and distorted, those terminating the branches blunt or tapering.

Bermuda, Florida, North Carolina, and Massachusetts; in the cell walls of various algae, particularly Rhodophyceae; early autumn.

REFERENCES: Phyc. Bor.-Am. 1626. Collins 1906b, p. 123, 1909b, p. 279 (as *Endoderma viride*).

Entocladia wittrockii Wille

Plants filamentous, the filaments simple or irregularly branched, with tapering ends and branches sometimes laterally united; cells nearly cylindrical, 5–10 μ diam., and 1.0–1.5 diameters long; cells at the tips of the branches about 6 μ diam., to 26 μ long.

Connecticut to Maine; in the cell walls of various marine algae, particularly Phaeophyceae; common in summer and autumn.

REFERENCES: Phyc. Bor.-Am. 265, 1469. Collins 1891, p. 340, 1900b, p. 44, 1909b, p. 279 (all as *Endoderma wittrockii*); Hazen 1902, p. 226.

Entocladia flustrae (Reinke) Batters

Filaments irregularly branched, the cells short-cylindrical or irregular, 5–10 μ diam., forming a disk which may eventually develop a congested pseudoparenchymatous central portion, where the cells are polygonal and 7–12 μ diam.; chromatophores single parietal plates each with a pyrenoid.

New York to Maine; growing in the chitinous membranes of sertularians, which become green; a plant of spring and summer.

REFERENCES: Phyc. Bor.-Am. 160. Collins 1900b, p. 44 (as *Epicladia flustrae*), 1909b, p. 287.

Entocladia testarum Kylin

Filaments interlaced, near the center of the colony the cells often fused, oval or irregular to spherical when to 10 μ diam., but peripherally of a single layer of free filaments, the cells about 3.5 μ diam., 3–8 diameters long; usually with 2 or 3 pyrenoids; sporangial cells cylindrical, clavate or rounded, 7–13 μ diam., with a pronounced neck.

Southern Massachusetts; in dead mollusk shells in a salt marsh during the summer.

REFERENCE: Thivy 1943, p. 259.

Pseudendoclonium Wille, 1900

Plants forming an irregular crust of short and closely branched filaments, some of which are rhizoidal in character; suberect filaments short, irregularly branched, the branches arising from the middle of the supporting cells; colorless hairs absent; cell structure showing a lateral platelike chromatophore with a single pyrenoid; asexual reproduction by zoöspores with four flagella; akinetes produced from liberated branch cells, and may become thick-walled resting cells enclosed in jelly.

Pseudendoclonium submarinum Wille

Plants forming a superficial or somewhat penetrating layer; granular in appearance, subparenchymatous in structure, distinctly filamentous in essential arrangement; cells 6–7 μ diam.

Rhode Island to Maine; on woodwork near high-water mark; summer.

REFERENCES: Phyc. Bor.-Am. 1124. Collins 1909b, p. 284.

Ochlochaete Thwaites in Harvey, 1849

Plants forming more or less irregular epiphytic disks which are monostromatic at the margin, with more or less alternately branched filaments, but by superposition becoming several cell layers thick in the center, where the cells extend outward into unseptate hairs without swollen bases; cell structure showing one irregularly lobed parietal chromatophore with one to three pyrenoids; asexual reproduction by zoöspores borne in enlarged, projecting sporangial cells near the center of the disk, numerous quadriflagellate zoöspores being formed in each.

KEY TO SPECIES

1. Epiphytic on marine plants; cell diameter of peripheral filaments 3.5–7.5 μ, or more in the center of the colony...... O. ferox
1. Growing on stones and shells; cell diameter 5–12 μ in peripheral filaments .... O. lentiformis

Ochlochaete ferox Huber

Plants of filaments radiating from a center to form a disk, with the strands more or less closely united and branching laterally, in parts often two layers thick; cells rounded or angular, showing a parietal chromatophore with one pyrenoid; setae often numerous.

Southern Massachusetts; on Zostera in warm shallow water; summer.

REFERENCES: Phyc. Bor.-Am. 1521. Rosenvinge 1893, p. 931; Collins 1908c, p. 157, 1909b, p. 285.

Ochlochaete lentiformis Huber

Plants forming lentiform colonies about 50–800 μ diam., filamentous, pulvinate, the filaments in the center more or less fused to a firm pluristromatic tissue, marginally the filaments a little free; cell size generally 5–13 μ, or occasionally to 23 μ, near the colony margin often somewhat elongate; setae present, not numerous; sporangia with a short neck; chromatophores with 1–3 pyrenoids.

Southern Massachusetts; on dead shells and stones, in shallow water; during the summer.

REFERENCE: Thivy 1943, p. 262.

Pringsheimiella v. Hoehnel, 1920

Plants small monostromatic epiphytic disks with marginal growth, the central cells somewhat taller than those at the edge, which are radially elongate; cells in young plants sometimes bearing long colorless hairs; protoplast showing a large platelike chromatophore with a single pyrenoid; asexual plants compact, with external walls somewhat gelatinously thickened, the cells near the center of the disk producing numerous quadriflagellate zoöspores; sexual individuals more diffuse, with intercellular spaces, walls of equal thickness, the cells near the center of the disk producing many small quadriflagellate gametes.

Pringsheimiella scutata (Reinke) Schmidt et Petrak

Disklike plants 1–2 mm. diam.; cells variable in size and shape, to 12 μ diam., the marginal ones radially elongate, the central distinctly taller than broad; zoösporangia oval to subpyriform, 15–22 μ diam., 28–38 μ high; zoöspores 15 μ diam., gametes 4 μ diam.

Bermuda, Connecticut to Maine; growing on Zostera and on various algae; summer.

REFERENCES: Phyc. Bor.-Am. 1524. Collins 1891, p. 340, 1900b, p. 45, 1908c, p. 157, 1909b, p. 288 (all as *Pringsheimia scutata*).

Protoderma Kützing, 1843, *emend.* Borzi, 1895

Plants forming small patches on the substratum, one cell layer in thickness, most of the disk appearing quite parenchymatous, the margin showing its filamentous character in short radiating series of little-differentiated cells; hairs and setae absent; cell structure showing a single simple platelike chromatophore with one pyrenoid; asexual reproduction by biflagellate zoöspores produced four to eight in a sporangial cell, or by aplanospores.

Protoderma marinum Reinke

Plants forming a considerable, thin, light green layer; structure subparenchymatous, the angular cells 6–12 μ diam., irregular in shape and arrangement except at the margin, where they are in rather indistinct, radiating series.

Bermuda, Florida, Connecticut to Maine; growing on stones and dead shells in tide pools and near low-water mark.

REFERENCES: Phyc. Bor.-Am. LIII. Collins 1900b, p. 45, 1909b, p. 217; Taylor 1928a, p. 59.

Tellamia Batters, 1895

Plants endozoic; branching irregular, mostly creeping, often congested, pseudoparenchymatous, with short, often irregular cells; scattered erect branches of few cells present, the tips pointed; hairs absent; cell structure showing a single parietal chromatophore with no pyrenoid; asexual reproduction by numerous zoöspores produced in sporangia.

Tellamia contorta Batters

Plants yellowish green or brown, the basal layer of densely and irregularly branched filaments, which are sometimes falcate or coiled, and often anastomosed, bearing erect filaments or short branches often united laterally, the tips acute; cells 6–9 μ diam., 3–10 μ long, ovoid or ellipsoid, or in the horizontal branches sometimes inflated to 20 μ diam.

Southern Massachusetts and Maine; growing in the periostracum, not the firm shell, of Littorina.

REFERENCE: Collins 1909b, p. 280.

CHAETOPELTIDACEAE

Plants unicellular or more or less filamentous, seldom aggregated to form a disk, the cells usually bearing one or more prominent simple or basally sheathed setae; cells with one nucleus and usually one parietal chromatophore, with one to two pyrenoids.

Diplochaete Collins, 1901

Cells solitary or in vague gelatinous filaments, globose or nearly so, bearing two or more long sheathless setae; chromatophore single, parietal; reproduction unknown.

Diplochaete solitaria Collins

Plants epiphytic, cells 25–30 μ in diameter, the walls 5–8 μ thick, hardly lamellose; setae two, arising from near the lower margin of the cell, tapering from a base which may be 4–6 μ thick.

From the tropics; southern Massachusetts, on *Polysiphonia.*

REFERENCES: Collins 1909b, p. 227; Thivy 1943, p. 263.

GOMONTIACEAE

Plants filamentous, penetrating shells, the filaments branched, septate, the cells with parietal chromatophores, sometimes multinucleate; sporangia formed from cells near the matrix surface, becoming detached from the vegetative filaments, large, embedded, with rhizoidal projections extending toward the outside.

Gomontia Bornet et Flahault, 1888

Plants perforating in shells or wood, usually reported as of creeping, branched filaments, the cells irregular, often crowded, or, where more deeply penetrating, more regular and slender; protoplast with a parietal lobed or reticulate chromatophore and one to several nuclei; asexual reproduction by biflagellate (or quadriflagellate?) zoöspores formed in sporangia which are developed near the surface of the substrate, and which enlarge, developing thick walls with specially indurated projections reaching from the surface.

Gomontia polyrhiza (Lagerheim) Bornet et Flahault

Pl. 1, figs. 13–14; pl. 7, fig. 4

Plants coloring the inhabited shell grass-green, the filaments widely and irregularly branching, 4–8 μ diam.; sporangia 30–125 μ diam., 150–250 μ long; nucleus single in vegetative cells; zoöspores of two sorts, one 10–12 μ long by 5–6 μ diam., the other about 3.5 μ diam., 5 μ long; aplanospores 4 μ diam.

North Carolina, Connecticut to Nova Scotia, Ellesmere Island and Devon Island; growing in old dead shells along the shore, and occasionally in live Spirorbis shells and barnacles, sometimes mixed with Mastigocoleus and Hyella.

REFERENCES: Phyc. Bor.-Am. 315. Collins 1900b, p. 44, 1909b, p. 370; Rosenvinge 1926, p. 16; Taylor 1928a, p. 58; Kylin 1935, p. 3.

From culture experiments Kylin (1935, p. 3) concludes that the filamentous stages usually found with these sporangia belong to other plants, a view which, if confirmed, would justify his association of this plant with Codiolum. The illustrations given (*op. cit.*) by Bornet and Flahault, as well as common observations, certainly suggest that under some circumstances filamentous phases are produced.

ULVACEAE

Plants capillary to broad, usually tubular or membranaceous, occasionally reduced to one or two rows of cells; attached or becoming free-floating; forming tubes or sheets one or two cells in thickness, the cells showing one or two large lateral chromatophores with, as a rule, single pyrenoids and a single nucleus; sexual and asexual plants morphologically indistinguishable; reproductive cells unaltered or slightly enlarged; asexual reproduction by bi- or quadriflagellate zoöspores which germinate directly; sexual reproduction by biflagellate gametes.

KEY TO GENERA

1. Adult plants tubular, at least in part 2
1. Adult plants not tubular 4

2. Cells closely placed .... ENTEROMORPHA, p. 61
2. Lateral cell walls thick; protoplasts markedly separated 3

3. Cells in longitudinal rows; walls lamellose .... CAPSOSIPHON, p. 69
3. Cells not in rows; walls not lamellose .... MONOSTROMA, p. 70

4. Plants broadly membranous 5
4. Plants narrowly filamentous 6

5. Membrane of two united cell layers .... ULVA, p. 73
5. Membrane one cell thick .... MONOSTROMA, p. 70

6. Filaments not of consistent width .... ENTEROMORPHA, p. 61
6. Filaments generally biseriate .... PERCURSARIA, p. 61

Percursaria Bory, 1828

Plant filamentous, the filaments slender, generally of two rows of cells more or less symmetrically placed side by side; cells rectangular, the chromatophores single parietal plates.

Percursaria percursa (C. Agardh) J. Agardh Pl. 1, fig. 8

Plants infrequently proliferating, the filaments mostly simple, to several centimeters long, flexuose and twisted, light green; filaments of 1–4, most generally of two rows of cells, each 10–15 μ wide, the cells to 28 μ long, walls rather thick.

New Jersey to the lower St. Lawrence; forming floating masses with other algae in brackish salt-marsh pools, or stranded on rocks.

REFERENCES: Algae Exsic. Am. Bor. 219 (as *Tetranema percursum*); Phyc. Bor.-Am. 469. Collins 1884b, p. 131 (as *Ulva percursa*), 1900b, p. 44, 1903a, p. 26, 1909b, p. 197 (as *Enteromorpha percursa*).

Enteromorpha Link, 1820 [1]

Plants capillary to ample, simple or chiefly alternately branched, tubular or with the branches showing uniseriate filamentous tips; at first attached, sometimes later free-floating; adult holdfasts formed by downgrowths from cells of the stalklike portion; cells usually close-placed; chloroplast single, a lateral plate more or less covering the outer or lateral face of the cell, generally with one pyrenoid; nucleus solitary; reproduction by zoöspores and gametes produced on similar plants.

KEY TO SPECIES

1. Plants of non-tubular branched strands......... E. cruciata, p. 62
1. Plants forming flat blades, minutely tubular only at the margins and near the base........................ E. linza, p. 68
1. Plants in most parts tubular................................. 2

[1] Recent studies seem to show that little reliance can be placed on the forms of the plants in this genus for interspecific distinctions. The work of Bliding and others is summarized for several of our species by Kylin (1949, p. 19). Since it is not clear to the writer how these data can be adapted to the needs of a systematic flora he is not adopting them completely, hoping that with time and confirmatory studies a practicable morphological system will appear.

2. Plants simple; infrequently and sparingly proliferous............ 3
2. Plants clearly branched.. 6

3. Cells in the lower part of the plant in groups of 2–4, above scattered; plants subfiliform-cylindrical.... E. groenlandica, p. 69
3. Cells marginally in longitudinal rows; plant subfiliform-compressed .... E. marginata, p. 65
3. Cells irregularly placed; plants broader....................... 4

4. Plants relatively large; cells in surface view 9–15 μ diam. E. intestinalis, p. 66
4. Plants usually small, less than 1 dm. tall; cells in surface view 7 μ diam. or less.. 5

5. Membrane less than 15 μ thick; cuticle not specially thickened E. minima, p. 67
5. Membrane more than 15 μ thick; cuticle usually thicker on the inside of the membrane.................. E. micrococca, p. 67

6. Branches relatively few, similar to the main axis................ 7
6. Branches abundant, tapering from base to apex.................. 8

7. Branches restricted toward the base of the plant; cells irregularly placed .... E. compressa, p. 64
7. Branches scattered; cells in the younger parts always in longitudinal rows .... E. prolifera, p. 65

8. Branches ending in long uniseriate tips......... E. plumosa, p. 63
8. Branches not having long uniseriate tips....................... 9

9. Branches biseriate, sharply distinct from the entangled pluriseriate main axes............................ E. torta, p. 63
9. Branches not thus delimited................................ 10

10. Main axes bearing slender erect pluriseriate branches, not regularly repeatedly forked....................... E. erecta, p. 64
10. Branches repeatedly redivided................ E. clathrata, p. 63

Enteromorpha cruciata Collins

Plants filiform, branching, mostly of a single series of cells, but at the point of branching often of two or more series; branches issuing at right angles or nearly so, usually opposite but also irregular, simple, usually short and tapering; monosiphonous portions 20–30 μ diam., the cells about as long as broad and with a thick wall; in the irregular portions where several branches issue close together the cells are more rounded and may reach a diameter of 50 μ.

Maryland and Maine; in floating masses with other algae; summer.

REFERENCES: Phyc. Bor.-Am. 222. Collins 1896a, p. 3, 1900b, p. 44, 1903a, p. 27, 1909b, p. 198.

Enteromorpha torta (Mertens) Reinbold

Plants filiform, loosely entangled with other algae, the axes usually simple or sparingly divided, of rectangular cells usually in both longitudinal and transverse series, several cells broad, bearing long proliferous branches of but two rows of cells, the cells to 10 μ diam.

Maine; growing in a lagoon entangled with other algae.

REFERENCES: Collins 1900b, p. 44, 1903a, p. 26, 1909b, p. 198, 1912, p. 82.

Enteromorpha plumosa Kützing

Solitary or scattered, attached, 1–3 dm. tall, repeatedly, delicately, irregularly or sometimes in part oppositely branched, nearly capillary throughout, the branches cylindrical, soft, usually with long uniseriate tips; cells subrectangular, in the tips about 8–12 μ diam., below in longitudinal and often also in transverse series, 12–20 μ wide, 18–26 (–40) μ long, with chromatophores not covering the whole cell face.

Bermuda, Florida, New Jersey to Maine, Nova Scotia, and James Bay; growing on shells and stones, or coarse algae, in relatively quiet water; summer.

REFERENCES: Phyc. Bor.-Am. 463 (as *E. hopkirkii*). Harvey 1858a, p. 58; Farlow 1881, p. 44 (as *Ulva hopkirkii*); Collins 1900b, p. 44, 1903a, p. 27 (both as *E. hopkirkii*), 1909b, p. 198; Taylor 1928a, p. 56.

Enteromorpha clathrata (Roth) J. Agardh Pl. 3, fig. 1

Plants at first attached, later free-floating, light green, to about 4 dm. long; slenderly cylindrical and repeatedly branched, the main divisions reaching 0.5–2.5 mm. diam., the upper virgate, all tending to taper from the base but not ending in a prolonged single series of cells; cells irregularly rectangular, 10–28 μ diam., 13–38 μ long, in distinct longitudinal series, or somewhat irregularly disposed below in old plants.

Bermuda, New Jersey to Nova Scotia, Ile Miquelon, James Bay, Baffin Island, and Ellesmere Island; when young attached to rocks, woodwork, coarse algae or Zostera, later forming floating masses in warm and quiet and often brackish water; summer. The more delicate forms of this plant, especially those from brackish water, have generally been called *E. crinita* (Roth) J. Agardh, but it seems impossible to recognize two distinct groups of plants here.

REFERENCES: Algae Exsic. Am. Bor. 128; Phyc. Bor.-Am. LXXVIII; 460 (as *E. crinita*). Harvey 1858a, p. 57; Farlow 1881, p. 44; Hariot 1889, p. 155 (all as *Ulva clathrata*); Kjellman 1883, p. 287; Collins 1900b, p. 44, 1903a, p. 28, 1909b, p. 199; Collins 1896a, p. 2, 1900b, p. 44, 1903a, p. 27, 1909b, p. 199 (as *E. crinita*).

Enteromorpha erecta (Lyngbye) J. Agardh

Plants attached, soft and very slender, 1–2 dm. tall; the elongate axis distinct, cylindrical, tubular, laterally beset with many smaller erect branches; cells in the smaller branches often in transverse series, and in the whole plant in fairly distinct longitudinal rows, subrectangular or somewhat rounded, the radial walls rather heavy; the chromatophore occupying most of the face of the cell, which in the main axis may be 13–24 μ diam., 15–28 (–38) μ long; in section the cells are subequal or taller than broad, the whole membrane 18–28 μ thick.

Long Island to New Brunswick, and the lower St. Lawrence; growing on rocks, woodwork, and fuci in relatively exposed situations.

REFERENCES: Phyc. Bor.-Am. 461. Collins 1900b, p. 44, 1903a, p. 28, 1909b, p. 200.

Enteromorpha compressa (Linnaeus) Greville Pl. 3, fig. 3

Plants generally gregarious, attached, bright to dark green; to 3 dm. tall, tubular, more or less compressed or collapsed, below long tapering, and characteristically with several branches from the gradually contracted stalklike base which are entirely similar to the principal blade; above expanded, 2–20 mm. wide; cells small, 10–15 μ diam., rounded-subquadrate, in the adult plants irregularly placed, the walls not thickened; in section the cells vertically elongate, the whole membrane 13–20 μ thick.

New Jersey to the lower St. Lawrence, Hudson Bay, Baffin Island, Devon Island, and elsewhere in the American Arctic; a plant of rocks and woodwork in moderately exposed or somewhat protected situations.

REFERENCES: Harvey 1858a, p. 57; Farlow 1881, p. 43 (as *Ulva enteromorpha* v. *compressa*); Collins 1900b, p. 44, 1903a, p. 25; 1909b, p. 201.

Within our range there is also reported f. **subsimplex** J. Agardh (Phyc. Bor.-Am. 964): this shows the frond slenderly tapering above the stalk, linear or barely expanded at the tip; it is reported from Maine.

Enteromorpha marginata J. Agardh

Plants simple or a little proliferously branched, gregarious, very slender, compressed, rich green; length 2–3 cm., width usually 15–20 cells, the whole blade 12–100 μ thick, several blades often stranded together; cells in longitudinal series, particularly along the margins, with thick walls, rounded-subquadrate, 4–8 μ diam., the chromatophore covering the face of the cell.

Bermuda, New Jersey, Connecticut to Maine, and in the arctic at Resolution Island; growing on the stems and roots of Spartina and other objects in similar situations.

REFERENCES: Phyc. Bor.-Am. 466. Collins 1884b, p. 131, 1888a, p. 310 (as *Ulva marginata*), 1900b, p. 44, 1903a, p. 25, 1909b, p. 202.

Enteromorpha prolifera (Müller) J. Agardh Pl. 3, fig. 2

Plants solitary or tufted, generally remaining attached, to 6 dm. tall, more or less abundantly proliferously branched, with occasional branches to the second degree; 1.0–1.5 cm. diam., the membrane 15–18 μ thick, but firm; cells rounded-subangular, 10–19 μ diam., in the younger parts always in longitudinal series.

Bermuda, South Carolina to Prince Edward Island, the Gaspé,[1] and James Bay; on rocks and woodwork in sheltered localities near low-tide line; through the summer.

REFERENCES: Phyc. Bor.-Am. 470 (*p.p.*). Collins 1900b, p. 44, 1903a, p. 21, 1909b, p. 202; Taylor 1928a, p. 56.

Two minor varietal designations have been applied to plants from this area. V. **flexuosa** (Wulfen) Doty (1948, p. 259) shows simple thalli to 2 cm. tall, the cells rectangular, about 11 μ diam., with very thin walls. V. **tubulosa** (Kützing) Reinbold (Phyc. Bor.-Am. 471; Collins 1909b, p. 203) shows plants to 3 dm. tall, exceedingly slender, subsimple, the cells 8–16 μ diam. Proliferous *E. intestinalis* may be confused with this species unless examined microscopically.

Enteromorpha intestinalis (Linnaeus) Link

Pl. 3, fig. 7; pl. 4, figs. 4–5

Solitary or gregarious, at first attached, often becoming free-floating, subintestiniform, yellowish green; 1–20 dm. or even more in length, 1 mm. to 10 cm. wide; membranous; frond tapering below, above the stalk tubular, cylindrical, clavate, or generally inflated and bullate; simple or rarely branched from the very base, or proliferous; cells rounded-polyhedral, not arranged in any order in the inflated portion, in surface 9–15 μ diam., the whole membrane 20–40 μ thick, the cells in section vertically rounded-oblong, about 14–17 μ deep, generally with the wall rather thicker on the inner side.

Bermuda, Florida, North Carolina to the lower St. Lawrence, Ile St. Pierre, Hudson Bay, James Bay, Baffin Island, Devon Island, and Ellesmere Island; attached to shells, stones, and woodwork in the lower intertidal zone and a little below low-tide line, or in tide pools, in somewhat protected places; through most of the year.

REFERENCES: Phyc. Bor.-Am. 464. Harvey 1858a, p. 57; Farlow 1881, p. 43 (as *Ulva enteromorpha* v. *intestinalis*); Collins 1900b, p. 44, 1903a, p. 23, 1909b, p. 204 (with forms); Le Gallo 1947, p. 301.

This species is quite variable and, although common, is frequently reported by error for other species in the genus. F. **clavata** J. Agardh (Pl. 3, fig. 4; Phyc. Bor.-Am. 966), with a frond which

[1] See footnote on p. 46.

becomes claviform-dilated and irregularly swollen, has a length of 1–20 dm., diam. 0.5–10.0 cm. or more. F. **cylindracea** J. Agardh (Pl. 3, figs. 5–6; Phyc. Bor.-Am. 465) has a frond which above the stalk portion is throughout cylindrical, about 3–10 mm. diam., and very long. In f. **maxima** J. Agardh the plants when fully developed are floating, and irregularly inflated, reaching extreme dimensions. All are reported within our range, and some forms may occur in brackish or quite fresh water.

Enteromorpha minima Nägeli

Plants small, gregarious, attached, the base a minute disk of crowded, branched filaments, the blades 1–10 cm. tall, simple or slightly proliferously branched, yellowish green and rather soft; blade linear, dilated sharply above the stalk, width to 1–2 (–5) mm., tubular or compressed, the membrane 8–10 μ thick or somewhat more, cells 5–7 μ diam., angular, arranged in no definite order, the walls rather thin; in section the cells appearing nearly cubical, the walls on the inner and the outer faces about equally thick.

Bermuda, Connecticut, southern Massachusetts, New Brunswick, and Newfoundland; growing on rocks and shells in the lower part of the littoral zone; appearing during the summer.

REFERENCES: Phyc. Bor.-Am. 468. Collins 1896b, p. 458, 1900b, p. 44, 1903a, p. 24, 1909b, p. 201; Kylin 1949, p. 30 (as *Blidingia minima*).

Within our range there is also reported f. **glacialis** Kjellm. (Phyc. Bor.-Am. 1183; Kjellman 1883, p. 291; Collins 1909b, p. 201); it grows in numerous close tufts to form an expanded layer, the membrane 9–15 μ thick, the cells 5–8 μ diam.; this form was found in northern Massachusetts, New Brunswick, Nova Scotia, and Prince Edward Island, especially favoring sites on rocks covered only at high tide by salt water, being wet with fresh water from above at low-tide periods.

Enteromorpha micrococca Kützing

Plants small, gregarious, attached, 1–5 cm. tall, simple or slightly proliferous, rather firm and bright green; dilated sharply above the stalk, tubular or compressed, becoming much twisted, the membrane

15–20 (–30) μ thick, cells 3–7 μ diam., angular, in no definite order, with the walls particularly thick, especially on the inner side of the membrane; chromatophore rather smaller than the face of the cell.

Northern Massachusetts to New Brunswick and Hudson Bay; on rocks, such as cliff faces, and particularly where the plant is moistened by the drip of fresh water from above during low tidal periods.

REFERENCES: Phyc. Bor.-Am. 66. Collins 1891, p. 336, 1900b, p. 44, 1903a, p. 20, 1909b, p. 204. Kylin 1949, p. 30, considers this a mere variant of *E. minima.*

Within our range there is also reported f. **subsalsa** Kjellm. (Phyc. Bor.-Am. 467, 1068; Collins 1900b, p. 44, 1909b, p. 204); this form shows a compressed blade with many spreading branches issuing from the margin and is particularly twisted; it grows in northern Long Island and Massachusetts lagoons and marshes.

Enteromorpha linza (Linnaeus) J. Agardh Pl. 3, fig. 8

Plants simple, usually gregarious, the blades with a short stalk and tapering base, the blade flat, linear to lanceolate, or with a crisped margin, yellowish green, to 37 cm. tall, 30 mm. wide; cells of the stalk longitudinally seriate, those in the blade angular, in vague linear series or in no definite order, 10–15 (–20) μ diam., about 12 μ tall in section; tapering base of the blade hollow, flat portion of the blade of two layers united by their inner cuticula to a combined thickness of 35–50 μ, except near the margins, where they are separated by a definite tubular space.

Bermuda, South Carolina to the lower St. Lawrence, and Ile St. Pierre; growing on rocks and woodwork in the upper intertidal zone, often in quite exposed situations; late spring and summer.

REFERENCES: Phyc. Bor.-Am. 16. Harvey 1858a, p. 59 (as *Ulva linza*); Farlow 1881, p. 43 (as *U. enteromorpha* v. *lanceolata*); Collins 1900b, p. 44, 1903a, p. 23, 1909b, p. 206; Le Gallo 1947, p. 302.

V. **oblanceolata** Doty (1947, p. 19, 1948, p. 258), with oblanceolate blades to 13 cm. tall and 5 cm. broad toward the distal end, has been reported from southern Massachusetts.

Enteromorpha groenlandica (J. Agardh) Setchell et Gardner

Plants in dense tufts, filiform, 2–5 (–15) cm. long, capillary toward the base, above expanded to about 1 mm. diam., persistently tubular, the membrane 25–35 μ thick, the cells in section radially 2–4 times as long as broad, in surface view loosely in 2–4-celled groups below, above not grouped.

Northern Massachusetts, the lower St. Lawrence, Hudson Bay, and Newfoundland; with other filamentous algae on rocks and woodwork in the lower littoral zone; a plant of the spring flora in New England, but in its northern range persisting into the summer.

REFERENCES: Algae Exsic. Am. Bor. 216 (as *Monostroma collinsii*); Phyc. Bor.-Am. 13; Phyk. Univ. 442 (both as *M. groenlandicum*). Rosenvinge 1893, p. 954; Collins 1900b, p. 44, 1909b, p. 208 (all as *M. groenlandicum*).

Capsosiphon Gobi, 1879

Plants attached, unbranched and tubular, brownish, gelatinous; the cells mostly in twos and fours in longitudinal series which are somewhat loosely connected laterally; chromatophores single, parietal, green, with one pyrenoid; walls thick, yellowish brown, gelatinous, the mother-cell walls persistently evident after cell division; reproduction by quadriciliate zoöspores.

Capsosiphon fulvescens (C. Agardh) Setchell et Gardner

Plants tufted, gregarious, attached, yellowish brown; 5–20 (–100) cm. long, simple, subcylindrical, 2–30 mm. diam. above, tapering to the base, the thallus wall to 15 μ thick; cells round to oval, 6.5–8.3 μ diam., single or 2–4 in a group, occasionally cells long remaining undivided, reaching 18 μ diam.

New Jersey to New Brunswick; on rocks in the intertidal zone, during the summer, especially where there is some seeping fresh water; perhaps local and uncommon.

REFERENCES: Algae Exsic. Am. Bor. 212 (as *Enteromorpha aureola*); Phyc. Bor.-Am. 264 (as *Ilea fulvescens*). Collins 1884b, p. 131 (as *Ulva aureola*), 1900b, p. 44, 1902a, p. 175, 1903a, p. 30, 1909b, p. 206 (as *I. fulvescens*).

Monostroma Thuret, 1854

Plants at first saccate, later usually splitting into broad flattened blades or narrow segments; at first attached, later in some species becoming free; walls in the saccate stage one cell in thickness, the expanded stage likewise of one cell layer, the cell walls generally thin, but sometimes gelatinous; cells usually with a singe platelike chromatophore which nearly surrounds the protoplast, one pyrenoid and one nucleus; asexual reproduction by the formation in each cell of several quadriflagellate zoöspores; sexual reproduction by the similar formation of many biflagellate gametes.

KEY TO SPECIES

1. Plants dark green, becoming brownish or drying, not adhering well to paper, the thallus membrane to 65–72 μ thick M. fuscum v. blyttii, p. 70
1. Plants light green, not discoloring, the thallus membrane much thinner 2
2. In surface view the walls between the cells very thin, the cells sharply angular 3
2. In surface view the walls thicker, sometimes very thick and gelatinous, the cells with rounded angles or altogether rounded... 4
3. Thallus membrane thin, generally less than 10 μ in thickness, with the cells somewhat quadrate; in surface view the cells 4.5–10 μ diam., in fairly distinguishable rows M. leptodermum, p. 71
3. Thallus membrane somewhat thicker, generally 15–25 μ in thickness, with the cells much broader than tall; in surface view the cells 13–15 μ diam., not in regular rows...... M. grevillei, p. 71
4. Thallus usually early divided into long lanceolate or oblanceolate segments, generally conspicuously ruffled; in surface view the walls between the cells of moderate thickness M. pulchrum, p. 72
4. Thallus divided into broad segments; in surface view the walls between the cells thick to very thick....... M. oxyspermum, p. 72

Monostroma fuscum (Postels et Ruprecht) Wittrock f. **blyttii** (Areschoug) Collins

Plants large, to about 1–2 dm. diam., at first tubular, soon and irregularly splitting, eventually forming very broad irregular dark green lobes, turning brown and staining the paper on drying; membrane rather firm, 65–72 μ thick, cells angular, closely set, in sectional view vertically elongate, 55–60 μ tall, with two chromatophores.

Rhode Island, northern Massachusetts to the lower St. Lawrence, Newfoundland, Labrador, Baffin Island, and Ellesmere Island; a characteristic Ulva-like plant of the northern New England shores.

REFERENCES: Algae Exsic. Am. Bor. 98; Phyc. Bor.-Am. 715 (as *M. fuscum*). Farlow 1881, p. 41; Hariot 1889, p. 155 (both as *M. blyttii*); Collins 1900b, p. 44 (as *M. fuscum*), 1903a, p. 13, 1909b, p. 213.

Monostroma leptodermum Kjellman

Plants with a light green cuneate-obovate frond, or divided into similar segments, very soft and delicate, to 10 cm. long; membrane 7–12 μ thick, the cells in section 4–5 μ high and rounded-quadrate; in surface view the cells in somewhat curving longitudinal and transverse series, markedly quadrangular, 4.5–12 μ broad; lateral walls rather thin.

Long Island and northern Massachusetts; growing in pools or below low water on Zostera, etc.; early spring.

REFERENCES: Phyc. Bor.-Am. 1272. Rosenvinge 1893, p. 944; Collins 1900b, p. 44, 1903a, p. 15, 1909b, p. 213.

Monostroma grevillei (Thuret) Wittrock

Plants attached, often persistently saccate, eventually opening and splitting to the base, the segments broad, or more or less elongate, 0.2–1.5 dm. tall, thin and slippery but firm, light green, the membrane 15–25 perhaps to 35 μ thick, the cells in section 12–14 μ high, transversely oval; in surface view the cells angular, generally 13–15 μ, occasionally to 22 μ diam.; sporangial cells forming marginal bands, becoming vertically elongate, the walls shortly persisting after discharge; when dried on paper remaining rather dull.

Long Island to New Brunswick, Quebec, Baffin Island, and Ellesmere Island, but the records from south of Maine open to question; growing in rock pools or near low watermark; a plant of the spring flora.

REFERENCES: Phyc. Bor.-Am. 15, 1467, also 217, 1271 as *M. lactuca.* Rosenvinge 1893, p. 946, 1894, p. 149; Collins 1900b, p. 44, 1903a, p. 12, 1909a, p. 23, 1909b, p. 209 (including *M. lactuca*); Kylin 1949, p. 14.

Within our range there is also reported v. **vahlii** (J. Agardh) Rosenvinge (Rosenvinge 1893, p. 949; Collins 1909b, p. 209); plants elongate, to 32 cm. tall, more slender, the segments to 2 cm. broad, more persistently tubular, the wall or membrane 15–25 μ thick, the cells showing some longitudinal seriation; reported from northern Massachusetts, in salt marshes, appearing in March and disappearing in April.

Monostroma pulchrum Farlow

Plants attached, light green, soft, 4–16, perhaps to 30 cm. long, only saccate in the earliest stages of growth, soon divided into ovate or more characteristically lanceolate or oblanceolate segments, attenuate at the base, 1–4, perhaps to 5 cm. wide, which have an elegantly ruffled margin; blade to 20–32 μ thick; cells irregularly placed, rounded, with rather thin walls, 7.5–19 μ, generally 11–15 μ in surface diam., in section broader than tall, the surface membranes to 8–9 μ thick on each face of the blade.

Connecticut to Nova Scotia and Newfoundland; growing attached to coarse algae on rocky shores; appearing in the spring.

REFERENCES: Algae Exsic. Am. Bor. 217 (as *M. lactuca*); Phyc. Bor.-Am. 658. Farlow 1881, p. 41; Hariot 1889, p. 155; Collins 1900b, p. 44, 1903a, p. 14, 1909b, p. 211.

Monostroma oxyspermum (Kützing) Doty Pl. 4, figs. 1–3

Plants attached, ultimately in dense tufts, soft, light green, to about 3–10 cm. tall, or in protected pools detached and much larger, even to 60 cm., the initial sac early split, the very broad segments of the membrane becoming plane or irregularly ruffled; cells in surface view irregularly placed or somewhat grouped 2–4 together, generally 7–18 μ, infrequently to 26 μ diam., rounded-angular to round or oval, the intervening walls moderately thick to very thick and gelatinous in appearance; the thallus membrane 20–40 μ, occasionally to 60 μ thick, with the cells in section rounded rectangular and about 15–21 μ tall.

Bermuda, Florida, South Carolina, New Jersey to New Brunswick, and Newfoundland, on woodwork, stones, shells, marine grasses etc.; in shallow, protected localities, especially salt-marsh ditches and where the water is somewhat brackish; a plant of spring and summer.

REFERENCES: Algae Exsic. Am. Bor. 174 (as *M. crepidinum*); Phyc. Bor.-Am. 220 (as *M. crepidinum*), 14, 1122 (as *M. latissimum*), 1575 (as *M. orbiculatum* v. *varium*), 406 (as *M. undulatum* v. *farlowii*). Farlow 1881, p. 42 (as *M. crepidinum*); Collins 1909b, p. 211 (as *M. crepidinum, M. latissimum* and *M. undulatum* v. *farlowii*), p. 212 (as *M. orbiculatum* v. *varium*); Taylor 1937c, p. 71 (as *M. latissimum* and *M. wittrockii*), p. 72 (as *M. orbiculatum* v. *varium*), p. 73 (as *M. crepidinum*); Doty 1947, p. 12.

In the first edition of this book the records of Monostroma with rather thick walls were chiefly accepted as published. During the intervening years numerous attempts to identify specimens with these species descriptions have been indecisive and unsatisfactory. Without implying that the names of European origin are synonyms, it seems best to consider that so far as American records are concerned plants determined as *M. latissimum* (Kützing) Wittrock, *M. undulatum* v. *farlowii* Foslie, and *M. wittrockii* Bornet are minor variants of one species ranging from the Maritime Provinces to Florida and Bermuda. Hamel (1930a–32) considers some of these as varieties of *M. oxycoccum* (Kützing) Thuret, but the criteria for separation are hardly more satisfactory at the varietal than at the species level. It seems probable that *M. orbiculatum* v. *varium* Collins and *M. crepidinum* Farlow are not independent entities and that the names should go into synonymy.

Ulva Linnaeus, 1753

Plants at first irregularly subfiliform, later expanded, marginally attached or substipitate, plane or crispate, orbicular to elongate-laciniate; of two cell layers, these in close approximation or separated by the gelatinous walls of the cells, which may be much thickened both internally and externally; cells showing single chromatophores usually on the external faces, with one or two pyrenoids; nuclei solitary, except that the elongate holdfast cell extensions become multinucleate; asexual reproduction by the production of four to eight quadriflagellate zoöspores in each sporangial cell; sexual reproduction by the dioecious production of biflagellate anisogametes, usually eight in each cell.

Ulva lactuca Linnaeus Pl. 4, fig. 6

Plants attached, foliaceous, bright green; the holdfast small, the stalk inconspicuous or apparently absent, the blade lanceolate to rounded, often somewhat lobed and undulate or folded, to 6 dm. long or more, and relatively broad; membrane thick near the base, the marginal portions somewhat thinner, cells usually about as high as broad in section, closely placed in surface view.

From the tropics, Bermuda, Florida to the lower St. Lawrence, Newfoundland, Hudson Bay, and James Bay; growing on woodwork, rocks, or coarse algae in moderately exposed situations through the year.

REFERENCES: Harvey 1858a, p. 60; Farlow 1881, p. 43; Collins 1900b, p. 45, 1903a, p. 8, 1909b, p. 214.

Within our range this plant is quite variable. V. **latissima** (Linnaeus) DeCandolle (Pl. 4, fig. 9; Phyc. Bor.-Am. LXXVI; Harvey 1858a, p. 59, as *U. latissima*; Farlow 1881, p. 43) occurs as plants of irregular outline, becoming detached and drifting, in sheets reaching 1–3 meters long and nearly as broad, pale green, often torn, not flat; thickness of the plant 35–40 μ, the cells in section nearly square; it grows in quiet harbors and salt-marsh pools. In the variety **mesenteriformis** (Roth) Collins (1900b, p. 45, 1909b, p. 215) the plants are solitary, ovate to oblong or longer, bullate, contorted, or plicate, often torn, forming masses loose on the bottom of marsh pools in Connecticut. V. **rigida** (C. Agardh) Le Jolis (Pl. 4, figs. 7–8; Phyc. Bor.-Am. 407; Farlow 1881, p. 42; Taylor 1928a, p. 57) appears as attached plants, firm and rather stiff, with a distinct holdfast and stalk, dark green when well developed, becoming somewhat cleft into broad lobes which are often folded; thickness of the plant 60–110 μ, the cells in sections of the blade higher than broad and cell walls thicker than in the other forms; it grows on rocks and woodwork on exposed shores.

PRASIOLALES

Plants filamentous to foliaceous; walls firm, cells often showing regular arrangements in groups in the broader forms; cells uninucleate, the radiating chromatophores with a central pyrenoid; multiplication by fragmentation; asexual reproduction by the production of akinetes either by direct conversion of vegetative cells or after one division across the plane of the blade.

PRASIOLACEAE

Characters of the Order.

Prasiola (C. Agardh) Meneghini, 1838

Plants small, foliaceous, firmly membranous, usually attached by a distinct short stalk or the edge of the membrane or in some forms eventually becoming free; cells dividing in groups of fours and multiples, the groups remaining more or less distinct.

Prasiola stipitata Suhr Pl. 1, figs. 4–7

Plants tufted, dark green, much curled and shrunken when dry; variable in size and form, 2–6 (–8) mm. long, the base narrow and stalklike, above expanding more or less, becoming in well-developed plants lanceolate to fan- or kidney-shaped, the apex truncate, the margins often incurled; cells 5–7 (–10) μ diam., seriate in the stalk, in the upper part crowded in regular blocks; akinetes formed at the truncate end of the plant, spherical, 10–12 μ diam.

Southern Massachusetts and Nova Scotia; growing on exposed rocks near or above the high-tide level, particularly where coated with the excrement of sea birds; through the year.

REFERENCES: Phyc. Bor.-Am. 2135. Lewis in Collins 1916, p. 90; Collins 1918a, p. 57.

CLADOPHORALES

Plants filamentous, usually with a distinct basal holdfast, simple or branched, the filaments uniseriate, the cells with a thin to thick firm wall, a large central vacuole, a generally multinucleate cytoplast with a greatly reticulate or perhaps ultimately fragmented chloroplast and one to many pyrenoids; reproduction basically showing alternation of similar generations; asexual reproduction by bi- to quadriflagellate zoöspores or by akinetes; sexual reproduction by heterothallic iso- or anisogametes.

CLADOPHORACEAE

Characters of the Order.

KEY TO GENERA

1. Filaments unbranched, or with few short, simple branchlets....... 2
1. Filaments progressively, often abundantly, branched.............. 5
2. Filaments attached by a basal end............................ 3
2. Filaments free, or attached by lateral holdfasts.................. 4
3. Cells very large, walls firm; plant not easily adhering to paper CHAETOMORPHA, p. 78
3. Cells moderately large, short, walls soft; plant adhering readily UROSPORA, p. 76
4. Filaments coarse, symmetrical, a mass not collapsing on removal from water, unbranched.................. CHAETOMORPHA, p. 78
4. Filaments more slender and irregular of contour, a mass collapsing on removal from water; unbranched or with a few lateral or rhizoidal spur branches................ RHIZOCLONIUM, p. 80
5. Filaments free or somewhat twisted together, but not united into cords by special hooked branchlets........ CLADOPHORA, p. 82
5. Filaments united by rhizoidal, hooked, or spinelike branchlets in the lower parts of the plant............ SPONGOMORPHA, p. 89

Urospora Areschoug, 1866

Plants filamentous, unbranched, basally attached; vegetative cells short, broad, with thick walls, a netlike parietal chromatophore containing many pyrenoids, many nuclei, and a large central vacuole; asexual reproduction by division of the contents of ordinary upper cells into many quadriflagellate zoöspores and also by akinetes; sexual reproduction by the formation of anisogametes.[1]

KEY TO SPECIES

1. Filaments above little broader than below...................... 2
1. Filaments to 10 times as broad above as below....... U. wormskjoldii
2. Filaments 50–180 μ diam., soft and slippery............ U. collabens
2. Filaments 30–60 μ diam., rather more firm.......... U. penicilliformis

[1] Recent work would suggest that the diploid cell formed after zygosis is a Codiolum or a Codiolum-like plant, which may reproduce by zoöspores to form Urospora filaments. The relations between our species of the genera Codiolum and Urospora not having been established, these plants are here handled separately.

Urospora penicilliformis (Roth) Areschoug

Plants tufted, deep green, moderately firm of texture; filaments 1–8 cm. tall, 30–60 μ diam., the vegetative cells nearly cylindrical, 0.3–2.0 diameters long; the fertile cells somewhat swollen, to 85 μ diam.

New Jersey, Connecticut to the lower St. Lawrence, and Newfoundland; growing on rocks and woodwork between tide marks in exposed places; a plant of the spring and summer.

REFERENCES: Phyc. Bor.-Am. 18 (as *Ulothrix isogona*). Harvey 1858a, p. 90 (as *Hormotrichum speciosum p. p.*); Farlow 1881, p. 45 (as *U. isogona*); Collins 1900b, p. 45 (as *Urospora penicilliformis*), 1909b, p. 368 (as *Hormiscia penicilliformis*), Taylor 1937c, p. 78 (as *H. penicilliformis*); Silva 1952, p. 270.

Urospora collabens (C. Agardh) Holmes et Batters

Plants forming bright green slippery tufts, the filaments soft, above moniliform, but somewhat more slender and more cylindrical below, variable in thickness, 50–180 μ diam.; the cells clearly contracted at the septa, 1–3 diameters long, the wall rather thin; fertile cells much transversely swollen.

Northern Massachusetts to Nova Scotia; on exposed rocky shores; appearing in the spring.

REFERENCES: Phyc. Bor.-Am. 970. Farlow 1881, p. 45; Collins 1900b, p. 45 (both as *Ulothrix collabens*), 1909b, p. 369 (as *Hormiscia collabens*); Taylor 1937c, p. 79 (as *H. collabens*).

Urospora wormskjoldii (Mertens) Rosenvinge

Plants large and coarse but slippery, filaments in tufts to 1–2 dm. long, cylindrical and slender near the base where 30–60 μ diam., the cells 3–10 diameters long, reaching 500–675 (–1000) μ diam. in the upper portion, but in the extreme tip portions of young and sterile filaments again somewhat more slender; cells in the thicker portions shorter than broad, or to 2 diameters long, much swollen and so narrower at the septa; in the fertile portions much inflated and the segments truncate spherical.

Maine, New Brunswick, and Devon Island; on rocks near low water in the spring, or at the northern station in the summer.

REFERENCES: Harvey 1858a, p. 91; Collins 1909b, p. 368; Taylor 1937c, p. 79 (all as *Hormiscia wormskjoldii*).

Chaetomorpha Kützing, 1845 [1]

Plants filamentous, unbranched, attached by a basal holdfast cell, cylindrical or usually somewhat broader above the base, or in the form of more or less entangled free filaments; usually with firm, often heavy, lamellose walls; asexual reproduction by quadriflagellate zoöspores formed in but slightly enlarged cells; sexual reproduction by biflagellate isogametes.

KEY TO SPECIES

1. Filaments 100 μ diam. or less.................. C. cannabina, p. 78
1. Filaments over 100 μ diam.................................. 2

2. Plants free, drifting, or entangled............................ 3
2. Plants basally attached.. 4

3. Plants usually light-green, cells mostly 2 diameters long, 150–300 μ diam.................................. C. linum, p. 78
3. Plants dark-green, cells mostly 3–4 diameters long, 300–380 μ diam. .... C. atrovirens, p. 79

4. Cells 1–2 diameters long, 150–350 μ diam........... C. aerea, p. 79
4. Cells 1.5–3.0 diameters long, 300–700 μ diam... C. melagonium, p. 80

Chaetomorpha cannabina (Areschoug) Kjellman

Plants filamentous, light green, filaments unattached, entangled, soft and relatively delicate, 75–100 μ diam., the width irregular in the same mass and even in the same filament; cells usually 500–600 μ long, being 3–8 diameters.

Maine.

REFERENCE: Collins 1909b, p. 325.

Chaetomorpha linum (Müller) Kützing Pl. 1, figs. 1–2

Plants composed of loosely entangled, unattached filaments, yellowish green, somewhat stiff and curled; cylindrical or the cells

[1] The plants named *Chaetomorpha olneyi* Harvey (1858a, p. 86; Farlow 1881, p. 48) and *C. longiarticulata* Harvey (1858a, p. 86; Farlow 1881, p. 48) are too poorly defined to place exactly, but they may well be *C. aerea* and *C. linum,* respectively.

slightly swollen; 100–375 μ diam., cells 1–2 diameters long, occasionally shorter or longer even to 5 diameters.

From the tropics, Bermuda, Florida, North Carolina, New Jersey to Nova Scotia; commonly drifted ashore in large masses, often entangled with coarse algae from moderate depths; throughout the year.

REFERENCES: Algae Exsic. Am. Bor. 175; Phyc. Bor.-Am. 22; Phyk. Univ. 469. Harvey 1858a, p. 87; Farlow 1881, p. 47; Collins 1900b, p. 43, 1909b, p. 325, 1918a, p. 79.

Chaetomorpha atrovirens Taylor

Plants forming loosely entangled mats, usually anchored among other and coarser algae; filaments curved, twisted, interwoven, the cells not at all or but slightly swollen, diameter at the middle of the cells 300–380 μ; cells usually 3 diameters long, but sometimes 1.5–5.0 diameters.

Connecticut to Nova Scotia, Prince Edward Island, and Newfoundland; growing in moderately deep water, entangled with the bases of Laminaria or Phyllophora plants, infrequent south of Cape Cod; summer, perhaps throughout the year.

REFERENCES: Phyc. Bor.-Am. 412. Harvey 1858a, p. 85; Farlow 1881, p. 47 (the two as *C. picquotiana*); Collins 1909b, p. 324 (as *C. melagonium* f. *typica*); Taylor 1937b, p. 227.

Chaetomorpha aerea (Dillwyn) Kützing Pl. 1, figs. 10–12

Plants gregarious, bright green, to 10–15 (–25) cm. tall, attached by a basal cell which is disklike below, but becomes quite elongated above, to 800 μ or more in length; filaments slender toward the base, above to 150–350 (–500) μ diam., stiff and straight, the cells 1–2 diameters long, or shorter, little constricted at the septa; zoöspores formed in the upper cells of the filament, which become cask-shaped to subglobose, 600–700 μ diam.

South Carolina to New Hampshire, and Prince Edward Island; growing on rocks in moderately exposed situations near low-tide mark; forming zoöspores in early summer.

REFERENCES: Phyc. Bor.-Am. 1526. Harvey 1858a, p. 86; Farlow 1881, p. 46; Collins 1900b, p. 43, 1909b, p. 324; Croasdale 1941, p. 213.

Chaetomorpha melagonium (Weber et Mohr) Kützing

Plants erect, solitary or gregarious, dark green; attached by basal disklike holdfasts, the lower cells elongate above the disk, about 240 μ diam. and 3.0–5.5 mm. long; above straight and stiff, 1–3 (–10) dm. long, 300–700 (–1050) μ diam., the cells rather long below, to 6.5 diameters, above usually 1.5–3.0 diameters long, cylindrical to slightly swollen or in the fertile part truncate-subspherical.

New Jersey to the lower St. Lawrence, Ile St. Pierre, Labrador, Hudson Strait, Melville Peninsula, Baffin Island, Devon Island, Ellesmere Island, and other stations in the American Arctic. Uncommon south of Cape Cod.

REFERENCES: Phyc. Bor.-Am. 413. Harvey 1858a, p. 85; Farlow 1881, p. 46 (as *Chaetomorpha melagonium*); Collins 1900b, p. 43, 1909b, p. 324 (as *C. melagonium* f. *rupincola*); Le Gallo 1947, p. 302.

Rhizoclonium Kützing, 1843

Plants filamentous, simple or sparingly branched, the tapering or rhizoidal branches spreading, of few cells; main filaments usually entangled, of moderately long cells somewhat unevenly articulated; cells with moderately to quite thick walls; chromatophores parietal, netlike, with numerous pyrenoids; nuclei one to few in each cell; reproduction by biflagellate zoöspores or by akinetes formed from little-modified cells.

KEY TO SPECIES

1. Forming erect tufts, branching only at the base, the branches in part similar to the axis....................... R. erectum, p. 82
1. Branches when present short and rhizoidal....................... 2

2. Filaments 10–14 μ diam., unbranched............. R. kerneri, p. 81
2. Filaments 15–30 μ diam., sometimes branched.... R. riparium, p. 81
2. Filaments 40–70 μ diam., unbranched........... R. tortuosum, p. 80

Rhizoclonium tortuosum Kützing

Plants forming dull, rather dark green entangled masses; filaments curled, rather stiff, 40–70 μ diam., cells 1–2 diameters long; branches absent.

Bermuda, North Carolina, New York to the lower St. Lawrence; forming dark crisped masses in the lower tide pools of exposed shores.

REFERENCES: Phyc. Bor.-Am. 23. Harvey 1858a, p. 85 (as *Chaetomorpha tortuosa*); Farlow 1881, p. 49; Collins 1900b, p. 45, 1909b, p. 328.

In addition to the typical form, this plant is known in f. **polyrhizum** Holden (Phyc. Bor.-Am. 625; Collins 1909b, p. 328), which has abundant short rhizoidal branches, similar to the axis, the terminal cells abruptly conical; it is reported from Connecticut to Maine.

Rhizoclonium riparium (Roth) Harvey Pl. 1, fig. 3

Plants entangled, yellowish green, forming considerable expanses on the substrate; the filaments curled, 20–30 μ diam., the cells 1–2 diameters long, the rhizoidal branches usually numerous.

Bermuda, North Carolina, New Jersey to the lower St. Lawrence, Hudson Strait, Labrador, and Baffin Island; forming mats on the ground or rocks in the littoral zone.

REFERENCES: Algae Exsic. Am. Bor. 213. Harvey 1858a, p. 92; Farlow 1881, p. 49; Rosenvinge 1893, p. 914, 1894, p. 127; Collins 1900b, p. 45, 1909b, p. 327; Whelden 1947, p. 52.

This plant is present within our range in two varieties. V. **implexum** (Dillwyn) Rosenvinge (Phyc. Bor.-Am. 266; Rosenvinge 1893, p. 915, as *R. implexum*) has the branches few or usually lacking; the plant ordinarily grows on mud or sand, and ranges from New Jersey to Labrador. V. **polyrhizum** (Lyngbye) Rosenvinge (Phyc. Bor.-Am. 24, as *R. riparium;* Rosenvinge 1893, p. 915) has frequent branches of one to a few cells in length and is found from Connecticut northward.

Rhizoclonium kerneri Stockmayer

Plants filamentous, filaments entangled, yellowish green, without branches; cells 10–14 μ diam., 3–7 diameters long.

Bermuda, and Massachusetts to Maine; in masses in tide pools.

REFERENCES: Phyc. Bor.-Am. 623. Farlow 1881, p. 49 (as *R. kochianum*); Collins 1900b, p. 45, 1909b, p. 329.

Rhizoclonium erectum Collins

Plants tufted, the tufts to 10 cm. tall; filaments below prostrate, thick-walled, cells irregular, 70–100 μ diam., 1–2 diameters long, or these prostrate filaments not evident; erect filaments to 30 cm. long when extended, curled and entangled, occasionally forked at the base, generally simple, sometimes with a few short spur branchlets near the top, 20–50 μ diam., the cells 3–6 diameters long.

Maine; in a tide pool exposed on a rocky shore; summer.

REFERENCES: Phyc. Bor.-Am. 975. Collins 1900b, p. 45, 1901b, p. 291, 1909b, p. 327.

Cladophora Kützing, 1843 [1]

Plants filamentous, sparingly to repeatedly branched, attached by rhizoidal extensions from the lower cells, these filaments spreading over the substratum and often giving rise to new erect shoots; growth primarily apical; cytoplasm peripheral around a large central vacuole, the chromatophore reticulate with many pyrenoids, or of separate disks, the nuclei numerous in each cell; asexual reproduction by numerous quadriflagellate zoöspores in the outer cells of the branches; sexual reproduction similar, by biflagellate isogametes.

KEY TO SPECIES

1. Low, matted tufts with creeping base........ C. magdalenae, p. 88
1. Plants essentially erect, sometimes eventually free-floating....... 2

2. Main filaments usually exceeding 150 μ diam.................... 3
2. Main filaments usually less than 150 μ diam.................... 4

3. Branchlets short and clustered............... C. hutchinsiae, p. 88
3. Branchlets not clustered and rather long.......... C. gracilis, p. 86

4. Main filaments distinctly flexuous or angled.................... 5
4. Main filaments straight or nearly so.......................... 14

5. Branchlets in clusters at the tips of the branches................ 6
5. Branchlets not distinctly clustered.......................... 7

[1] It is probably not possible to place *C. morrisiae* Harvey (1858a, p. 54) from the figure and the description given.

6. Plant floating, except at the earliest stages C. expansa v. glomerata, p. 86
6. Plant remaining attached during its growing cycle C. laetevirens, p. 87

7. Plant floating, except at the earliest stages.................... 8
7. Plant remaining attached during its growing cycle............... 11

8. Branchlets in long comblike series at the ends of the filaments.... 9
8. Branchlets scattered 11

9. Filaments 40–100 μ diam., cells 4–8 diameters long C. gracilis f. tenuis, p. 87
9. Filaments to 140 μ, rarely to 160 μ diam., cells 3–5 diameters long .... C. gracilis f. expansa, p. 87

10. Main filaments 100–150 μ diam................ C. expansa, p. 85
10. Main filaments 30–60 μ diam................. C. flavescens, p. 85

11. Plant mass of a spongy texture.......... C. flexuosa f. densa, p. 85
11. Plant mass not spongy....................................... 12

12. Cells long, to 20 diameters................ C. rudolphiana, p. 84
12. Cells not over 8 diameters long............................. 13

13. Branchlets long, in comblike series near the tips of the branches C. gracilis, p. 86
13. Branchlets not in comblike series............... C. flexuosa, p. 85

14. Branchlets curved 15
14. Branchlets straight or nearly so............................ 16

15. Plant mass of a soft, spongy nature..... C. albida v. refracta, p. 84
15. Plant relatively stiff and firm................. C. refracta, p. 87

16. Branches opposite or whorled.................. C. rupestris, p. 88
16. Branches neither opposite nor whorled....................... 17

17. Filaments not more than 30 μ diam.............. C. albida, p. 83
17. Filaments exceeding a diameter of 30 μ....................... 18

18. Branchlets 20–30 μ diam., acute at the tips; main filaments not more than 75 μ diam................. C. glaucescens, p. 84
18. Branchlets not acute; main filaments 80–140 μ diam. C. crystallina, p. 86

Cladophora albida (Hudson) Kützing

Plants light green, soft and dense, to 10 cm. tall; filaments below loosely dichotomous, above much and irregularly branched, 20–

30 (–60) μ diam., the cells 2–6 diameters long, little contracted at the septa; branchlets spreading, somewhat unilateral, often curved.

New Jersey to southern Massachusetts, and Ile St. Pierre.

REFERENCES: Harvey 1858a, p. 80; Farlow 1881, p. 51; Collins 1900b, p. 43, 1902b, p. 119, 1909b, p. 336; Le Gallo 1947, p. 302.

There is also reported v. **refracta** (Wyatt) Thuret (Phyc. Bor.-Am. 720; Harvey 1858a, p. 79, as *C. refracta;* Collins 1909b, p. 336); in this the upper branches and branchlets are very much recurved; it is known from New Jersey and Connecticut to Maine.

Cladophora glaucescens (Griffiths ex Harvey) Harvey

Plants 10–40 cm. tall, attached, soft, yellowish or grayish green; primary filaments elongate, plumose, and somewhat clustered, soft, the lower cells 50–75 μ diam., 4–6 diameters or more in length; branchlets long, alternate or sometimes unilateral, 25–40 μ diam., cells proportionally nearly as long as those below.

Florida, South Carolina (?), New Jersey to Labrador; growing in upper tide pools, marsh pools, and shallow bays; particularly in the northern part of its range a plant of spring and early summer.

REFERENCES: Algae Exsic. Am. Bor. 205; Phyc. Bor.-Am. 817(?). Harvey 1858a, p. 77; Farlow 1881, p. 52; Collins 1900b, p. 43, 1902b, p. 120, 1909b, p. 336.

Cladophora rudolphiana (C. Agardh) Harvey

Pl. 5, figs. 3–4; pl. 6, fig. 2

Plants attached, tufted, yellowish green, slippery, reaching a meter in length; main filaments 40–60 (–100) μ diam., alternately to oppositely branched, the branches spreading, flexuous; branchlets somewhat unilaterally placed, elongate, tapering, 20–40 μ diam., the cells 5–20 diameters long, the branchlets, particularly if short, blunt at the apex.

New Jersey to Maine; growing in the upper sublittoral, particularly in warm shallow bays.

REFERENCES: Phyc. Bor.-Am. 267, 817 *p. p.* Harvey 1858a, p. 80; Farlow 1881, p. 54; Collins 1900b, p. 44, 1902b, p. 120, 1909b, p. 336.

Cladophora flavescens (Roth) Kützing

When fully developed, floating in yellowish-green masses, widely branching, the flexuous filaments 30–60 μ diam., cells 6–10 diameters long; branchlets scattered, straight, nearly as thick, tapering to blunt apices.

New York to Maine; from tidal and spray-filled rock pools.

REFERENCES: Phyc. Bor.-Am. 1077, 1229 (as *C. fracta* f. *flavescens*). Collins 1909b, p. 338.

Cladophora flexuosa (Dillwyn) Harvey

Plants attached, 10–20 cm. tall, light green; main filaments somewhat stiff, irregularly flexuous, cells 80–120 (–160) μ diam., about 6 diameters long, with alternate flexuous branches 40–80 μ diam.; branchlets alternate or unilateral, curved and sometimes recurved, the cells about 2 diameters long.

Florida, North Carolina, New Jersey to Prince Edward Island, and Newfoundland; growing in rock pools near low-water mark.

REFERENCES: Algae Exsic. Am. Bor. 206; Phyc. Bor.-Am. 1076, 1527. Harvey 1849, p. 202, 1858a, p. 78; Farlow 1881, p. 54; Collins 1900b, p. 43, 1902b, p. 121, 1909b, p. 339.

From Rhode Island there is reported f. **densa** Collins (Phyc. Bor.-Am. 979; Collins 1902b, p. 121, 1909b, p. 340); it branches densely in all orders, producing a spongy plant; other variants of this species approach *C. gracilis* rather closely.

Cladophora expansa (Mertens) Kützing Pl. 5, fig. 5

Plants light green, forming loose cushions or soft entangled masses of considerable extent; filaments much and loosely branched, the main branches angled-flexuous, elongate, (80–) 100–150 μ diam., the cells 3–6 (–12?) diameters long, divaricately or alternately divided, bearing smaller spreading secondary branches; branchlets somewhat unilateral, blunt, to 40 μ diam.

New Jersey to Nova Scotia and Newfoundland; a plant of marsh pools and lagoons, forming large masses in quiet warm water; summer.

REFERENCES: Algae Exsic. Am. Bor. 210; Phyc. Bor.-Am. 121, 977, 1280. Farlow 1881, p. 55; Collins 1900b, p. 43, 1902b, p. 123, 1909b, p. 340.

From Connecticut and southern Massachusetts there is also reported v. **glomerata** Thuret (Phyc. Bor.-Am. 1027; Collins 1902b, p. 123, 1909b, p. 341); it shows the branchlets rather closely set in tufts.

Cladophora crystallina (Roth) Kützing

Plants light yellowish green, soft and silky, 10–30 cm. tall, slightly matted; branching ditrichotomous, distant, erect or spreading, above sometimes whorled or alternately unilateral; main branches 80–140 μ diam., branchlets 25–40 μ diam., cells throughout 4–12 diameters long, cylindrical, hardly constricted at the septa.

Bermuda, Connecticut, and southern Massachusetts; growing on mud and clay flats between tide levels.

REFERENCES: Phyc. Bor.-Am. 1581. Collins 1909b, p. 342.

Cladophora gracilis (Griffiths ex Harvey) Kützing

Pl. 5, fig. 2; pl. 6, figs. 3–4

Plants tufted, bright grayish green, glossy, somewhat harsh in texture, to 3 dm. long or more; filaments flexuose, irregularly and angularly bent, the alternate branches spreading, 140–160 μ diam., the cells 3–5 diameters long; branchlets unilateral, comblike, moderately elongate and slender, 40–60 μ diam., the cells 3–5 diameters long, the tips acute.

New Jersey to Nova Scotia, and Newfoundland; variable in aspect, occurring in exposed or sheltered places.

REFERENCES: Algae Exsic. Am. Bor. 209; Phyc. Bor.-Am. 724, 726 (f., as *C. hirta*), 1528, 1529. Harvey 1858a, p. 80; Farlow 1881, p. 55; Collins 1900b, p. 44, 1902b, p. 121, 1909b, p. 342 (with varieties); Taylor 1928a, p. 63.

This species is quite variable, and the following forms have been reported within our range. F. **australis** Collins (Collins 1909b, p. 343) has the filaments less sharply angular, and the branches of all orders more spreading; it is reported to be common south of

Cape Cod. F. **elongata** Collins (Phyc. Bor.-Am. 725; Collins 1902a, p. 122) is grayish green, the branches distant and very erect, the plant stretching out over the surface of shallow water to a length of a meter or more; it occurs in Maine, on Zostera and other objects, in warm pools with a steady gentle current. F. **expansa** Farlow (Phyc. Bor.-Am. 981; Farlow 1881, p. 55) shows plants early detached, irregularly branched, forming particularly loose floating masses in tide pools on the coast of New Jersey, northern Massachusetts to New Brunswick. F. **subflexuosa** Collins (Phyc. Bor.-Am. 1530) exhibits plants shorter than in the typical form, the branches flexuous, the ultimate branching fasciculate. It is a reduced form of shallow rock pools without persistent current, occurring in Maine, but is perhaps not related to *C. gracilis*. F. **tenuis** Farlow (Farlow 1881, p. 55; Collins 1902b, p. 122, as *C. gracilis* v. *vadorum*), with filaments more slender than the typical plant, 40–100 μ diam., the cells 4–8 diameters long, forms loose floating masses in the sublittoral zone throughout New England.

Cladophora refracta (Roth) Kützing Pl. 6, fig. 1

Plants attached, tufted, dull to grayish green, 10–20 cm. tall, spongy in texture, the main filaments rather stiff, much branched, the branches erect, but their upper parts reflexed, 100–120 μ diam., the cells 2–3 diameters long; branchlets rather close, reflexed, often unilateral, 30–50 μ diam., blunt at the tips.

South Carolina, New Jersey, Connecticut to Nova Scotia; growing on coarse algae or rocks near low-tide line.

REFERENCES: Algae Exsic. Am. Bor. 207; Phyc. Bor.-Am. 573. Harvey 1858a, p. 79; Farlow 1881, p. 52; Collins 1900b, p. 43, 1902b, p. 124, 1909b, p. 344.

Cladophora laetevirens (Dillwyn) Harvey

Plants attached, rather stiff, bright green, to 10–20 cm. tall; main filaments much branched, 50–150 μ diam., the cells 2–6 diameters long; branching alternate to sometimes opposite, the branches erect; branchlets short, obtuse or subacute, densely fastigiate at the tips, the cells somewhat contracted at the septa, to 3 diameters long.

New York, northern Massachusetts to the lower St. Lawrence; growing in the lower littoral or the sublittoral.

REFERENCES: Harvey 1858a, p. 82; Farlow 1881, p. 53; Collins 1900b, p. 43, 1902b, p. 125, 1909b, p. 345.

Cladophora hutchinsiae (Dillwyn) Kützing

Plants attached, large, grayish green, stiff and coarse, to 4 dm. tall; main filaments flexuous, 120–300 (–400) μ diam., the cells 2–4 diameters long; sparingly branched, the branches elongate, flagelliform, 160–240 μ diam.; branchlets mostly on the ultimate and penultimate branches, rather short, scattered or in unilateral series, constricted at the septa, 90–100 μ diam., the cells 1–2 diameters long, the tips blunt.

Florida, New Jersey, and Anticosti Island, Quebec (?).

REFERENCES: Farlow 1881, p. 53; Collins 1900b, p. 43, 1902b, p. 126, 1909b, p. 345.

V. **distans** (C. Agardh) Kützing (Harvey 1858a, p. 83) has the main filaments long, secondary branches few, branchlets few and long, cells longer than in the typical form, nodes not constricted; it has been reported from Rhode Island, New Jersey, and northern Massachusetts, but the records are not well confirmed.

Cladophora rupestris (Linnaeus) Kützing Pl. 5, fig. 1

Plants tufted, dark green, densely branched and stiff, to 5–20 cm. tall; primary branches 90–185 μ diam., cells 3–9 diameters long; branches alternate to usually opposite or in whorls of four, erect; branchlets numerous and close, short, suberect, to 65–85 μ diam., the cells 2–5 diameters long, the tips blunt or a little pointed.

Long Island to southern Massachusetts (all these of dwarfed plants or the records doubtful); northern Massachusetts to Nova Scotia, the lower St. Lawrence, and Newfoundland; growing on rocks in the littoral, especially where covered by fuci, and on exposed coasts; throughout the year.

REFERENCES: Phyc. Bor.-Am. 728. Harvey 1858a, p. 74; Farlow 1881, p. 51; Collins 1900b, p. 44, 1902b, p. 125, 1909b, p. 346.

Cladophora magdalenae Harvey

Plants forming spreading mats, dark green, coarse, and stiff; branching dichotomous to unilateral, spreading, flexuous, the main filaments 60–100 μ diam., the cells 2–4 diameters long; branchlets few, irregular, curved.

Connecticut to southern Massachusetts; forming mats among other algae in the littoral zone; appearing in late autumn and winter.

REFERENCES: Phyc. Bor.-Am. 572. Farlow 1881, p. 56; Collins 1900b, p. 43, 1909b, p. 348.

Spongomorpha Kützing, 1843

Plants filamentous, abundantly branched, attached by rhizoidal branches from the lower cells; several erect axes usually twisted together, joined by short recurved, hooked, or spinelike branchlets, or by descending rhizoidal ones; intercalary cell division frequent; cells with a reticulate chromatophore, pyrenoids numerous, nuclei single to several in each cell; sporangia intercalary, seriate, forming numerous biflagellate zoöspores.

KEY TO SPECIES

1. Spinous or hooked branches present........... S. spinescens, p. 90
1. Such branches absent, rhizoidal branches present................ 2

2. Filaments 200–500 μ diam. at the tips............. S. hystrix, p. 90
2. Filaments less than 150 μ diam................................ 3

3. Filaments 30–40 μ diam......................... S. lanosa, p. 89
3. Filament tips 60–100 μ diam...................... S. arcta, p. 90

Spongomorpha lanosa (Roth) Kützing Pl. 5, figs. 6–7

Plants forming bright green or eventually faded tufts 1–5 cm. tall, at first attached, later becoming free; erect filaments united into ropelike masses, several growing together and forking, these covered with less matted free branches and united by rhizoidal branches 16–30 μ diam.; principal filaments to 30–40 μ diam., the cells 2–3 diameters long, uninucleate; when branching dividing below the summit of the supporting cell.

Connecticut to Nova Scotia, Newfoundland, and Baffin Island; growing on various coarse algae, particularly in the spring, and becoming detached later.

REFERENCES: Algae Exsic. Am. Bor. 204; Phyc. Bor.-Am. 661. Harvey 1858a, p. 76; Farlow 1881, p. 51; Collins 1900b, p. 43, 1902b, p. 118 (all as *Cladophora lanosa*), 1909b, p. 358.

This species is also represented throughout the same range on rocks and in tidal rock pools by v. **uncialis** (Müller) Kjellman (Phyc. Bor.-Am. 77, as *Cladophora lanosa* v. *uncialis;* Harvey 1858a, p. 77, as *C. uncialis;* Farlow 1881, p. 54, as *C. lanosa* v. *uncialis;* Collins 1909b, p. 359); these plants form irregular dense tufts which do not become detached, but remain in a coarse bleached state.

Spongomorpha arcta (Dillwyn) Kützing[1] Pl. 6, figs. 5–6

Plants forming dark green hemispherical to globose tufts 3–15 cm. tall; filaments erect, rather stiff, upper cells 4–6 diameters long; 60–100 (–150) μ diam. at the enlarged tips, the terminal cells obtuse; vegetative branches erect, hooked branches relatively infrequent or more usually lacking, descending rhizoidal branches numerous below, 40–60 μ diam., the cells 2–6 diameters long.

New Jersey to the lower St. Lawrence, Newfoundland, northern Labrador, Hudson Strait, and Baffin Island; growing on exposed rocky coasts; common in spring and early summer.

REFERENCES: Phyc. Bor.-Am. 224, 815. Harvey 1858a, p. 75; Farlow 1881, p. 50; Hariot 1889, p. 154 (as *Cladophora arcta*); Collins 1900b, p. 43, 1902b, p. 115 (both as *C. arcta*), 1909b, p. 359.

Spongomorpha hystrix Strömfelt

Plants forming large, dense dark green tufts to 10–15 cm. tall; erect filaments mostly straight, 100–300 μ diam. below, the cells 0.5–2.0 diameters long, the tip cells 200–500 μ diam., to 5 diameters long; rhizoidal branches fairly common in the older parts, 40–70 μ diam., the cells 3–10 diameters long.

Northern Massachusetts to Newfoundland, Devon Island, and Ellesmere Island; growing in tide pools on exposed shores.

REFERENCES: Phyc. Bor.-Am. 982. Collins 1900b, p. 43, 1902b, p. 116 (both as *Cladophora hystrix*), 1909b, p. 358.

Spongomorpha spinescens Kützing

Plants tufted, to 10 cm. tall; main filaments about 80 μ diam. below, the cells 0.5–1.0 diameter long; tip cells to 100 μ diam., the

[1] *Cladophora rhizophora* Kützing (De-Toni 1889, p. 340), described from Newfoundland, may be this species.

upper cells to 2 diameters long; branches abundant, erect, obtuse at the tips; spreading acute spinelike or hooked branchlets present, associated with rhizoidal filaments in uniting the filaments into ropelike tufts.

Northern Massachusetts to Nova Scotia; growing in rock pools on Ascophyllum and various objects.

REFERENCES: Phyc. Bor.-Am. 721 (as *Cladophora arcta* v. *centralis*). Farlow 1881, p. 51; Collins 1900b, p. 43 (as *C. arcta* v. *centralis*), 1902b, p. 117 (as *C. spinescens*), 1909b, p. 360.

SIPHONALES

Plants coenocytic; simply filamentous, or these filaments branched or aggregated to more or less elaborate shapes, free or calcified, in the less simple forms usually showing differentiation of rhizoidal and assimilative portions, and often of stoloniferous portions as well; coenocyte wall sometimes braced by extensions of the membrane across the cavity, and in some species at certain points with ringlike thickenings which almost divide the cavity into separate chambers; nuclei and small chromatophores very numerous, pyrenoids absent or present; multiplication by fragmentation; asexual reproduction by zoöspores formed in more or less enlarged sporangia, by aplanospores, or by akinetes; sexual reproduction by anisogametes or by oögamy.

KEY TO FAMILIES

1. Plant microscopic 2
1. Plant macroscopic 3

2. Endophytic .... CHAETOSIPHONACEAE, p. 91
2. In the matrix of shells................. PHYLLOSIPHONACEAE, p. 94

3. Erect axes abundantly branched............. BRYOPSIDACEAE, p. 93
3. Erect axes subsimple or sparingly branched.... DERBESIACEAE, p. 92

CHAETOSIPHONACEAE

Plants forming cushion-shaped to subcylindrical and branched microscopic coenocytes, vegetative cross walls being absent, although the filamentous types are often deeply constricted; long, unseptate hairs present; cytoplasm peripheral, with many disklike chromatophores containing pyrenoids and with many nuclei; sporangia segregated by a cross wall or involving entire coenocytes, the zoöspores with two or four flagella.

Blastophysa Reinke, 1888

Plants microscopic, epi- or endophytic, the coenocytic cells rounded, cushion-shaped, or lobed; wall thick, often locally particularly heavy, lamellose, and bearing long, colorless, unseptate hairs; multiplication by the formation of long outgrowths from the cells which swell at the tips and develop from each of these portions a new cell separated from the then empty outgrowth by a wall; cell division, possibly by constriction, also sometimes present; sporangia little modified from the form of vegetative cells, producing many quadriflagellate zoöspores.

Blastophysa rhizopus Reinke, f. Pl. 2, figs. 6–8

Plants of oval or oval-acuminate cells, scattered or joined by slender colorless tubes, and bearing 1–3 long colorless hairs with a very slightly enlarged base; coenocytes 25–52 (–120) μ diam., 55–90 (–130) μ long, with numerous rounded or angular chromatophores with single pyrenoids.

Southern Massachusetts; in the tissues of coarser algae; summer.

REFERENCE: Collins 1912, p. 99.

DERBESIACEAE

Plants matted, with a creeping rhizomatous portion bearing attaching holdfasts and erect simple or branched assimilative filaments; asexual reproduction by zoöspores with an anterior circle or flagella, borne several together in lateral sporangia on the erect filaments; sexual reproduction unknown.

Derbesia Solier, 1847

Plants of unicellular filaments, primarily erect and simple or sparingly branched, below stoloniferous and attached by lobed holdfasts; zoösporangia lateral on the erect filaments, segregated by a pair of transverse walls; zoöspores round-ovoid, with an anterior ring of flagella.

Derbesia vaucheriaeformis (Harvey) J. Agardh

Plants forming erect dense brushlike tufts, the filaments dichotomously forked, 4–5 cm. tall, 20–50 μ diam.; walls sometimes present in pairs across the segments above a fork, about 30–35 μ apart; sporangia replacing branches, ovoid or broadly pyriform, 100–130 μ diam., 190–300 μ long, pedicellate, the pedicels about 15 μ diam., 50–100 μ long with two partitions near the base enclosing a cell 2–4 diameters long; zoöspores large, about 16 formed in each sporangium.

Bermuda, New Jersey (?), and southern Massachusetts, where on stone walls at the entrance to a protected harbor, but very rare in our range.

REFERENCES: Phyc. Bor.-Am. 318. Harvey 1858a, p. 30 (as *Chlorodesmis vaucheriaeformis*); Farlow 1881, p. 60 (as *Derbesia tenuissima*); Collins 1909b, p. 406.

BRYOPSIDACEAE

Plants generally erect from a more or less extended rhizoidal holdfast, branching, the branches alternate, with more or less constricted bases; asexual reproduction unknown; sexual reproduction by the conversion of the branchlets into gametangia, or the lateral production of gametangia on these, with the completion of a wall across the base, and the formation of many small biflagellate anisogametes; after liberation of the gametes the fertile branchlets drop off.

Bryopsis Lamouroux, 1809

Plants erect, tufted or spreading, bushy, with more or less evident main axes and more or less pyramidally or uni- or bilaterally arranged branchlets formed successively from the apex; plans diploid, reproducing after the segregation of entire branchlets by a basal membrane, the contents undergoing meiotic divisions so that these branchlets eventually function as heterogametic gametangia, the smaller gametes brownish, the larger yellowish green.

KEY TO SPECIES

1. Branchlets not sharply delimited from the lesser branches, radially placed, forming an indistinct tuft.............. B. hypnoides

1. Branchlets distinct from the lesser branches, roughly bilaterally placed, the longest forming a more or less plane triangular blade .... B. plumosa

Bryopsis hypnoides Lamouroux

Plants more or less erect, tufted, to 10 cm. high, dull and rather dark green, the axis usually much branched, branches irregularly placed, progressively smaller, without very marked difference between the lesser branches and the rather slender branchlets.

Bermuda, Long Island, and Massachusetts; very infrequent, growing in pools in warm, shallow, yet not stagnant water.

REFERENCES: Harvey 1858a, p. 32; Collins 1906b, p. 124, 1909b, p. 403.

Bryopsis plumosa (Hudson) C. Agardh Pl. 7, figs. 1–3

Plants erect, tufted, to 10 cm. tall, light to olive-green, the axis often naked below except for a few rhizoids, but above sparingly to much branched, branching generally pyramidal; branchlets in size distinct from the axis, roughly bilaterally attached to the lesser branches, sharply constricted at the base, obtuse at the apex, simple or very infrequently forked, not very slender, being 65–100 μ diam., grading rather evenly from the longest at the base of the branch to the initials at the apex, so that the whole appears as a somewhat flat, triangular blade.

Bermuda, Florida, the Carolinas, New Jersey to Maine, Nova Scotia, and Smith Sound (?); growing in tide pools and on woodwork in moderately protected locations, particularly south of Cape Cod, becoming rare northward.

REFERENCES: Phyc. Bor.-Am. 227. Harvey 1858a, p. 31; Farlow 1881, p. 59; Collins 1900b, p. 43, 1909b, p. 403.

PHYLLOSIPHONACEAE

Plants endophytic or endozoöic, of large oval cells or branched filaments without cross walls, coenocytic, with many nuclei and many disklike chromatophores lacking pyrenoids; reproduction by aplanospores produced in great numbers.

Ostreobium Bornet et Flahault, 1889

Plants filamentous, of much-branched and apparently anastomosing coenocytic filaments of very irregular form and variable diameter; aplanospores formed in the swollen ends of branches.

Ostreobium quekettii Bornet et Flahault Pl. 2, fig. 3

Plants of slender filaments, mostly 4–5 μ diam., much twisted in the tissues, with considerable local inflations to 40 μ diam., or more moderate vermiform enlargement; at the tips of the branches reduced to 2 μ diam.

Connecticut, Massachusetts, Devon Island, and Ellesmere Island; in old shells of oysters, etc.

REFERENCES: Collins 1909b, p. 408; Rosenvinge 1926, p. 18.

XANTHOPHYCEAE

HETEROSIPHONALES

Plants coenocytic, balloon-shaped to filamentous, with differentiation of rhizoidal and assimilative portions; chromatophores peripheral, small, sometimes with pyrenoid-like structures; food reserves of oil; asexual reproduction by zoöspores or aplanospores; sexual reproduction by iso- or anisogametes, or oögamous.

VAUCHERIACEAE

Plants filamentous, irregularly or dichotomously branched; filaments without regular wall formation except to segregate the reproductive organs; asexual reproduction by zoöspores produced singly in sporangia formed at the tips of branches, multiflagellate, the flagella in pairs over the whole surface; sexual reproduction oögamous, dioecious or monoecious, when fruiting organs of both sexes are sometimes borne together on a single stalk; the antheridia cylindrical, generally curved, producing many small biflagellate antherozoids; oögonia single or in groups, producing single eggs, which are fertilized in place and develop thick walls.

Vaucheria De Candolle, 1801

Plants of branched coenocytic filaments somewhat tufted or loosely entangled, or forming considerable mats, particularly in shallow water, the filaments without normal vegetative cross walls or constrictions; antheridia more or less cylindrical, sessile, or stalked, usually near the oögonia; oögonia single or grouped, sessile or on a short or an elongated stalk; oöspore when mature often with much stored oil.

KEY TO SPECIES

1. Antheridia attached directly to the oögonia.................... 2
1. Antheridia attached to the filament elsewhere.................... 3

2. Filaments generally less than 15 μ diam.; oöspore ovoid
V. minuta, p. 98

2. Filaments generally much more than 15 μ diam.; oöspore spherical 8
3. Antheridium not separated from the supporting filament by an empty cell or space.......... 4
3. Antheridium separated from the filament by an empty cell or space 5
4. Oöspores globose, 150–180 μ diam.......... V. thuretii, p. 97
4. Oöspores elliptical, 60–70 μ diam., 80–100 μ long V. arcassonensis, p. 97
5. Oögonium separated from the filament by an empty cell V. litorea, p. 99
5. Oögonium not separated from the filament by an empty cell....... 6
6. Plant dioecious; oöspores spherical, 125–150 μ diam. V. compacta, p. 98
6. Plants monoecious 7
7. Oöspores spherical with a crownlike lobed papilla at the tip V. coronata, p. 100
7. Oöspores depressed or lenticular with a simple terminal pore V. piloboloides, p. 98
8. Antheridia straight, with one terminal and one lateral pore; oögonia obovoid, to 110 μ long.......... V. intermedia, p. 99
8. Antheridia curved, with 3–6 pores; oögonia clavate, over 200 μ long .... V. sphaerospora, p. 100

Vaucheria arcassonensis Dangeard

Filaments about 35–55 μ diam.; plants monoecious; the antheridia usually attached to the filament close to the oögonia, or between two of them, very short-stalked, strongly curved, discharging terminally; oögonia very short-stalked, rarely sessile, ovoid, cylindrical or a little curved, the pore terminal; oöspore thick-walled, elliptical, 60–70 μ diam., 80–100 μ long.

Southern Massachusetts and Maine; fruiting from April to early May.

REFERENCE: Blum and Conover, 1953, p. 395.

Vaucheria thuretii Woronin

Filaments 30–120 μ diam., generally 80 μ; monoecious, the antheridia sessile, lateral, spreading, oblong-ovate to ovoid with an apical pore; oögonia sessile or on short lateral branches, single, obovoid or pyriform, oblique, 200 μ diam., 250–300 μ long; oöspores globose, 150–180 μ diam.; aplanospores formed in short lateral branches, ovoid, 80 μ diam., 100–120 μ long.

New Jersey to New Brunswick; forming dense green patches in muddy brackish ditches near the shore, fruiting in January.

REFERENCES: Phyc. Bor.-Am. 1029. Farlow 1881, p. 104; Collins 1900b, p. 45, 1909b, p. 424; Brown 1929, p. 95.

Vaucheria minuta Blum et Conover

Filaments 8.5–15, seldom to 17 μ diam.; antheridia and oögonia on a fruiting branch 200–800 μ long, the antheridium terminal, separated from the oögonium immediately below it by an empty space, or the oögonium terminal in the absence of an antheridium; antheridia cylindrical, 13–17 μ diam., 45–65 μ long, opening by a single lateral pore; oögonium ovoid to cylindric, 50–55, seldom to 65 μ diam., 100–125, seldom to 170 μ long, opening by a wide terminal pore; oöspore thick-walled, ovoid to cylindric, 50–53 μ, seldom to 60 μ diam., 68–92 μ long.

Southern Massachusetts and Maine; fruiting from January to early May.

REFERENCE: Blum and Conover 1953, p. 399.

Vaucheria compacta (Collins) Collins

Plants forming extensive matted tufts; dioecious, antheridia pedicelled, a very short branch terminating in the empty cell and the antheridium which it supports; oöspores spherical, 125–150 μ diam.

Massachusetts; salt marshes and salt-marsh creeks, on the mud at low tide, fruiting in January.

REFERENCES: Phyc. Bor.-Am. 477 (as *V. piloboloides* v. *compacta*). Collins 1900a, p. 13, 1900b, p. 45, 1909b, p. 429 (all as the variety); Taylor 1937b, p. 226; Christensen 1952, p. 173; Taylor and Bernatowicz 1952, p. 78.

Vaucheria piloboloides Thuret

Filaments sparingly branched, 40–60 (–80) μ diam.; monoecious; antheridia usually terminal on short branches, separated from the filament by empty cells, cylindrical to spindle-shaped with a terminal and one or two lateral conical projections which become discharge pores; diam. 30–45 μ, length 150–200 μ; oögonia terminal or on short branches near the antheridia, clavate, hemispherical at the top, 140–210 μ diam., 320–500 μ long; oöspores lenticular, 105–145 μ diam., 75–110 μ thick, with a somewhat thickened membrane, attached at the distal end of the oögonium; aplanospores sometimes formed at the ends of branches, 80 μ diam., 250 μ long.

Bermuda and Connecticut; growing on muddy and sandy shores below low-water mark.

REFERENCES: Phyc. Bor.-Am. 476. Collins 1900b, p. 45, 1909b, p. 429; Brown 1929, p. 101; Taylor and Bernatowicz 1952, p. 76.

Vaucheria intermedia Nordstedt

Filaments 15–35, seldom to 50 μ diam.; plants monoecious, the fruiting branches short, the antheridia attached to the stalk of the oögonium, or to the oögonium itself, by an intermediary sterile or suffultory cell; antheridia nearly straight, cylindrical or a little tapered, 12–20, seldom to 35 μ diam., 70–110, seldom as little as 50 or as much as 125 μ in length, usually with single terminal and lateral pores at the tips of papillae; oögonia obovoid, short, 50–90 μ diam., 75–110 μ long, the spherical moderately thick-walled oöspore, which is 50–60 μ diam., nearly filling the oögonium.

Southern Massachusetts; on mud about high-tide line, fruiting in January.

REFERENCE: Blum and Conover 1953, p. 397. European records show oögonia to 100–120 μ diam.

Vaucheria litorea C. Agardh

Filaments long, loosely tufted, 70–95 μ diam.; dioecious, the antheridia on the ends of branches of various lengths, isolated by an empty cell, cylindrical with 2–4 short lateral projections, obtuse, 54–64 μ diam.; oögonia at the extremity of reflexed branches, isolated by single empty cells, clavate to obovoid, 190–205 μ diam., 300–450 μ long; oöspore subglobose, with a membrane to 12 μ thick, occupying the upper part of the oögonium.

New Jersey to Massachusetts; growing on mud and gravel near low-water mark.

REFERENCES: Phyc. Bor.-Am. 166. Farlow 1881, p. 105; Collins 1900b, p. 45, 1909b, p. 430; Brown 1929, p. 101.

Vaucheria coronata Nordstedt

Filaments 30–45 μ diam.; plants monoecious, antheridia terminal on short successive branches developed below an oögonial initial, supported on an empty cell, cylindrical to oval, discharging through 1–2 lateral papillae, 26–39 μ diam., 46–64 μ long; oögonia oval to subspherical, sessile or nearly so, 85–114 μ diam., 107–171 μ long, the coroniform papilla at the tip developing the receptive pores; oöspore spherical, 78–107 μ diam.

Southern Massachusetts and Maine; on mud about high-tide line, fruiting from March to early May.

REFERENCE: Blum and Conover 1953, p. 396.

Vaucheria sphaerospora Nordstedt

Filaments 28–50 μ diam.; plants monoecious, one oögonium and one antheridium occurring on each fruiting branchlet; antheridia curved toward and closely adjacent or appressed to the oögonia, each supported on a suffultory cell 54–56 μ long which is attached to one side of an oögonium; antheridia fusiform, 38–57 μ diam., 128–157 μ long, with 3–6 pores through papillae 14–33 (–45) μ long; oögonia 214–386 μ long, cylindrical below, swollen above, where 89–178 μ diam.; oöspores spherical, free from the oögonial wall, 86–141 μ diam.

Northern Quebec (Ungava Bay), in a river estuary, fruiting in early September.

REFERENCE: Wilce *in herb.*

PHAEOPHYCEAE

ECTOCARPALES

Plants generally filamentous, branched and uniseriate with apical, trichothallic, or dominantly intercalary growth; sporophyte and gametophyte phases of similar appearance, reproducing by zoöspores and by iso- or anisogametes.

ECTOCARPACEAE

Plants filiform, subsimple, or more or less branched, from a creeping, penetrating, or disciform base; generally uniseriate, occasionally some cells in the lower part with one or two longitudinal septa; cells uninucleate, chromatophores parietal, irregularly band-shaped, simple or branched, or often disk-shaped; reproductive organs lateral, replacing branchlets, or intercalary from transformed vegetative cells, both unilocular and plurilocular types known, usually but not always on separate individuals, and the plants in general presenting a diploid generation producing zoöspores after meiosis in unilocular sporangia, followed by a haploid generation of similar aspect producing gametes in plurilocular gametangia.[1]

KEY TO GENERA

1. Endophytic species, sometimes with short exserted filaments....... 2
1. External free filaments dominant; often large bushy plants........ 3

2. Sporangia differing little from vegetative cells. MIKROSYPHAR, p. 116
2. Sporangia and gametangia morphologically distinct organs STREBLONEMA, p. 113

3. Gametangia in dense racemose clusters.......... SOROCARPUS, p. 116
3. Gametangia scattered, seriate or fasciculate.................... 4

[1] The terminology of the reproductive organs of the Phaeophyceae offers several difficulties owing to the frequent differences in function of the swarmers from that which should be expected from the morphology of the organs producing them. Although it is relatively easy to select a terminology which is correct and adequate for the student of life histories, its use in systematic work becomes subject to interminable revisions as individual species are studied. Consequently the old terms "plurilocular gametangium" and "unilocular sporangium" are adopted in the species descriptions for the structures ordinarily so designated, but no implication of function should necessarily derive from this use of terms.

4. Sporangia and gametangia seriate, intercalary..... PYLAIELLA, p. 102
4. Sporangia and gametangia lateral or terminal.................. 5
5. Chromatophores generally band-shaped, occasionally platelike ECTOCARPUS, p. 103
5. Chromatophores discoid, numerous............... GIFFORDIA, p. 110

Pylaiella Bory, 1823 [1]

Plants filiform, sometimes large, primarily uniseriate, widely oppositely or irregularly branched, attached by rhizoidal filaments more or less coalesced to a holdfast; unilocular sporangia cask-shaped, formed in intercalary series from cells of the vegetative branches; plurilocular gametangia usually intercalary, oblong to irregularly cylindrical, formed by repeated longitudinal and transverse division of vegetative cells.

Pylaiella littoralis (Linnaeus) Kjellman Pl. 9, figs. 1–3

Plants erect and tufted to 5–50 cm. tall, generally much branched, with age the central parts of the tufts twisted into ropelike strands; branching scattered, opposite or irregular, filaments 25–70 μ diam. near the base, with occasional longitudinal walls, above more slender; cells with rounded disk-shaped chromatophores; hairs absent; plurilocular gametangia formed from one to several contiguous cells of lateral branchlets, ovoid to cylindrical, sometimes terminal but usually with the basal and distal parts of the branchlet unaltered; unilocular zoösporangia formed in a manner similar to the gametangia, 2–30 in a series, disk- or cask-shaped, or if terminal subspherical.

New Jersey to northern Labrador, Newfoundland, James Bay, southeastern Hudson Bay, Hudson Strait, Baffin Island, Devon Island, and Ellesmere Island; common on rocks, coarse algae, or other objects in shallow water; fruiting in the winter and spring in the southernmost part of its range, becoming rare later in the season.

REFERENCES: Phyc. Bor.-Am. 171, 414. Harvey 1852, p. 139 (as *Ectocarpus littoralis*); Farlow 1881, p. 73 (as *E. littoralis*); Hariot 1889, p. 155; Whelden 1947, p. 115; Wilce *in herb.*

[1] As *Pilayella.* Schuh's report (1933a) of *P. fulvescens* appears to have been based on a misidentification.

This plant is exceedingly variable with situation and season. V. **firma** (C. Agardh) Kjellman, with branching alternate and main filaments to 50 μ diam., the sporangia mostly intercalary, has been reported from Maine, and its f. **macrocarpa** (Foslie) Kjellman (Phyc. Bor.-Am. 733), tufted, much branched, the branchlets alternate to unilateral, erect spreading to right-angled, cells in the lower parts of the plant 18–25 μ diam., 2–6 diameters long, gametangia 180–1320 μ long, 24–36 μ diam., has been reported from New Hampshire. V. **fluviatilis** (Kützing) Hauck, forming tufts 1–3 dm. tall, softly flexuose, the flaccid filaments elongate, sparingly branched, the branches tapering to the apices, gametangia in the middle or the upper part of the branchlets, or largely terminal, to 1 mm. long, has been reported from northern Massachusetts in quiet, rather brackish water. V. **opposita** Kjellman (*E. brachiatus* C. Ag.) has the branching mainly opposite, the main filaments about 45 μ diam., the sporangia mostly intercalary; it has been recorded in our range with f. **rupincola** Kjellman, in which the main filaments are seldom over 30 μ diam. V. **robusta** Farlow (Phyk. Univ. 516; Algae Exsic. Am. Bor. 168, as *E. littoralis* v. *robusta;* Farlow 1876, p. 709, as *E. farlowii*), with filaments 0.75–1.0 dm. tall, densely branching, branches robust, opposite or irregular, the cells 30–50 μ diam., fertile branches short and rigid, often almost completely transformed into sporangia, which are stout and cylindrical, has been reported from Massachusetts and Maine. V. **varia** (Kjellman) Kuckuck (Phyc. Bor.-Am. 669), with tufts to 30 cm. long, generally entangled into ropelike strands, branches opposite, cells 45 μ diam., 2–3 diameters long, gametangia short, ovate, ellipsoid to globose, on a 1–4-celled stalk, 45 μ long, 30 μ diam., sporangia mostly solitary on short branches, has been reported from Connecticut to Maine. The species is so subject to seasonal and ecological variation that recognition of stable varieties is difficult; it is doubtful if this list adequately covers the range of the American plant, and also whether some of these records and descriptions could stand critical study.

Ectocarpus Lyngbye, 1819

Plants usually freely branched, from a decumbent, rhizoidal, or, infrequently, a penetrating base, the uniseriate branches often terminating in colorless hairs, the axis sometimes corticated by rhizoids near the base; cells with parietal band-shaped, perhaps sometimes discoid chromatophores; reproduction by sessile or short-stalked globose to ellipsoid unilocular sporangia which are scattered, replacing lateral branchlets, or by plurilocular gametangia which are varied in form, stalked or sessile, generally abundantly longitudinally and transversely divided, dehiscing by a pore; gametes isogamous.[1]

KEY TO SPECIES

1. Plants very minute, seldom over 2 mm. tall. E. subcorymbosus, p. 109
1. Plants small, usually less than 2 cm. tall 2
1. Much larger, freely branched species 3

2. Filaments much entangled; gametangia linear
E. tomentosoides, p. 108
2. Filaments not conspicuously entangled; gametangia ovoid to ellipsoid .... E. paradoxus, p. 109

3. Filaments densely entangled into ropelike strands held together by recurved branchlets ....E. tomentosus, p. 108
3. Filaments free, or loosely entangled 4

4. Ultimate branchlets somewhat clustered in fascicles 5
4. Not fasciculate 6

5. Fascicles congested; gametangia ovate-acuminate to subulate, 70–150 μ long .... E. fasciculatus, p. 107
5. Fascicles loose; gametangia elongate-conical to thick subulate, to 250 μ long .... E. penicillatus, p. 107

6. Gametangia acute to often hair-tipped .... E. siliculosus, p. 105
6. Gametangia usually blunt 7

7. Gametangia usually terminal on the branchlets, very slender, to 250 μ long, 10–15 μ diam. .... E. dasycarpus, p. 105
7. Gametangia usually lateral, commonly shorter, 20–35 μ diam.
E. confervoides, p. 106

[1] The following key does not take into account the various dwarf or otherwise atypical varieties of large species, which are very common; some which have been recognized in the literature are mentioned under the typical forms. Inadequately described by Harvey (1852), *E. dietziae, E. hooperi,* and *E. landsburgii* cannot be recognized. Likewise, *E. lutosus* Harvey is of doubtful value; the specimens on which Farlow's revised description is based seem to be but a phase of *Pylaiella littoralis.* His records of *E. longifructus* Harvey from Maine and *E. amphibius* Harvey from New York have not been confirmed.

Ectocarpus dasycarpus Kuckuck

Plants tufted, not entangled, brown, 5–7 cm. tall; branching loosely pseudodichotomous, with small lateral branchlets; cells of the main filaments 20–40 μ diam., 2–3 diameters long, little constricted at the septa; plurilocular gametangia sometimes sessile, generally terminal on one- to several-celled lateral branchlets, cylindrical, 10–15 μ diam., the length variable, but to 250 μ, the tip not prolonged into a hair.

Rhode Island and southern Massachusetts; growing on various algae; fruiting in early summer.

REFERENCE: Collins 1900b, p. 46.

Ectocarpus siliculosus (Dillwyn) Lyngbye Pl. 8, figs. 4–5

Plants tufted, yellowish to brownish olive, at first attached and up to 3 dm. tall, later often free-floating, more or less entangled in the center; branching below at times pseudodichotomous, but above the branches alternate or unilateral, ascending; cells in the lower part of the main axes 40–60 μ diam., 4–5 diameters long, above the cells as long as broad or a little shorter, chromatophores irregularly band-shaped, often oblique; plurilocular gametangia sessile or usually very short-stalked, typically conico-subulate, rarely longer or shorter, 50–600 μ long, 12–25 μ diam., sometimes slightly curved, usually ending in a hair; unilocular sporangia sessile or on short stalks, ellipsoid, 30–65 μ long, 20–27 μ diam., often on the same plant as the gametangia.

Bermuda, the Carolinas, New Jersey to the lower St. Lawrence, Newfoundland, James Bay and southeastern Hudson Bay, Baffin Island, and Ellesmere Island; growing on coarse algae or other objects; fruiting in summer.

REFERENCES: Phyc. Bor.-Am. 319, 1386. Harvey 1852, p. 139, 140 (as *E. viridis*); Farlow 1881, p. 71 (as *E. confervoides* v. *siliculosus*).

F. **hiemalis** (Crouan) Kuckuck (Phyc. Bor.-Am. 372; Farlow 1881, p. 71, as *E. confervoides* v. *hiemalis*) has plurilocular gametangia 300–600 μ long by 23–30 μ diam., with a wide base but a very short terminal hair; it is reported from southern Massachusetts, fruiting in winter. In the absence of hair tips on the gametangia it is often hard to distinguish *E. confervoides* from the present species; since the gametangia are usually more slender toward the extremities of the plant, mature organs from the lower portions of the fruiting tufts should be observed.

Ectocarpus confervoides (Roth) Le Jolis Pl. 8, figs. 1–3

Tufts attached, more or less loosely entangled below, dark brown, to 5–30 (–60) cm. tall; branching irregular, pseudodichotomous or chiefly alternate but not opposite, the branches tapering, often forming numerous rhizoids, particularly near the base of the plant, and rather sparingly forming hyaline hairs at the ends; lower cells of the main branches 20–50 (–65) μ diam., 1–2 occasionally 3 diameters long; chromatophores ribbon-shaped, often forked; plurilocular gametangia sessile or with short stalks, scattered, not ending in hairs, short-subulate to fusiform, 60–150 (–250) μ long, 20–35 μ diam.; unilocular sporangia ovoid, sessile or on short stalks, 35–50 μ long, 20–40 μ diam.

Bermuda, Florida, North Carolina, New Jersey to the lower St. Lawrence, Newfoundland, and Baffin Island; attached to coarse algae and other algae and other objects; fruiting in summer.

REFERENCES: Phyc. Bor.-Am. 871. Farlow 1881, p. 71.

This species varies much in stature and general aspect, particularly with habitat and season. Some of the forms and varieties which have been recognized within our area may be worthy of retention. V. **brumalis** Holden (Phyc. Bor.-Am. 576; Collins 1900a, p. 13), having fronds 1–3 cm. long, the main filaments 25 μ in diam., the branches 12–20 μ, gametangia cylindrical to short-conical, to over 150 μ long, 15–25 μ diam., often curved, sessile or terminal on one- to several-celled branches, is a winter form from Spartina roots in Connecticut. F. **irregularis** Collins (Phyc. Bor.-Am. 670; Collins 1906a, p. 107), described as having fronds 15 cm. long, somewhat twisted below, the branching rather erect above, the main filaments to 45 μ diam., about half as thick in the branches, cells 1–3 diameters long in the main axis, to 8 diameters in the branches, slightly constricted at the nodes, gametangia variable, 20–35 μ diam., 50–150 μ long, cylindrical or tapering from the base, usually curved, the tip blunt, has been reported from Maine, fruiting in summer. F. **pygmaeus** (Areschoug) Kjellman, forming soft tufts 3–12 mm. long, or forming cushions, erect filaments 7–36 μ diam., tapering to base and apex, simple or with scattered branches, gametangia 60–75 μ long, 25–30 μ diam., few, generally pedicellate and frequently terminal on the branches, sporangia pedicellate, 17–32 μ long, 14–22 μ diam., has been reported from Rhode Island and southern Massachusetts, fruiting in the spring.

Ectocarpus penicillatus (C. Agardh) Kjellman

Plants forming attached tufts, somewhat rusty brown, to 10 cm. tall; branching without distinct primary axes, in the upper portion these apparently dichotomous with short radial branchlets aggregated into fascicles; cells with irregular, often branched, bandlike chromatophores; plurilocular gametangia sessile or short-stalked, elongate-conical or thick-siliquose, 50–200 (–250) μ long, 18–35 μ diam.; unilocular sporangia sessile or short-stalked, short-ellipsoid, 35–50 μ long, 25–30 μ diam., erect-appressed or divergent.

Connecticut to southern Massachusetts, and New Hampshire; growing on stones or large algae; fruiting in early summer.

REFERENCES: Phyc. Bor.-Am. 479. Croasdale 1941, p. 214.

Ectocarpus fasciculatus (Griffiths) Harvey

Plants in erect tufts from a disklike or somewhat penetrating holdfast (probably of rhizoidal origin), to 2–10 (–20) cm. tall, olive to brown, the axes sometimes much twisted together below and consolidated by numerous rhizoids, but free above; main filaments 30–50 μ diam., the cells subequal or to 1.5 diameters long; primary branches long, alternate, erect, characteristically with arcuate or zigzag branchlets, the ultimate shorter, repeatedly unilateral in congested fascicles, generally ending in hairs, the cells 8–15 μ diam., 1.5–3.0 diameters long; plurilocular gametangia ovate-acuminate to subulate, sessile or rarely short-stalked, unilateral on the inner side of the lesser branchlets, generally 70–150 μ long by 18–25 μ diam.

New Jersey to the lower St. Lawrence, but particularly north of Cape Cod; usually growing on Laminariaceae; an annual, fruiting in early summer.

REFERENCES: Phyc. Bor.-Am. 730, 1232. Harvey 1852, p. 141.

V. **abbreviatus** (Kützing) Sauvageau (Phyc. Bor.-Am. 731; Collins 1901a, p. 133), the plants 1–4 cm. tall, the cells short, gametangia often on main axes as well as on branchlets, sessile, erect-appressed, sometimes fasciculate, ovate-oblong, and obtuse, has been reported from northern Massachusetts and Maine and perhaps occurs through the range of the species, forming a dense coating on Rhodymenia and Saccorhiza. F. **polyrhizus** Collins (Phyc. Bor.-Am. 1535), from the Collins manuscript and the exsiccata largely spreading in the form of a dense disk from which filaments may grow into the host tissues, and from which erect subsimple filaments with abundant rhizoidal downgrowths arise, with gametangia on the spreading filaments as well as on the erect ones, 60–70 μ long, 20–25 μ diam., has been reported from Maine, on or among other Ectocarpaceae, in summer. V. **refractus** (Kützing) Ardissone consists of plants with the terminal branchlets reflexed or recurved and is reported from Maine on Porphyra, in summer.

Ectocarpus tomentosus (Hudson) Lyngbye Pl. 8, figs. 6–8

Plants forming erect dark brown spongy strands reaching a length of 2.5–20 cm.; filaments repeatedly branched, much entangled, extending freely from the surface of the ropelike cords, below 8–12 μ diam., the cells 2–4 diameters long, the lesser branches alternate, spreading or recurved, the cells 2–3 diameters long; cells with 1–2 bandlike chromatophores in each cell; plurilocular gametangia principally on the free branches, sessile or short-stalked, ovoid to more usually linear-obtuse, erect, incurved, or refracted, 25–100 μ long, 6–17 μ diam.; unilocular sporangia scattered on the branches, short-stalked, ovoid to almost spherical, 20–45 μ long, 15–30 μ diam.

New Jersey to Prince Edward Island; growing on various coarse algae; fruiting in spring or early summer, but soon thereafter largely stripped of free branches.

REFERENCES: Phyc. Bor.-Am. 478. Harvey 1852, p. 141; Farlow 1881, p. 70; Kylin 1947, p. 11, fig. 5 (as *Spongonema tomentosum*).

Ectocarpus tomentosoides Farlow

Plants epiphytic, ramifying on or near the surface of the host, eventually sending filaments upward to 5–10 mm. above it in pulvinate fashion; free filaments densely entangled, sparsely and irregularly branched, 6–8 μ diam., cells short, to 2 diameters long; plurilocular gametangia borne on numerous short, straight or slightly falcate branches, spreading from the filaments, sessile, linear-cylindrical, 60–80 μ long, 6–7 μ diam., usually simple, but occasionally forked.

Northern Massachusetts to Maine, and Devon Island; forming a coating on Laminaria; fruiting in the spring.

REFERENCES: Phyc. Bor.-Am. 230. Farlow 1889, p. 11; Rosenvinge 1893, p. 890; Kylin 1947, p. 6 (as *Laminariocolax tomentosoides*).

Ectocarpus paradoxus Montagne

Plants tufted, the tufts 5–20 mm. tall, yellowish olive, abundantly branched, the lower branches opposite to verticillate, the upper dichotomous-fastigiate, in size ranging from 30–60 μ diam., the cells 1.5–3.0 diameters long, or even shorter; gametangia on stalks of 1–2 cells, rounded-ovate, 50–130 μ long; unilocular sporangia short-pedicellate, rounded-ovate.

Maine.

REFERENCE: Schuh 1933b, p. 107.

Ectocarpus subcorymbosus Farlow, *emend.* Holden

Pl. 7, figs. 5–6

Plants small, tufted, to 1–3 mm. tall; bases crowded, disciform, of radiating filaments laterally conjoined, bearing 1–few-branched erect filaments; branches alternate or more rarely opposite, in the latter case the main filament or one of the branches may be represented by a long hair, or hairs may terminate the branches; the cells 12–15 μ diam., 2–4 diameters long, chromatophores disk-shaped; plurilocular gametangia terminal or lateral, when usually on a short stalk of 1–3 cells, sometimes crowded near the ends of branches, cylindrical, the apices bluntly rounded, 30–140 μ long, 18–25 μ diam., the segments 8–10 μ long, transversely 1–3-celled.

Maryland, Connecticut, and Massachusetts; growing on the leaves of Ruppia in coves and salt-marsh pools; fruiting in summer and autumn.

REFERENCES: Algae Exsic. Am. Bor. 197; Phyc. Bor.-Am. 415, 630. Collins 1905, p. 227. In view of the type of chromatophore found here, perhaps to be removed from the genus Ectocarpus.

Giffordia Batters 1893

Plants bushy; branching alternate or often opposite, the branchlets sometimes terminating in short hairs without basal growth; cells with numerous small discoid chromatophores peripherally disposed; reproductive organs sessile, often seriate on the branchlets, consisting of unilocular sporangia, or of plurilocular gametangia which may be of two or three types in the same species, with loculi of contrasting sizes, producing anisogametes.

KEY TO SPECIES

1. Small species hardly 2 cm. tall; filaments bound together by rhizoids below; gametangia acute................ G. ovata, p. 110
1. Much larger species; filaments free or loosely entangled........... 2

2. Branching not ordinarily opposite............................. 3
2. Branching opposite above, but the ultimate branchlets generally unilateral; gametangia subovate; sporangia sessile, globose G. granulosa, p. 112

3. Gametangia short, obliquely ovoid, not more than half longer than broad .... G. secunda, p. 112
3. Gametangia relatively longer.................................. 4

4. Cells of the main axis 2–3 diameters long; ultimate branching alternate, somewhat close but spreading; gametangia usually cylindrical, 20–30 μ diam......................G. mitchellae, p. 111
4. Cells of the main axis one-half as long as broad, to subequal; ultimate branching unilateral; gametangia usually long-ovoid, 15–35 μ diam............................. G. sandriana, p. 111

Giffordia ovata (Kjellman) Kylin

Tufted plants, to 2–6 cm. tall, the axes conjoined by rhizoids below; repeatedly distantly oppositely or more usually alternately branched, in the lower portions the cells 30–46 μ diam., about 2 diameters long, the cells somewhat shorter and cask-shaped above, with many small discoid chromatophores, the branchlet tips terminated by hairs; plurilocular gametangia scattered or crowded on the branches, often paired and opposite to each other, sessile, ovoid but acute, more rarely elongate-ovoid, 28–52 μ long, 18–28 μ diam.; unilocular sporangia generally opposite, subglobose, 20–46 μ long, 19–36 μ diam.

Rhode Island, southern Massachusetts, and Maine; growing on larger algae; fruiting in midsummer.

REFERENCES: Collins 1896b, p. 458 (as *Ectocarpus ovatus*); Kylin 1947, p. 9, fig. 3A, B; Taylor 1937c, p. 114 (as *E. ovatus*).

Giffordia mitchellae (Harvey) Hamel

Plants tufted, soft, and feathery, to 20 cm. tall; repeatedly branched, the branches spreading, delicate, alternate, more closely set above, at the tips attenuate or hair-tipped; the cells of the main axes about 50 μ diam., 2–3 diameters long, of the branchlets 30 μ diam., 1.5 diameters long, chromatophores discoid, numerous; plurilocular gametangia sessile in series on the upper side of the branches, elliptic-oblong to linear-cylindrical, 50–150 μ long, 20–30 μ diam., the meiosporangia with cells about 6–7 μ wide, the megasporangia (not reported from New England) with cells 10–17 μ wide.

From the tropics, Bermuda, Florida, North Carolina, Rhode Island, and southern Massachusetts; growing on various objects; fruiting in autumn.

REFERENCES: Phyc. Bor.-Am. 321. Harvey 1852, p. 142; Farlow 1881, p. 72; Collins 1891, p. 337; Taylor 1937c, p. 111 (all as *E. mitchellae*).

The v. **parvus** Taylor (in Lewis and Taylor 1921, p. 254) is described as 8–12 mm. tall, with the erect filaments 22 μ diam., sparsely branched; plurilocular gametangia sessile, erect on the upper side of the secondary branches, cylindrical, obtuse, averaging 60 μ long by 16 μ diam., 1–2 cells broad and so possibly megasporangial; it is recorded from southern Massachusetts, growing affixed to the carapace of a marine turtle and to floating timbers, and fruiting in summer.

Giffordia sandriana (Zanardini) Hamel

Plants attached by rhizoidal downgrowths, bushy, to 4–12 cm. tall; branching alternate below, above densely corymbose, in the ultimate branchlets unilaterally comblike, the main filaments 30–100 μ diam., cells short or to 2 diameters long, the branchlets 10–20 μ diam., the cells about as broad as long; plurilocular gametangia small, sessile, usually numerous and unilaterally disposed along the inner side of the lateral branches, long-ovoid, 45–60 μ long, 15–35 μ diam.

Southern Massachusetts; on various algae at 2 meters depth over a muddy bottom; fruiting in winter.

REFERENCES: Phyc. Bor.-Am. 320 (as *E. elegans*). Taylor 1937c, p. 112 (as *E. sandrianus*); Kylin 1947, p. 10, fig. 3C, D.

Giffordia secunda (Kützing) Batters

Plants tufted, to 2–8 cm. tall; filaments corticated by rhizoids below, with few vegetative branchlets or with these often recurved; main branching alternate, the cells 60–90 μ diam., 1–2 diameters long with many discoid chromatophores, the upper branching unilateral; plurilocular gametangia dimorphic, the macrogametangia blunt-fusiform to short-conical or ovoid, with an obliquely truncate base, 80–90 μ long, 50–60 μ diam., the cells 8–10 μ wide, the microgametangia ovoid to subspherical, 60–80 μ diam., the cells about 2 μ wide, becoming orange at maturity, and losing the chambered aspect of a plurilocular structure.

Rhode Island to New Hampshire.

REFERENCES: Harvey 1852, p. 142 (as *E. durkeei*); Farlow 1881, p. 70 (as *E. granulosus* v. *tenuis*); Taylor 1937c, p. 112 (as *E. secundus*).

Giffordia granulosa (J. E. Smith) Hamel Pl. 7, figs. 7–9

Plants 3–20 cm. tall, olivaceous, somewhat stiff, the primary filaments being corticated by rhizoids; repeatedly widely oppositely branched, the main axes 50–100 μ diam., cells shorter than long or subequal; secondary branches sometimes recurved at the tips, the short ultimate branchlets opposite or more often comblike on one side, the cells to 1.5 diameters long, chromatophores discoid, numerous; plurilocular gametangia generally numerous, unilateral on the upper branchlets, broadly ovoid, often asymmetrical, sessile on a broad base, 60–80 μ long, 40–60 μ diam.; unilocular sporangia sessile, asymmetrical, subglobose.

New Jersey to Maine, but rare north of Cape Cod; growing on other and coarser algae; fruiting in the winter and occasionally into early summer.

REFERENCES: Harvey 1852, p. 141; Farlow 1881, p. 70; Taylor 1937c, p. 113 (all as *E. granulosus*); Kylin 1947, p. 11.

Streblonema Derbès et Solier, 1851

Plants creeping or penetrating the host, of irregularly branched filaments, not forming a regular disk, the cells with a few lenticular or somewhat elongated chromatophores; erect filaments absent, or short and not exserted, not different in character from the spreading ones; colorless hairs present and becoming exserted; plurilocular gametangia irregular, erect, cylindrical to ovoid or branched, uni- to pluriseriate; unilocular sporangia oval to spherical.

KEY TO SPECIES

1. Short external filaments, 4–8 cells long, extending outside the host .... S. parasiticum, p. 116
1. External filaments not regularly present........................ 2
2. Sporangia crowded in circumscribed eruptive sori S. aecidioides, p. 115
2. Fruiting area not sharply defined............................ 3
3. Unilocular sporangia present............................... 4
3. Plurilocular gametangia present............................. 5
4. Sporangia oval, to 140 μ long................ S. chordariae, p. 113
4. Sporangia subpyriform to spherical, to 60 μ long. S. sphaericum, p. 114
5. Gametangia usually fasciculate, often branched. S. fasciculatum, p. 114
5. Gametangia simple, scattered................................ 6
6. Filaments seldom to 10 μ diam.; gametangia uniseriate or biseriate .... S. oligosporum, p. 114
6. Filaments seldom less than 10 μ diam.......................... 7
7. Gametangia uniseriate .... S. sphaericum, p. 114
7. Gametangia pluriseriate .... S. effusum, p. 115

Streblonema chordariae (Farlow) De Toni

Plants endophytic, the freely branching filaments penetrating among the superficial tissues of the host; the branches nodose, about

20 μ diam., the spreading filaments bearing erect hairs and branchlets, the former penetrating the surface of the host; unilocular sporangia stalked, solitary, or clustered oval, 140 μ long, 70 μ diam.

Rhode Island to Maine and Nova Scotia; endophytic in various plants.

REFERENCE: Farlow 1881, p. 69 (as *Ectocarpus chordariae*).

Streblonema sphaericum Derbès et Solier

Plants endophytic, the branching filaments ramifying among the outer tissues of the host; filaments irregular, branched, moniliform, the cells 10–14 μ diam., 1–4 diameters long, each with a few discoid chromatophores; erect filaments absent; colorless hairs 10–14 μ diam.; plurilocular gametangia sessile, lateral on the branches or terminal, cylindrical to subfusiform, generally uniseriate, 30–65 μ long, 12–16 μ diam.; unilocular sporangia sessile, solitary, or often clustered, subpyriform becoming nearly spherical, 40–60 μ diam.

Southern Massachusetts; endophytic in soft algae, such as Eudesme and Nemalion.

Streblonema oligosporum Strömfelt Pl. 11, fig. 6

Basal filaments spreading over or in the superficial matrix of the host plant or among the cortical cells, 5–10 μ diam., the erect filaments short, tufted, the longer erect branches sometimes a little exceeding the cortex of the host, below of cells with 2–4 discoid chromatophores, 5–10 μ diam., about as long as broad, but above terminating in colorless hairs; plurilocular gametangia ovoid-lanceolate to obtuse-linear, uniseriate or, in part, biseriate, 25–40 μ long, 8–15 μ diam., sessile on the decumbent filaments or terminal on the short erect branches.

Maine; endophytic in Gloiosiphonia; fruiting in summer.

REFERENCES: Phyc. Bor.-Am. 1336. Collins 1906b, p. 125; Kylin 1947, p. 20, fig. 16 (as *Entonema oligosporum*).

Streblonema fasciculatum Thuret

Plants endophytic, the filaments ramifying among the outer tissues of the host; filaments irregularly branched, 8–12 μ diam., the cells cylindrical-ellipsoid, 1–3 diameters long; colorless hairs 8 μ diam., terminating short branchlets; plurilocular gametangia stalked, solitary or fasciculate, simple, spindle-shaped or sparingly digitately branched, 65–200 μ long, each portion 10–20 μ diam.

Reported from Maine; endophytic in Eudesme; fruiting in summer.

REFERENCE: Collins 1896a, p. 3.

Streblonema aecidioides (Rosenvinge) Foslie

Plants minute, appearing as brown spots on the surface of the host; composed of vegetative filaments generally 6–8 μ diam. growing deeply between the host cells and below the epidermal layer, irregularly and widely branching; erupting locally through the epidermis in crowded sori of sporangia and hairs, sometimes also with chromatophore-bearing filaments 6–10 μ diam., cells 1.5–3.5 diameters long, the filaments to 50, rarely 150 μ long; plurilocular gametangia and unilocular sporangia in separate sori, formed laterally on vegetative filaments or short-stalked and 2–few together; gametangia linear, 35–80 μ long, 6.0–7.5 μ diam., with 8–14 uniseriate loculi; sporangia obovate, 28–62 μ long, 17–28 μ diam., abruptly contracted at the base.

Connecticut to Maine; endophytic in Laminaria; fruiting in spring.

REFERENCES: Phyc. Bor.-Am. 373. Rosenvinge 1893, p. 894; Collins 1896a, p. 4; Kylin 1947, p. 21, fig. 18 (as *Entonema aecidioides*).

Streblonema effusum Kylin, prox. Pl. 11, fig. 15

Plants microscopic, with filaments chiefly endophytic, ramifying between those of the host; filaments cylindrical to irregular or submoniliform, the cells 9.5–19.5 μ diam.; erect filaments more regularly cylindrical, the cells 9.5–11.5 μ diam., 15.0–17.0 μ long; gametangia sessile on the creeping filaments, or on short 1–3-celled stalks, or on the ends of short erect filaments, broadly lanceolate to ovate, the apex subacute, 15–26 μ diam., 38–75 μ long.

Southern Massachusetts; endophytic in Nemalion; fruiting in summer.

REFERENCE: Kylin 1947, p. 20 (as *Entonema effusum*).

Streblonema parasiticum (Sauvageau) De Toni

Plants primarily wide-spreading in the superficial tissues of the host, the cells 2–10 μ diam., 8–30 μ long; external filaments erect, crowded, simple, obtuse or hair-tipped, 4–8 cells long, the cells 6–8 μ diam., 6–12 μ long, with single platelike chromatophores; plurilocular gametangia sessile, or on 1–2-celled stalks, subcylindrical, attenuate at the tips, to 50 μ long, 9–10 μ diam.

Maine; endophytic in Cystoclonium; fruiting in summer.

REFERENCES: Phyc. Bor.-Am. 1337. Collins 1906b, p. 125; Taylor 1937c, p. 116 (as *Ectocarpus parasiticus*).

Mikrosyphar Kuckuck, 1895

Thallus consisting of uniseriate branched filaments creeping within the host membrane, sometimes free, sometimes united into a pseudoparenchymatous tissue; vegetative cells generally twice as long as broad, with one or two platelike chromatophores; hairs present or absent; reproduction by plurilocular gametangia and unilocular sporangia formed below and discharging at the surface of the host membrane.

Mikrosyphar porphyrae Kuckuck — Pl. 11, figs. 3–5

Plants minute, endophytic, forming brown spots on the host, composed of radiating filaments ramifying between and over the cells in the membrane, uniseriate, sometimes subparenchymatously aggregated, branching irregular, cells 3–5 (–8) μ diam., 6.4–9.6 μ long in the center of the colony to 19–23 μ at the margin, with 1–2 platelike chromatophores; branch tips, hairs, and sporangia penetrating the cuticle of the host; unilocular sporangia formed on the ends of the very short 1–2-celled branchlets, spherical, 6.5 μ diam.; plurilocular gametangia lateral, sessile, erect, of 3–5 cells, blunt-cylindrical, about 5 μ diam., 16 μ long.

Southern Massachusetts; not infrequent in old Porphyra blades; fruiting during the latter part of the summer.

REFERENCES: Phyc. Bor.-Am. 2293. Collins 1918b, p. 143.

Sorocarpus Pringsheim, 1862

Plants of erect, bushy habit; repeatedly and widely branched, the branches uniseriate, bearing lateral and terminal hairs; plurilocular gametangia in small densely crowded racemose clusters at the bases of branchlets or hairs, producing anisogametes; unilocular sporangia unknown.

Sorocarpus micromorus (Bory) Silva Pl. 9, fig. 6

Plants tufted, to 20 cm. tall; branching irregularly alternate, the filaments about 50 μ diam. below and 20 μ diam. above, the cells 1.5–3.0 diameters long, with a few discoid chromatophores; branches terminating in, or bearing laterally, colorless hairs; plurilocular gametangia oval, clustered, the clusters on 1–2-celled stalks or subsessile, near the bases of branchlets, branches, or hairs, the gametangia 20–25 μ long, 12–15 μ diam.

Rhode Island, southern Massachusetts, Maine, and Ile Miquelon; rare, growing on larger algae; probably fruiting in the spring.

REFERENCES: Hariot 1889, p. 155; Collins 1896b, p. 458; Schuh 1933f, p. 347 (all as *S. uvaeformis*); Silva 1952, p. 272.

SPHACELARIALES

Plants generally filamentous, branched and polysiphonous; growth from large apical cells, the segments from which generally divide longitudinally in a regular and characteristic fashion; sporophyte and gametophyte phases of similar appearance, reproducing by zoöspores and by iso- or anisogametes.

SPHACELARIACEAE

Plants with a stoloniferous base or erect from the holdfast; branches sometimes similar to the axis, naked or corticated by rhizoids, but in other cases the axis and its main divisions clothed with short branchlets of limited growth; unilocular sporangia and plurilocular gametangia on branchlets derived from the vegetative axis or from the cortex.

KEY TO GENERA

1. Plants 0.5–3.0 cm. tall, usually without distinctive heavier axes SPHACELARIA, p. 118
1. Plants larger, the main axes distinctly thicker than the determinate branches 2

2. Vegetative branchlets short, whorled, rather closely covering the axis .... CLADOSTEPHUS, p. 123
2. Vegetative branches alternate or opposite....................... 3
3. Habit congested, branches clustered; uncorticate branchlets several times divided........................ HALOPTERIS, p. 123
3. Habit open, plane; uncorticate branchlets simple or once divided CHAETOPTERIS,[1] p. 122

Sphacelaria Lyngbye, 1819

Plants small, forming subglobose tufts or spreading mats, attached by basal disks or stolons ramifying on or penetrating the support; erect axes subsimple to bushy, generally abundantly laterally determinately branched, the ultimate divisions irregular or two-ranked; indeterminate branches infrequent, like the main axis; hairs present in some cases; segments from the apical cell transversely once divided before longitudinal segmentation, secondary transverse divisions present or absent; pedicellate, stalked, usually triradiate propagula of various shapes occur on many species; plurilocular gametangia, alike or heteromorphic, or unilocular sporangia occur on simple or forked lateral stalks.

KEY TO SPECIES

1. Base filamentous; disklike holdfasts little or not developed...... 2
1. Basal disks distinct... 3
2. Plants forming a dense felt, supporting erect filaments; propagula unknown .... S. britannica, p. 119
2. Plants forming loose tufts; propagula biradiate.. S. furcigera, p. 119
3. Basal disks connected by stolons; plants forming dense mats S. radicans, p. 119
3. Stolons absent, the individuals free.......................... 4
4. Ultimate branching producing flat, evenly bilateral blades........ 7
4. Ultimate branching at most irregularly bilateral................ 5
5. Surface cells becoming transversely divided...... S. racemosa, p. 120
5. Surface cells without transverse divisions..................... 6
6. Propagula contracted at the base; branching freely, somewhat bilaterally................................ S. cirrosa, p. 121
6. Propagula not contracted at the base; branching sparse S. fusca, p. 120

[1] *S. plumigera* may be confused with small *Ch. plumosa.*

7. Rhizoids only originating in the plane of the branching; stalked sporangia borne on the primary branches.... S. plumigera, p. 122
[7. Rhizoids probably originating without restriction; sporangial branchlets borne on the cortex only.......... Chaetopteris, p.122]

Sphacelaria britannica Sauvageau

Plants arising from a basal layer of felted creeping stoloniferous filaments to about 1 cm. in height; erect filaments irregularly branched, 14–30 μ diam., secondary segments nearly or as long as broad (10–16 μ), not divided transversely, but simple or longitudinally divided by 1–2 walls; propagula and plurilocular gametangia unknown; unilocular sporangia on 1–3-celled stalks, oval to spherical, 40–46 μ diam.

Maine and southern Nova Scotia; forming considerable mats in crevices of rocks or on the bottoms of tide pools.

REFERENCE: Collins 1911a, p. 270.

Sphacelaria furcigera Kützing

Plants tufted, arising from a basal layer of stoloniferous filaments which are superficial or somewhat penetrating, to a height of as much as 3 cm.; erect filaments 16–45 μ diam., branching irregular, segments as long as broad or longer, with 1–3 longitudinal septa; propagula pedicellate, the stalks attenuate at the base but nearly cylindrical, at the apex equally biradiate, cylindrical or distally attenuate, the cells simple or with one longitudinal wall; plurilocular gametangia on separate plants, short-stalked, heteromorphic, the microgametangia cylindrical, 45–65 μ long, 24–28 μ diam., with cells about 3 μ diam., the macrogametangia more irregular in form, 30–60 μ long, 28–40 μ diam., with cells 5–7 μ diam.; unilocular sporangia sometimes associated with propagula on the same plant, on one-celled stalks, globose, 50–70 μ diam.

Bermuda, Florida, and Massachusetts; on scallop shells and other objects.

Sphacelaria radicans (Dillwyn) C. Agardh

Plants small, forming a low turf 1.0–1.5 cm. tall, the basal parts attached by small disklike holdfasts which give rise to stolons bearing new disks and filaments; erect axes little branched, the branches like the primary axis, 35–55 μ diam., divided into secondary segments which are as long as broad or longer toward the base of the filaments; upper secondary segments often containg one undivided tanniferous cell; hairs found on sterile plants; rhizoids numerous; propagula unknown; plurilocular gametangia not well substantiated; unilocular sporangia sessile along the axis or the main branches, or with the base somewhat immersed or, more rarely, short-stalked, often opposite, 24–64 μ long, 30–52 μ diam.

Long Island, Connecticut, and southern Massachusetts, with much doubt as to all records; more reliably reported from northern Massachusetts to Nova Scotia, southeastern Hudson Bay, and Newfoundland; forming dense turfs in tide pools, often becoming buried in sand.

REFERENCES: Phyc. Bor.-Am. 322(?). Harvey 1852, p. 137; Farlow 1881, p. 76.

Sphacelaria fusca (Hudson) C. Agardh

Plants arising from a compact one-layered disk of radiating filaments, the erect filaments 1–3 cm. tall, sparingly branched, 60–80 μ diam.; propagula pedicellate at the base, stalked, the stalk increasing in girth from base to apex, triradiate, the arms not constricted at their bases, projecting at about 90° from each other, cylindrical or tapering somewhat to the tips; stalk with a buttonlike cell at the summit, but no terminal hair between the arms; gametangia and sporangia unknown.

Massachusetts, Maine.

REFERENCES: Phyc. Bor.-Am. 1392. Collins 1908c, p. 157.

Sphacelaria racemosa Greville v. **arctica** (Harvey) Reinke

Plants to 4–5 cm. tall, somewhat tufted with clustered branches, the primary filaments capillary, below corticated by rhizoids which arise first on the bases of the branchlets; branchlets rather erect, radially alternate, less often opposite or (in a form) pinnate, cortical segments generally transversely divided; plurilocular gametangia distributed along the branchlets on 1–2-celled stalks, elongate but obtuse conical to clavate, 45–55 μ diam., 120–200 μ long; unilocular sporangia similarly disposed, 40–60 μ diam., 50–75 μ long, 2–20 together, rarely single; propagula unknown.

Baffin Island and Ellesmere Island; growing on coarse algae and stones in exposed localities; fruiting in winter and spring. Records for New England are apparently in error.

REFERENCES: Farlow 1879, p. 169; Kjellman 1883, p. 274 (as *S. arctica*); Rosenvinge 1926, p. 19.

Sphacelaria cirrosa (Roth) C. Agardh Pl. 17, figs. 1–6

Plants forming soft rounded tufts, to 0.5–2.0 cm. tall; color dull olivaceous brown, texture a little stiff; basal portion of filaments radiating in a simple compact disk; lower branches occasionally recurved and rhizoidal at the tip or forming a secondary attaching disk, but true stolons absent; erect filaments 40–100 μ diam., the secondary segments about as long as broad; branching usually abundant, alternate, opposite, or irregular, the ultimate determinate branchlets short, sometimes somewhat spreading in a plane; propagula frequent, to about 500–600 μ long, the stalk portion sharply contracted to the point of attachment on the parent branch, the arms usually 3 in number, subcylindrical or generally fusiform, and terminating the stalk between the arms a hair which may drop off; plurilocular gametangia very rare, often in the same tufts with sporangia, unilateral on side branches, subtruncate-cylindrical, 70–80 μ long, 60–65 μ diam.; unilocular sporangia uncommon, solitary on one-celled stalks, globose, 75–100 μ diam.

New York to the lower St. Lawrence, Newfoundland, and Baffin Island; epiphytic on Fucaceae and sometimes on other algae and Zostera, and in a small form on stones, shells, and barnacles; fruiting in autumn and winter, forming propagula abundantly in summer.

REFERENCES: Algae Exsic. Am. Bor. 164; Phyc. Bor.-Am. 416. Harvey 1852, p. 137; Farlow 1881, p. 76.

In addition, there is recognized from our range f. **meridionalis** Sauvageau, which has shorter tufts that are less noticeably pinnate and stiffer, the arms of the propagula are more spindle-shaped, and when young the apical cells of the arms are somewhat inflated; Connecticut, southern Massachusetts, and New Hampshire; usually on rocks.

Sphacelaria plumigera Holmes

Plants erect, from a basal disk of some millimeters in diameter, to several centimeters tall, the ultimate branches spreading in a plane; erect shoots pinnately branched; 70–100 μ diam., in section in the lower part showing a core of larger cells derived from the axis by segmentation, surrounded by a small-celled cortex composed of densely matted rhizoids; lateral branches borne on the superior secondary segments, 30–40 μ diam., usually corticated below; ultimate branchlets 20–40 μ diam., infrequently corticated about the base, structurally showing a central cell and a layer of peripheral cells; hairs generally absent; rhizoids produced from the cells of the branches in the general plane of branching only, descending and enveloping the axis to form a thick, complete cortication; propagula and plurilocular gametangia unknown; unilocular sporangia rounded or ovoid, 50–68 μ long, 40–52 μ diam., borne by the primary branches on scattered simple or occasionally branched stalks which are generally pinnately arranged.

Southern Massachusetts; small plants dredged off a stony bottom in relatively shallow water; sterile in late summer.

REFERENCE: Lewis and Taylor 1933, p. 151.

Chaetopteris Kützing, 1843

Plants bushy, erect, the ultimate branching in plane pinnate blades; holdfast disklike, all axes polysiphonous, the erect percurrent main axis with irregular principal branches formed from the upper initial segments; the main branches with a pseudoparenchymatous outer cortex formed by rhizoids originating apparently at random around the axis; primary branchlets two-ranked, with a few irregular pinnules near their distal ends; reproductive organs stalked, on short branchlets derived from the outer rhizoidal cortex of older, otherwise naked branches.

Chaetopteris plumosa (Lyngbye) Kützing

Plants from a disklike base, height to 4–8 cm.; main axes slender and wiry, ultimate branching bilateral, flattened; the axes naked of branchlets but heavily corticated by rhizoids below, 250–500 μ diam., less corticated and with numerous simple determinate branchlets above, 45–60 μ diam., forming lanceolate plumose blades; unilocular sporangia and plurilocular gametangia borne on different individuals, in 1–2 lateral rows or irregular on the concave side of special cortical branchlets mostly on the lower portion of the plant, the sporangia broadly oval, on 1–4-celled stalks, to 46 μ long, 28–40 μ diam.

Nova Scotia, Prince Edward Island, Newfoundland, Labrador, Hudson Bay, James Bay, Hudson Strait, Baffin Island, and Ellesmere Island.

REFERENCES: Harvey 1852, p. 136; Farlow 1881, p. 77; Rosenvinge 1926, p. 19; Wilce *in herb.*

Halopteris Kützing, 1843

Plants freely branched, the branches polysiphonous, becoming closely invested by rhizoidal outgrowths producing a pseudoparenchymatous cortex; branches formed by division of the rather large apical cells; fertile branchlets arising in placental areas in the axils of the branches, forming clusters of unilocular sporangia or plurilocular gametangia.

Halopteris scoparia (Linnaeus) Sauvageau

Plants erect, often several from the same base, attached by rhizoids; tufts to 15 cm. tall, with a thick axis which becomes very densely tomentose below by growth of the corticating rhizoids; above densely branched, fasciculate, sometimes producing obconic clusters, the ultimate 65–100 μ diam., bearing regularly two-ranked, spreading spinuliform branchlets 0.3–2.0 mm. long; plurilocular gametangia rare and incompletely known, not on sporangial plants, short-stalked, 100–110 μ long, 90–100 μ diam., pluriseriate; unilocular sporangia in groups on simple stalks in the axils of short branches, elliptical to globular, 60–80 μ diam.

Northern New Brunswick, Nova Scotia, Prince Edward Island, and Hudson Strait.

REFERENCE: Phyc. Bor.-Am. 986.

Cladostephus C. Agardh, 1817

Plants bushy, branched, the branches of tomentose aspect, dark brown and a little stiff in habit; primary cortex becoming covered externally by meristematic activity at the surface, and eventually enveloped in turn by rhizoidal investment, developing a pseudoparenchymatous aspect in section; branches beset with short branchlets in whorls, these developed from the upper primary segments; unilocular sporangia and plurilocular gametangia developed on short branchlets produced by the secondary cortex, always on separate plants.

Cladostephus verticillatus (Lightfoot) Lyngbye

Pl. 17, figs. 9–11

Plants bushy, from small leathery disks which may be associated in expansions of 1–3 cm., the tufts erect to 5–25 cm., dark brown, tomentose and somewhat firmly spongy; primary branching ditrichotomous, erect-spreading, somewhat curved, below sometimes extensively denuded, but above retaining variably crowded whorls of upcurved determinate branchlets, about 25 in a whorl, their bases contracted and tapering from the middle toward the apex, bearing 1–few smaller branchlets; denuded lower portions with much thickened cortex, developing many crowded, irregularly placed, and smaller cortical branchlets, which in season bear ovoid to cylindrical plurilocular gametangia or globose unilocular sporangia on different individuals; gametangia usually lateral on the little branchlets, short-stalked, 50–90 μ long, 25–30 μ diam.; sporangia borne similarly, 55–80 μ long, 35–55 μ diam.

New York to southern Massachusetts, perhaps rarely north of Cape Cod; growing on rocks and shells at moderate depths in exposed localities; fruiting chiefly during the winter.

REFERENCES: Algae Exsic. Am. Bor. 165; Phyc. Bor.-Am. 30, 275. Harvey 1852, p. 135 (incl. *C. spongiosus*); Farlow 1881, p. 78.

TILOPTERIDALES

Plants filamentous, branched; growth trichothallic or localized intercalary, the segments usually ultimately undergoing limited longitudinal division; sporophyte and gametophyte phases of similar appearance, the asexual generation apparently reproducing by uninucleate monospores and the sexual generation involving microgametes and motionless cells which may represent macrogametes.

TILOPTERIDACEAE

Plants attached by rhizoids, filamentous, repeatedly branched, polysiphonous below, especially in the main axis; cells with numerous small disklike chromatophores; asexual reproduction in some cases by uninucleate spores, in others by quadrinucleate spores, both formed singly in sporangia; sexual reproduction probably by small biciliate microgametes produced in tubular microgametangia, in some cases perhaps acting as antherozoids and associated with large motionless macrogametes.

KEY TO GENERA

1. Plant tufted or entangled, radially branched, or the ultimate branches tending to lie in a plane; unilocular sporangia lateral
   HAPLOSPORA, p. 125
1. Plant tufted, bilaterally pinnate, the ultimate branchlets lying in a plane; unilocular sporangia axial, usually paired
   TILOPTERIS, p. 126

Haplospora Kjellman, 1872

Plants filiform, monosiphonous above, more or less polysiphonous below; branches issuing irregularly from all sides of the main axis, in growth trichothallic; asexual reproduction by non-motile quadrinucleate spores formed singly in terminal, stalked, sessile or, rarely, intercalary sporangia; uninucleate monosporangia (oögonia?) partly immersed in the branches of sexual plants; sexual reproduction by intercalary tubular microgametangia.

Haplospora globosa Kjellman

Plants tufted, finely filamentous, sometimes somewhat entangled, 2.5–30 cm. tall; yellowish to rich brown, attached by rhizoids; filaments densely irregularly unilaterally or oppositely branched, below 50–135 μ diam., above more slender, the secondary branches often short, recurved, and hair-tipped; quadrinucleate monosporangia on 1–several-celled stalks, mostly borne laterally, infrequently sessile, spherical, 85–115 μ diam.; uninucleate monosporangia (oögonia?) usually several on the distal branches, spherical, developed by the bipartition of an axial cell, one of the halves forming the sporangium and the other remaining sterile, the base consequently usually immersed, 45–60 μ diam.; microgametangia intercalary, tubular, 75–150 (–430?) μ long, 30–40 (–75?) μ diam.

Southern Massachusetts and southeastern Hudson Bay; reported a plant of relatively deep water or shallow bays; sexual material was obtained from the Massachusetts station in midwinter, but monospores have been reported in Europe in the summer.

REFERENCE: Farlow 1882, p. 67 (as *Scaphospora kingii*).

Tilopteris Kützing, 1849

Plants filamentous, with plane branching; filaments polysiphonous below by longitudinal division of the initial segments, uniseriate above; bipinnate, the ultimate branchlets opposite with very short segments; sporangia formed by the transformation of one to several original branch segments, forming single uninucleate protoplasts at maturity; reproduction also reported to occur by quadrinucleate membrane-covered monospores; microgametangia tubular, intercalary, often associated with a few cells which are much larger and produce non-flagellate uninucleate cells which may be eggs.

Tilopteris mertensii (J. E. Smith) Kützing

Plants tufted, pale olivaceous, to 15–25 cm. tall; primary axes dominant, the lateral indefinite axes much shorter, pinnate throughout, 100–150 μ diam., bearing many branches mostly 1–5 mm. long; branchlets opposite and equal, or alternately longer and shorter; chromatophores small, disklike; sporangia (including oögonia?) 40–68 μ diam., to 60–75 μ long, generally in pairs in branchlets 30–40 μ diam.; microgametangia formed in similar branchlets, cylindrical or somewhat larger below, hollow, sometimes with a few very large divisions producing large and non-flagellate cells which may be eggs.

Southern Massachusetts; on a stony bottom in shallow water; with sporangia in the summer.

REFERENCES: Lewis and Taylor 1933, p. 151; Taylor 1941, p. 73.

CHORDARIALES

Plants of filamentous construction, the filaments with apical, trichothallic, or intercalary growth, the strands single and becoming corticated or more often grouped into branches of the multiaxial type, or habit reduced and the plants more or less pulvinate or disciform; plants as ordinarily described reproducing by unilocular sporangia and often with plurilocular gametangia or gametangium-like structures, but also bearing functional gametangia and isogametes on a small filamentous phase developed from zoöspores originating in the unilocular sporangia.

KEY TO FAMILIES

1. Plants large, tuberiform to usually erect and branched, corticated . . 2
1. Plants small to minute, or if crustose sometimes several centimeters in diameter 3

2. Plants slenderly branched, uniaxial, becoming corticated, growth at least in young tips trichothallic . . . . Acrothricaceae, p. 149
2. Plants slenderly branched, with four or five axial filaments, becoming corticated, growth apical . . . . Stilophoraceae, p. 150
2. Plants tuberiform to slenderly branched, multiaxial Chordariaceae, p. 144

3. Plants disklike, crustose or pulvinate 4
3. Plants small, tufted from a spreading or penetrating base, the erect portion of free filaments with intercalary growth Elachisteaceae, p. 138

4. Plants minute, pulvinate or punctiform, of spreading filaments sometimes bearing short erect branches . . Myrionemataceae, p. 127
4. Plants minute, or to several centimeters in diameter, disk- or crustlike 5

5. Sporangia of transformed surface cells . . . . Lithodermataceae, p. 137
5. Sporangia lateral to paraphyses in sori . . . . Ralfsiaceae, p. 133

MYRIONEMATACEAE

Small cushionlike plants, consisting of radiating branched filaments closely laterally approximated, bearing vertical filaments of limited growth, sporangia, gametangia, hairs, and paraphyses on the upper surface.

KEY TO GENERA

1. Erect filaments free 2
1. Erect filaments enclosed in jelly . . . . Microspongium, p. 128

2. Cylindrical or clavate-vesicular paraphyses present Ascocyclus, p. 132
2. Paraphyses absent 3
3. Basal layer simple.......................... Myrionema, p. 130
3. Basal part of two cell layers................. Hecatonema, p. 128

Microspongium Reinke, 1888

Plants in the form of small gelatinous cushions or elevated disks; basal portion of radiating filaments, at first of one layer, later distromatic; erect filaments arising from the basal portion, surrounded by a gelatinous investment, branched, bearing hairs; plurilocular gametangia and unilocular sporangia lateral on the vegetative filaments, gametangia filiform, uniseriate, the sporangia ovoid or clavate.

Microspongium gelatinosum Reinke

Plants small, somewhat cushionlike; erect filaments at first simple, with conspicuous hairs, later becoming more branched when producing the reproductive organs; basal layer distromatic on the plants with plurilocular gametangia, reported unistratose on those with unilocular sporangia; gametangia lateral on the erect filaments, uniseriate, 20–40 μ long, 5 μ diam.; sporangia usually single, lateral, ovoid or broadly clavate, 20–120 μ long.

Maine; growing on fuci and Alaria; fruiting in late summer.

REFERENCE: Collins' manuscript.

Hecatonema Sauvageau, 1897

Plant a compact disk of radiating filaments, the filaments commonly becoming divided once in the plane of the disk, so that a more or less completely bistratose layer results, which may produce a few rhizoids below, and above scattered erect assimilatory filaments which, like the basal layer, may bear pluriseriate sporangia and hairs.

KEY TO SPECIES

1. Gametangia stalked, lateral as well as terminal on the erect filaments 2
1. Gametangia sessile on the basal filaments or terminal on short erect pedicels .... H. reptans, p. 130

2. Erect filaments up to 2 mm. tall, the cells to 4 diameters long
H. terminalis, p. 129
2. Erect filaments much shorter, the cells to 1.5 diameters long
H. maculans, p. 129

Hecatonema terminalis (Kützing) Kylin

Plants inconspicuous, tufted; bases disklike, of irregularly entangled primary basal filaments about 10–18 μ diam., cells 8–24 μ long, the erect filaments to 2 mm. tall, infrequently branched, cylindrical, or a little attenuate, 8–12 μ diam., cells 8–48 μ long; plurilocular gametangia terminal and lateral on the erect filaments, on short stalks, ovoid to oblong, generally somewhat distorted, not rostrate, 48–120 μ long, 16–18 μ diam., the cells 4–7 μ diam.; unilocular sporangia little known, reported as ellipsoid or oval, 40–52 μ long, 24–30 μ diam., terminal on the branches.

New Jersey, Connecticut, Rhode Island, northern Massachusetts to Maine; on piling and other objects; fruiting in winter.

REFERENCES: Phyc. Bor.-Am. 126. Taylor 1937c, p. 115 (as *Ectocarpus terminalis*); Kylin 1947, p. 15.

Hecatonema maculans (Collins) Sauvageau

Plants disciform, epiphytic, of radiating filaments 12–14 μ diam.; producing erect vegetative filaments which are simple or branched below, 10–16 μ diam., the cells 1.0–1.5 diameters long, often terminating in a colorless hair; plurilocular gametangia sessile on the spreading filaments, or on shorter or longer stalks, or more generally lateral or terminal on the longer erect vegetative filaments (which may be 3 or 4 times the length of the gametangia and may have 1 or 2 branches), ovate-lanceolate, 50–115 μ long, 18–25 μ diam.

Massachusetts to Maine; epiphytic on Rhodymenia and other large algae, particularly in tide pools; fruiting during the latter part of the summer.

REFERENCES: Phyc. Bor.-Am. 274. Collins 1896b, p. 459 (as *Phycocelis maculans*).

There have been distinguished within our range a f. **sauvageauii** Collins (Phyc. Bor.-Am. 1536), with the erect filaments simple and the plurilocular gametangia, borne on the basal layer, 35–60 μ long, 11–13 μ diam., and a f. **soluta** Collins (Phyc. Bor.-Am. 1038; Collins

1906a, p. 108), with the basal layer open owing to the distinct separation of the filaments; both are reported from Maine.

Hecatonema reptans (Reinke) Sauvageau

Plants minute; the primary filaments decumbent, mainly superficial, forming a disk which may become two-layered, the cells 8–12 μ diam., 8–16 μ long; the erect filaments dispersed over the disk, 120–250 (–1000) μ long, unbranched, the cells 6–10 μ diam., 8–24 μ long; plurilocular gametangia on the basal layer of primary filaments or terminating the erect filaments, sessile or on 1–5-celled stalks, oblong to linear-oblong, 30–80 μ long, 12–30 μ diam.

Long Island to northern Massachusetts, and Nova Scotia; epiphytic on Petalonia, Dictyosiphon, and other coarse algae.

REFERENCES: Farlow 1881, p. 69; Taylor 1937c, p. 116 (both as *Ectocarpus reptans*); Kylin 1947, p. 15.

Myrionema Greville, 1827

Plant a minute, flattened cushion or disk, round or elongated; below a unistratose layer of radiating crowded filaments, the filaments differing in habit and dimensions according as gametophyte or sporophyte individuals are concerned; forming above short erect assimilators and colorless hairs; plurilocular gametangia usually pluriseriate, elongate, more or less stalked; unilocular sporangia on separate plants, ovoid or pyriform.

KEY TO SPECIES

1. Reproductive organs arising from the basal layer, not from free filaments 2
1. Reproductive organs produced by transformation of parts of free filaments 4

2. Ovoid unilocular sporangia and cylindrical plurilocular gametangia sessile or short-stalked . . . . M. strangulans, p. 132
2. Unilocular sporangia unknown 3

3. No free filaments; plurilocular gametangia sessile, simple, usually pluriseriate . . . . M. foecundum, p. 132
3. Free filaments present; plurilocular gametangia sessile or short-stalked, simple or branched, normally uniseriate M. corunnae, p. 131

4. Plurilocular gametangia formed in the upper part of simple free filaments .... M. balticum, p. 131
4. Plurilocular sporangia formed in branches of free filaments M. globosum, p. 131

Myrionema globosum (Reinke) Foslie

Plants forming small globose to hemispherical cushions 0.5–1.0 (–3.0) mm. diam., with a basal disk bearing erect branched filaments 5–6 μ diam.; each vegetative cell with 1–2 platelike chromatophores; plurilocular gametangia cylindrical-filiform, uniseriate, terminal or lateral on the erect filaments.

Rhode Island and Massachusetts; growing on Chaetomorpha.

REFERENCE: Collins 1900b, p. 47.

Myrionema balticum (Reinke) Foslie

Plants in the form of small disklike spots on the surface of the host; basal portion near the center producing erect assimilatory filaments, these progressively smaller toward the growing margin; erect filaments 100–125 μ long, 4–6 μ diam., each cell with 1–2 platelike chromatophores; plurilocular gametangia intermixed with the assimilators, formed from the upper parts of these filaments and so on stalks 2–4 cells long, cylindrical, uniseriate, or a few cells longitudinally divided.

Northern Massachusetts; growing on Laminaria.

REFERENCE: Collins 1900b, p. 47.

Myrionema corunnae Sauvageau

Plants disciform, forming small brown spots to 2–3 mm. diam. on the surface of the host; basal layer of filaments 4.5–7.0 μ diam., bearing hairs about 5 μ diam., and forming erect filaments which become 100–140 μ long, 6–7 μ diam., eventually all becoming converted into plurilocular gametangia, so that those toward the center have stalks 2–4 cells long, the gametangia filiform, uniseriate, or with a few longitudinal divisions, simple or sparingly branched, 25–120 μ long, 5–7 μ diam., the cells as long as broad or shorter.

Rhode Island, southern Massachusetts, and Maine; epiphytic on the blades of Laminaria.

REFERENCES: Phyc. Bor.-Am. 1234. Collins 1906c, p. 158.

In addition to the typical plant there is reported v. **filamentosa** Jónsson, which has its basal filaments nearly or quite free laterally, in Maine found intergrading with the typical condition.

Myrionema strangulans Greville Pl. 11, figs. 13–14

Plants forming small disks; basal filaments closely or sometimes a little loosely placed, cells 5.0–8.5 μ diam., 1–3 diameters long; hairs 8–13 μ diam.; assimilators rather densely crowded, their basal cells often in lateral contact below, the filaments 50–100 μ long, somewhat clavate-moniliform, the cells cylindrical below, to subspherical and shorter above, 6–11 μ diam., with several discoid chromatophores in each cell; reproductive organs on the basal layer or on the basal cells of the assimilators; plurilocular gametangia obtuse-cylindrical, sessile or on stalks of 1–2 cells, 15–50 μ long, 7–11 μ diam.; unilocular sporangia ellipsoid to obovoid, 35–65 μ long, 20–35 μ diam.

North Carolina, New Jersey to Maine, and in James Bay; epiphytic on Ulva, Rhodymenia, and other algae; fruiting throughout the year, but particularly in the summer.

REFERENCES: Phyc. Bor.-Am. 32, 1639. Harvey 1852, p. 132; Farlow 1881, p. 79; Taylor 1937c (all as *M. vulgare*).

Myrionema foecundum (Strömfelt) Foslie

Plants forming suborbicular disks to 0.5–3.5 mm. diam., or these confluent; erect filaments simple, short, composed of 3–5 short cells, in the older parts of the thallus all eventually changed into gametangia; gametangia crowded, sessile, cylindroconical with obtuse apices, of 1–2 rows of cells, 35–40 (–70) μ long, 7–12 μ diam.

Long Island Sound, Rhode Island, and Connecticut; epiphytic on large algae, especially Laminaria; fruiting during the summer.

REFERENCE: Collins 1900b, p. 47.

Ascocyclus Magnus, 1874

Plants minute, in the form of round epiphytic disks composed of radiating branched filaments; from the disks arise colorless hairs with basal growth, short, clavate vegetative filaments, elongated hyaline unicellular paraphyses, and cylindrical plurilocular gametangia.

KEY TO SPECIES

1. Erect filaments frequent; basal filaments biseriate.... A. distromaticus
1. Erect filaments absent; basal filaments uniseriate...... A. orbicularis

Ascocyclus distromaticus Taylor Pl. 11, figs. 7–12

Plants of filaments which are readily separated by pressure, 7.0–8.5 μ diam., the cells 1–2 diameters long, at first or sporadically remaining uniseriate, but generally dividing more or less irregularly in the plane of the support, so that the disk is in most part distromatic; erect vegetative filaments blunt at the end, cylindrical, or slightly contracted at the septa, 3–5 (–10) cells long, the cells 7.7–10.0 μ diam., 7.0–14.0 μ long; long hairs frequent, 5.5–9.5 μ diam., the upper cells to 120 μ long; paraphyses sessile or on one-celled stalks, at first obovate and brown, about 15 μ diam. and 30 μ long, becoming thick-walled and hyaline, subcylindrical or a little tapering, 7.5–10.0 μ diam., 70–100 μ long; plurilocular gametangia generally sessile or on one-celled stalks, sometimes on stalks of 2–3 cells, subcylindrical, obtuse, 7–10 μ diam., 17–56 μ long.

Southern Massachusetts; epiphytic on Rhodymenia; summer.

REFERENCE: Taylor 1937b, p. 228.

Ascocyclus orbicularis Magnus

Plants disklike, 1–3 (–10) mm. diam., formed of radiating filaments; erect vegetative filaments absent or rare, somewhat clavate; hairs 15–20 μ diam.; paraphyses sessile or on one-celled stalks, at first ovate and brown, later much elongate, truncate and saccate, hyaline, 8–12 (–25) μ diam., to 170 μ long; plurilocular gametangia sessile or on one-celled stalks, cylindrical or slightly clavate, obtuse, uniseriate, about 20–30 (–75) μ long, (5–) 8–12 μ diam.

Bermuda, Connecticut to Maine; fruiting during the summer.

REFERENCE: Phyc. Bor.-Am. 173. Levring 1940, p. 40.

RALFSIACEAE

Crustose, mostly perennial plants, composed of systems of radial filaments laterally united to form a horizontal layer, from which arise erect series of cells to form a more or less pseudoparenchymatous thick crust, and also colorless hairs; plurilocular gametangia and unilocular sporangia are formed in superficial sori on different plants, often associated with paraphyses.

Ralfsia Berkeley, 1831

Plants rarely minute, often one to several centimeters in diameter, forming expanded, more or less attached, usually somewhat brittle crusts, which are at times multiple by the overgrowth of overlapping lobes; of two layers, the basal of radiating filaments usually with rhizoids on the under surface, and the upper layer of parallel ascending assimilatory filaments, the erect assimilatory cell series compact, laterally united, usually with dark-colored cell walls, and with the chromatophores concentrated in the cells near the surface of the thallus; sometimes in marginal areas free from the substratum and with assimilatory layers on both sides; hairs inconspicuous, in tufts; unilocular sporangia and plurilocular gametangia in superficial sori on different plants, the sporangia lateral at the bases of the free paraphysal filaments, the gametangia terminal on erect filaments.

KEY TO SPECIES

1. Small species; cell rows arising at right angles from the basal layer to form the assimilatory layer.......................... 2
1. Larger species; curving rows of cells arising from the basal layer to form the assimilatory layer.......................... 3

2. On rocks and shells, thalli to 20 mm. diam.......... R. clavata, p. 136
2. On algae and Zostera, minute, punctiform.......... R. pusilla, p. 136

3. Closely adherent, the basal layer forming an assimilatory layer above only 4
3. Marginally free, with zonate imbricate lobes, layers of cells in curved rows formed on both sides of the basal layer R. fungiformis, p. 134

4. Paraphyses slender below, the usually colorless lower cells 20 times as long as broad; gametangia pluriseriate.. R. borneti, p. 135
4. Paraphyses with lower cells differing little from the others; gametangia of 1–2 cell series only, infrequent.. R. verrucosa, p. 135

Ralfsia fungiformis (Gunner) Setchell et Gardner

Plants large, perennial, of rounded imbricate lobes 2–4 (–6) cm. diam., 250–470 μ thick, attached to rocks by brown rhizoids 13–18 μ diam. over the central area, marginally free, overlapping at times to masses as much as 2 cm. thick, smooth above, superficially concentrically and radially marked; cortical cell rows 5.5–9.5 μ diam., firmly attached laterally; sori of paraphyses on the upper surface, the paraphyses 5–6 μ diam.; reproduction not known.

Northern Massachusetts to New Brunswick, Newfoundland, Labrador, James Bay, Hudson Strait, and Baffin Island; spreading over rocks in tide pools and other shallow situations.

REFERENCES: Phyc. Bor.-Am. 419 (as *R. deusta*). Farlow 1881, p. 87 (as *R. deusta*); Kützing 1869, p. 32 (as *Peyssonelia imbricata*, type seen); Wilce *in herb.*

Ralfsia verrucosa (Areschoug) J. Agardh Pl. 11, figs. 1–2

Plants forming olive-brown crusts, at first attached, but later in the older parts becoming loose and brittle; the smooth margins concentrically zonate, the older parts verrucose-roughened; reaching 0.5–10.0 cm. diam., becoming 1–2 mm. thick and the erect filaments many cells in length, 5.5–9.5 μ diam.; often showing several superposed thalli, which have been overgrown in turn; fertile elevations formed of short-stalked clavate paraphyses, 90–170 μ long, 4.0–5.5 μ diam. at the base, the lowest cell 2–3 (–5) diameters long, about 9–11 μ diam. at the top; unilocular sporangia pyriform or obovoid, 56–120 μ long, 22–43 μ diam.; plurilocular gametangia congested, of 1–2 vertical series of cells with the uppermost cells sterile, 7–8 μ diam., without paraphyses, infrequent.

New Jersey, Connecticut to the lower St. Lawrence, Hudson Strait, and Baffin Island; perennating in tide pools or on intertidal rocks or shells, occasionally on large algae; fruiting in summer and autumn.

REFERENCES: Algae Exsic. Am. Bor. 113; Phyc. Bor.-Am. 325, 1740. Farlow 1881, p. 87; Whelden 1947, p. 116.

Ralfsia borneti Kuckuck

Plants forming moderately thick rounded disks, the vegetative filaments 9.5–13.0 μ diam.; paraphyses about 150 μ long, cylindrical to slenderly clavate, about 6 cells in length, the upper short, but the lowest (immediately above the sporangial attachment) elongate, to

20 diameters long; unilocular sporangia lateral at the base of the paraphyses, elongate-pyriform, 75–100 μ long, 16–25 μ diam.; plurilocular gametangia longitudinally and transversely multicellular, to 75 μ long.

Connecticut and Rhode Island, New Hampshire, and Maine; growing on mussel shells, lithothamnia, and occasionally on stones; fruiting in the spring and summer.

REFERENCES: Phyc. Bor.-Am. 54. Collins 1900a, p. 12; Croasdale 1941, p. 214.

This species is not readily to be distinguished from *R. verrucosa* unless it is in fruit.

Ralfsia clavata (Carmichael) Crouan *sensu* Farlow

Thalli forming thin, closely adherent crusts with difficulty separated from the substratum, 2–20 mm. diam., 150–210 μ thick; vegetative cell rows 9–12 μ diam., with several short cells in each; paraphyses subcylindrical, 75–100 μ long, 5–6 μ diam. at the base, 10–12 μ diam. at the summit, the basal cell 10–42 μ and 2–8 diameters in length; unilocular sporangia borne on the bases of the paraphyses, ovoid to pyriform, 50–85 μ long, 20–32 μ diam.; gametangia blunt, subcylindrical, 1–2 cells broad, to 60 μ long, the cells about 3.5 μ diam.

New Jersey to the lower St. Lawrence; growing on stones, shells, or woodwork between tide levels; fruiting in spring and early summer.

REFERENCES: Phyc. Bor.-Am. 418. Farlow 1881, p. 88.

There may be recognized a f. **laminariae** Collins (Phyc. Bor.-Am. 1390, Taylor 1937c, p. 125): Fronds minute, seldom confluent; hairs frequent; unilocular sporangia occasional; plurilocular gametangia with a terminal sterile cell; on Laminaria stipes, Maine.

Ralfsia pusilla (Strömfelt) Batters

Plants appearing as very small disks, punctiform, and epiphytic; formed of a single layer of radiating filaments bearing short erect filaments, the component cells 7.5–15.0 μ long and broad, the filaments 85 μ long, or perhaps more, very readily separated from each other; paraphyses clavate, 100–125 μ long, 6 μ diam. at the base, 10–15 μ diam. at the summit; unilocular sporangia oval to sub-pyriform, 20–30 μ diam., 47–90 μ long.

Rhode Island, northern Massachusetts, and Maine; epiphytic; fruiting in summer.

REFERENCE: Phyc. Bor.-Am. 830.

LITHODERMATACEAE

Plants encrusting, strongly adhering by the under surface, growing peripherally; structure parenchymatous below; plurilocular gametangia originating as uniseriate lateral outgrowths of filaments which grow up from the surface, or by transformation of outgrowths from the surface cells; unilocular sporangia formed by transformation of surface cells.

KEY TO GENERA

1. Unilocular sporangia generally oval to nearly globose; chromatophores several in each cell........................ LITHODERMA
1. Unilocular sporangia generally somewhat elongated and pyriform; chromatophores single in each cell............... SORAPION

Lithoderma Areschoug, 1875

Thalli blackish olive, crustose, completely adherent by the under surface, thin, rounded, becoming thickened by overgrowth of successive crusts; the lower layer bearing a surface stratum of erect conjoined filaments, the cells with several small disciform chromatophores; plurilocular gametangia lateral on free surface filaments grouped in sori, possibly in some terminating the vertical rows of the thallus, linear-oblong or elongate-elliptical; unilocular sporangia formed by transformation of the end cells of the vertical filaments, oval to nearly globose.

Lithoderma extensum (Crouan) Hamel

Plants forming large rounded lobed crusts, to as much as 3 dm. diam. and 1 mm. thick, strongly adherent, above smooth and in water somewhat shiny; erect filaments 12–18 μ broad, 8–12 cells long, the cells shorter than their diameter, to subquadrate; the gametangia subcylindric to elliptical, blunt, 4–7 cells long, 1–2 cells broad, 1–4 together, lateral upon a supporting filament the sterile end cell of which is enlarged and subtruncate, but also reported as solitary to crowded, erect, terminating the erect cell rows of the thallus; sporangia about 20–45 μ long, 14–22 μ diam.

Northern Massachusetts (?), Devon Island, and Ellesmere Island; on small stones; fruiting in the spring.

REFERENCES: Collins 1906c, p. 158; Rosenvinge 1926, p. 20; Taylor 1937c, p. 126 (all as *L. fatiscens*).

Sorapion Kuckuck, 1894

Plant at first composed of one layer of cells forming a crust with marginal growth, which later develops from the base vertical branched filaments that apparently become united laterally; cells with single disklike chromatophores; unilocular sporangia formed by direct transformation of the end cells of the erect filaments, prominent, and aggregated into sori.

Sorapion kjellmani (Wille) Rosenvinge

Plant epiphytic, forming a crust, structurally of vertical filaments 5–7 cells long, the cells 11–17 μ diam. and as long as broad; unilocular sporangia in scattered sori, prominent on the thallus, large, 20–25 (–37) μ diam., and pyriform.

West side of Ellesmere Island; growing in shallow water on Chaetomorpha; fruiting in summer.

REFERENCE: Rosenvinge 1926, p. 21.

ELACHISTEACEAE

Tufted plants, showing a basal portion of densely intertwined filaments and long simple free filaments with intercalary growth as assimilators, which below often bear short lateral branchlets; plurilocular gametangia and unilocular sporangia borne among the short branchlets, but also in some genera on the assimilators.

KEY TO GENERA

1. Lower portion of the plant primarily superficial 4
1. Lower portion of the plant forming a penetrating, often firm and massive holdfast 2

2. Short paraphysal filaments accompanying the reproductive organs at the bases of the assimilators......... ELACHISTEA, p. 139
2. Paraphysal filaments ill defined or absent; reproductive organs present on the assimilators as well as below.................. 3

3. Paraphysal filaments absent; ours minute plants.. MYRIACTULA, p. 141
3. Paraphysal filaments ill defined; macroscopic species SYMPHORICOCCUS, p. 141

4. Free assimilators becoming pluriseriate............ GIRAUDIA, p. 143
4. Free filaments uniseriate.................................... 5

5. Gametangia in superficial patches on the assimilators HALOTHRIX, p. 143
5. Gametangia scattered in the upper parts of the assimilators LEPTONEMA, p. 142

Elachistea Duby, 1830

Plants minute, brush- or cushion-like, the basal portion of densely matted colorless filaments which often penetrate the host tissues, the vegetative filaments branched below only, uniseriate, of two sorts, above only large free straight assimilatory filaments, but below these crowded with often curved and moniliform short filaments or paraphyses; plurilocular gametangia and unilocular sporangia borne among the paraphysal filaments.

KEY TO SPECIES

1. Paraphyses moniliform, decidedly curved; upper cells of the assimilators nearly cylindrical................. E. fucicola, p. 140
1. Paraphyses little constricted at the septa...................... 2

2. Penetrating basal holdfast conspicuously developed; paraphyses straight or nearly so; upper cells of the assimilators nearly cylindrical .... E. lubrica, p. 140
2. Penetrating holdfast inconspicuous, the upper cells of the assimilators swollen .... E. chondri, p. 139

Elachistea chondri Areschoug

Plants tufted, solitary or confluent, 6–10 mm. tall; assimilators sharply tapering at the base, gradually tapering toward the free apex, the cells in the lower part shorter than their diameter, much swollen, 13–28 μ diam. at the base, above somewhat cask-shaped,

38–47 μ diam., 50–65 μ or 1.5–3.0 diameters long, the wall not conspicuously thickened, near the apex the cells again shorter, 28–34 μ diam., 11–18 μ long, the end cell rounded; paraphyses subcylindrical, the cells 2–3 diameters long; unilocular sporangia elliptical or subovate.

Rhode Island; on Chondrus in tide pools; fruiting in early summer.

REFERENCES: Phyc. Bor.-Am. 773. Collins 1901a, p. 133.

Elachistea fucicola (Velley) Areschoug Pl. 10, figs. 1–3

Plants tufted, the basal part a firm hemispherical cushion penetrating the host below, above bearing the free filaments in a dense cluster 1.0–1.5 cm. long; filaments from the cushion densely branched below, many of the divisions terminating in paraphyses or sporangia, but others in long brown assimilatory filaments which grow from near the base; assimilators tapered at the base, larger and submoniliform above it, cylindrical in the upper part, the apex obtuse if intact; cells shorter below, to longer and 1.5–2.0 diameters long above, 20–70, mostly 40–50 μ diam., the walls rather thick, chromatophores small, discoid, and numerous; paraphyses distinctly curved, moniliform above, the lower cells 10–23 μ diam., 20–60 μ long, the upper ones 13–30 μ diam., 22–48 μ long; unilocular sporangia obovate to pyriform, 60–100 (–200) μ long, 30–90 μ diam.

New Jersey to the lower St. Lawrence, southeastern Hudson Bay, and Ellesmere Island; epiphytic on fuci and other coarse algae; in good vegetative condition during the summer, but fruiting in autumn and winter, often when the assimilators have largely disappeared.

REFERENCES: Phyc. Bor.-Am. 535. Harvey 1852, p. 131; Farlow 1881, p. 81.

Elachistea lubrica Ruprecht

Plants tufted, the tufts separate or gregarious, with marked firm penetrating basal holdfasts; assimilatory filaments sharply tapered below, nearly cylindrical above, the cells in the lower part 0.3–0.5 diameter long, above to 35–65 μ diam., 2–3 diameters long; paraphyses but slightly curved, slightly club-shaped; unilocular sporangia ovate to somewhat pyriform 130–210 μ long, 75–100 μ diam.; plurilocular gametangia also known.

New Hampshire, Maine, the lower St. Lawrence, Newfoundland, southeastern Hudson Bay, James Bay, and the arctic; epiphytic on Ascophyllum, Halosaccion, and other algae; fruiting in the latter part of the summer.

REFERENCES: Phyc. Bor.-Am. 480. Collins 1891, p. 339, Croasdale 1941, p. 214.

Symphoricoccus Reinke, 1888

Small tufted filiform plants from a clearly delimited basal layer; assimilators gradually tapered to the end and sharply tapered near the base, with intercalary meristematic growth; plurilocular gametangia sometimes in loose sori, sometimes scattered on the assimilators, borne on the axial cells, seldom aggregated in a basal cluster; unilocular sporangia generally on the basal filaments, more seldom inserted on the assimilators in the same fashion as the gametangia.

Symphoricoccus stellaris (Areschoug) Kuckuck

Plants tufted, 2–5 mm. tall; the basal layer scanty, the assimilators loose, tapering sharply to the base, but very little toward the free end, with the lower segments as long as broad, the upper 2–4 diameters long, the diameter 25–50 μ, cells containing irregular platelike chromatophores; paraphyses ill marked, few, cylindric-clavate; plurilocular gametangia short-cylindrical or a little siliquose, blunt, in clusters on the upper part of the assimilators or at their bases; unilocular sporangia ellipsoid-pyriform, usually on the basal layer but smaller ones commonly scattered on the assimilators, about 50–100 μ long, 15–45 μ diam.

New Jersey and southern Massachusetts; growing on other algae.

REFERENCES: Silva 1952, p. 283; Collins' manuscript.

Myriactula Kuntze, 1898

Basal portion of the plant of colorless filaments, densely entangled, penetrating the host tissues; above forming a minute tuft of branches which fork densely below, bearing colorless hairs and unbranched assimilators; reproductive organs consisting of plurilocular gametangia and unilocular sporangia, either borne in clusters at the base of the assimilators or the gametangia also laterally on the upper part of the assimilators.

KEY TO SPECIES

1. Endophytic in the cryptostomata of Sargassum, the assimilators distally little tapered.................................. M. minor
1. Epiphytic, the assimilators distally much tapered........ M. chordae

Myriactula minor (Farlow) Taylor Pl. 10, figs. 8–9

Plants forming minute tufts, the basal portion slightly developed, giving off lateral filaments which penetrate the substratum; assimilatory filaments slightly curved, to 20–30 cells long, attenuated, especially at the base; somewhat moniliform, cells 7.5–18.0 μ diam. near the center, 2–3 diameters long; plurilocular gametangia very numerous, mostly clustered at the bases of the assimilatory filaments but sometimes lateral upon them, cylindrical, 7.6 μ diam., to 8–10 cells and 57 μ long; unilocular sporangia sometimes present.

Connecticut to Massachusetts; in the cryptostomata of Sargassum; fruiting in the summer.

REFERENCES: Phyc. Bor.-Am. 231 (as *Myriactis pulvinata* v. *minor*), 1592 (as *Elachistea minor*); Taylor 1937b, p. 229.

Myriactula chordae (Areschoug) Levring

Plants forming minute tufts; basal layer scanty, of filaments about 12 μ diam., bearing free assimilatory filaments which are relatively short, are much thicker above the base to which they are sharply contracted, and taper gradually toward the apex, the inflated cells being short, below as long as broad and above about 2 diameters long; sporangia subpyriform, about 20 μ diam., 50 μ long.

Southern Massachusetts; growing as an epiphyte on Stilophora; fruiting in late summer and early autumn.

REFERENCE: Taylor 1937c, p. 149 (as *M. pulvinata* f. *chordae*).

Leptonema Reinke, 1888

Plants in sparse tufts with a basal layer of contorted branching filaments, producing erect monosiphonous assimilatory filaments which may be slightly branched at the base; the unilocular sporangia ovoid, sessile, or short-stalked, lateral near the bases of the erect filaments; plurilocular gametangia in the upper part of the erect filaments, each formed by the lateral growth of a cell.

Leptonema fasciculatum Reinke v. **majus** (Reinke) Gran

Plants sparsely tufted, to 2 cm. tall, with spreading rhizoidal basal filaments bearing erect assimilators 22–28 μ diam., uniseriate, or with a few longitudinal walls; plurilocular gametangia formed in the assimilators, single or frequently many, whorled or crowded, extending at right angles to the axis, blunt-conical, 9–25 μ long, 9–13 μ diam.

Maine; growing on Dictyosiphon; fruiting in summer.

REFERENCES: Phyc. Bor.-Am. 678 (as *Elachistea fasciculata* v. *major*). Collins 1896b, p. 458.

Halothrix Reinke, 1888

Plant small, tufted, from a superficial holdfast portion giving off simple filaments with intercalary growth, the cells with several irregular disklike chromatophores; plurilocular gametangia formed by conversion of local regions of the assimilatory filaments into a sterile axis covered with small sessile crowded plurilocular gametangia.

Halothrix lumbricalis (Kützing) Reinke

Plants tufted, to 25 mm. tall, yellowish brown; erect filaments simple, 20–56 μ diam., the cells 85–95 μ long; in reproduction bearing radiating plurilocular gametangia about 4.0–5.5 μ diam. in continuous sori 45–200 μ or more long on the assimilators, generally completely encircling them, increasing the diameter to 47 –75 μ, each gametangium about 4–6 cells long.

Long Island to Maine; growing on the leaves of Zostera; appearing in the winter and persisting into the spring.

REFERENCES: Phyc. Bor.-Am. 31. Collins 1891, p. 337.

Giraudia Derbès et Solier, 1851

Plants forming densely packed clusters of filaments branching from the base, simple above, the main divisions monosiphonous at the base, polysiphonous above, tapering to the apex and bearing one or more hairs; short monosiphonous branches arising near the base; plurilocular gametangia of three kinds, resulting either from the conversion and enlargement of all exposed cells in slender parts of the thallus into laterally discharging gametangia of a few cells each, or the subdivision of one or a few contiguous cells in a broader polysiphonous region to form sori of simple plurilocular gametangia several cells long and one or two cells in diameter oriented laterally, or by the production of much-branched structures at the bases of the assimilators which form large digitately divided gametangia the lobes of which may be two to four cells wide.

Giraudia sphacelarioides Derbès et Solier

Plants tufted, yellowish brown, 5–15 mm. tall, attached by rhizoids, filamentous, the filaments 30–80 μ thick above, polysiphonous and segmented; plurilocular gametangia from the bases of the assimilators, 10–15 μ diam., to 120 μ long, or from the erect assimilators in sori when 25–40 μ long.

Southern Massachusetts; fertile in the winter.

REFERENCES: Farlow 1881, p. 75; Schuh 1900b, p. 206.

CHORDARIACEAE

Plants subspherical to filiform, simple or branched, sometimes hollow, of filamentous construction within a gelatinous matrix, a core of relatively colorless filaments giving off lateral assimilatory branches which constitute the cortex and contain most of the chromatophores; growth apical; colorless hairs usually abundant; unilocular sporangia and plurilocular gametangia immersed, formed from branchlets of the cortical filaments, or functionally present on the branches of a small filamentous gametophyte.

KEY TO GENERA

1. Plants subglobose, irregularly lobed.............. LEATHESIA, p. 148
1. Plants elongated, usually branched.............................. 2

2. Regular branching of 1–2 orders, the main axis subsimple, the final branches long; substance firm............ CHORDARIA, p. 148
2. Branching irregular; substance soft............................ 3

3. Plant large and bushy, slender, lubricous throughout; axes irregularly redividing, laterally beset with short branchlets; cortical filaments ending in subspherical cells SPHAEROTRICHIA, p. 146

3. Plant smaller and coarser; main axes sparingly divided, lateral branches few to many; cortical filaments of asymmetrical oval cells .... EUDESME, p. 145

Eudesme J. Agardh, 1882

Fronds cylindrical, branched, from a small disk holdfast, gelatinous; the axis solid or locally inflated, of longitudinal filaments laterally branched, large and slender intermixed, loosely conjoined; externally bearing a cortex of radiating peripheral assimilatory filaments, the assimilators subsimple, cylindrical below, above of asymmetrical oval cells, often surrounding the unilocular sporangia, which are ovoid to round; plurilocular gametangia formed by the conversion of the upper portion of the assimilators, often formed on the same plants as the unilocular sporangia.

KEY TO SPECIES

1. Coarse, the axis usually exceeding 5 mm. diam., pale brown, when well developed with large subsimple lateral branches E. virescens

1. Slender, not reaching 5 mm. diam., bright brown, the axis simple or with short, flexuous, or contorted, nodulose lateral branchlets .... E. zosterae

Eudesme virescens (Carmichael) J. Agardh Pl. 12, fig. 3

Plants of moderate size, 1–3.5 dm. long, pale brown to olivaceous, gelatinous, the base a small holdfast disk supporting a slender stalk which immediately expands into the thick axis; axis evident below, above usually dissolving into a few large smooth branches, each with occasional to frequent shorter spreading branchlets of 1, rarely 2 degrees, issuing at right angles, all usually 0.5–1.0 cm. diam. in mature plants, at first solid, becoming somewhat hollow; peripheral filaments slender and curved, cylindrical or only slightly moniliform, the cells ellipsoidal, 15–20 μ diam.; plurilocular gametangia conical, formed in rows by lateral outgrowth of upper cells of the assimilators, of 3–6 cells or a little larger; unilocular sporangia ovoid to nearly globular, 65–122 μ long, 35–75 μ diam., formed basally in the tufts of assimilators.

Long Island to Nova Scotia, Newfoundland, northern Labrador, and southeastern Hudson Bay (possibly occurring much farther down the Atlantic coast); growing on Zostera rhizomes, coarse algae, or stones; fruiting in early summer.

REFERENCES: Harvey 1852, p. 126 (as *Mesogloia virescens*); Farlow 1881, p. 85 (as *Castagnea virescens*); Taylor 1937c, p. 140 (as *Aegira virescens*).

Eudesme zosterae (J. Agardh) Kylin

Pl. 10, figs. 10–11; pl. 12, fig. 2

Plants relatively small and slender, 1 dm. to, rarely, 2 dm. tall, bright brown, gelatinous and slippery, the base a small disk from which the axis with a contracted base arises, the main portion usually 1–3 mm. diam., bearing few to several short secondary branches, simple, or less often 1–3 times redivided, all flexuous to contorted, the surface often nodulose; peripheral filaments erect, rather stout and rigid, cylindrical below, moniliform above, where the cells are spheroidal, 20–40 μ diam.; plurilocular gametangia cylindrical to obtuse-conical, formed in rows by lateral outgrowth of the upper cells of the assimilators, sometimes bilaterally placed, each usually 3–6 cells or 15–45 μ long, 1–2 cells or 7.5–18.0 μ diam.; unilocular sporangia formed in the tufts of assimilators at their bases, ovoid to nearly globular, 50–100 μ long, 25–70 μ diam.

Bermuda, Florida, North Carolina, Long Island, Massachusetts to Maine, Nova Scotia, the lower St. Lawrence, and Ile St. Pierre; growing on the leaves of Zostera or other plants; fruiting in summer and autumn.

REFERENCES: Algae Exsic. Am. Bor. 102. Harvey 1852, p. 127 (as *Mesogloia zosterae*); Farlow 1881, p. 86 (as *Castagnea zosterae*); Taylor 1928a, p. 112, 1937c, p. 141; Le Gallo 1947, p. 304 (all as *Aegira zosterae*); Kylin 1933, p. 57, 1947, p. 57.

Sphaerotrichia Kylin, 1940

Plants of moderate size, filiform, branched, gelatinous and slippery; of more or less indefinite growth, developing from a primary axial filament of limited distal growth terminated by a large subspherical cell, this filament having its initial segmentation distal to the point of formation of lateral filaments, and with advancing growth becoming obscure within the medulla; medulla essentially solid, of outwardly forked loosely conjoined filaments; lateral assimilatory filaments clavate, cylindrical below, inflated above with a very large terminal cell; unilocular sporangia ovoid, on the bases of the assimilatory filaments; zoöspores from these developing to filamentous plantlets with plurilocular gametangia.[1]

Sphaerotrichia divaricata (C. Agardh) Kylin
Pl. 12, fig. 1; pl. 14, fig. 1

Plants solitary or gregarious, moderately large, reaching 5 dm. or more in length, light to dark brown, exceedingly slippery and much branched; main axes not evident below, but subsidiary axes fairly evident in the upper parts of the plant, long and flexuous, subdichotomously or laterally freely branched, the branches spreading, the ultimate branchlets ordinarily lateral, all being slender; peripheral filaments 2–6 cells long, the lower cells subcylindrical, about 8–16 μ diam., 8–26 μ long, the terminal cells sharply inflated, subglobose, 22–44 μ diam., 22–40 μ long, with several disciform chromatophores; colorless hairs abundant, conspicuous in the living plant, 8–14 μ diam. at the base; unilocular sporangia generally single, stalked, with a peripheral filament on a common basal cell, rounded to obovate, 38–65 μ long, 21–38 μ diam.

New Jersey to the lower St. Lawrence, Newfoundland, and northern Labrador; common in the summer months on coarse algae.

REFERENCES: Algae Exsic. Am. Bor. 198; Phyc. Bor.-Am. 175. Harvey 1852, p. 124 (as *Chordaria divaricata,* possibly also p. 126 as *Mesogloia vermicularis*) ; Farlow 1881, p. 84; Collins 1904, p. 182; Taylor 1937c, p. 142 (all as *M. divaricata*), Wilce *in herb.*

The plant as found in Buzzards Bay in very sheltered harbors, particularly polluted ones, is sometimes much stiffer below than that of the open coast, of greater diameter, often hollow, not particularly slippery, bushy below, and only in the upper branches with the more ordinary slender, slippery character. There also seemed to be a fairly consistent difference in the length of the peripheral filaments. It may be possible to define this variety; it is ecologically rather distinct.

[1] *Mesogloia vermicularis* C. Ag., reported from Newfoundland by Farlow (1881, p. 85) and from other stations by later workers, is not currently known on our coast and may have been wrongly identified.

Chordaria C. Agardh, 1817

Plant slenderly cylindrical, branched; growth of axis and branches not indefinite; with a firm pseudoparenchymatous medulla of colorless longitudinal filaments outwardly of shorter cells, bearing a cortex of crowded radiating assimilatory filaments which are stalked, simply clavate, or branched only near their bases, embedded in firm jelly; unilocular sporangia borne at the bases of the assimilatory filaments.

Chordaria flagelliformis (Müller) C. Agardh

Pl. 12, fig. 6; pl. 14, fig. 4

Plants moderately large, 3–7 dm. long, solitary or often gregarious, very dark brown; the axes from a disklike holdfast, the plants firm in texture but slippery; 0.3–1.5 mm. diam. in all main branches, the branching widely lateral from a short and usually simple main axis, the upper branches longer than those near the base and the uppermost usually much exceeding the primary axis, the longer sometimes with a few small branches of the second order; assimilatory filaments 7–10 cells long, clavate, the upper cells progressively shorter, the terminal somewhat swollen, 10–16 μ diam.; colorless hairs with basal growth very abundant; unilocular sporangia pyriform or long-oval, 60–100 μ long, 20–38 μ diam.

New Jersey to the lower St. Lawrence, northern Labrador, James Bay, southeastern Hudson Bay, Hudson Strait, Baffin Island, and Ellesmere Island; growing on rocks and woodwork, more rarely on fuci; forming sporangia throughout the year.

REFERENCES: Phyc. Bor.-Am. 175, 324. Harvey 1852, p. 123; Farlow 1881, p. 83; Hariot 1889, p. 155; Croasdale 1941, p. 214; Wilce *in herb.*

V. **densa** Farlow is described as being 1–3 dm. long, with very short lateral branches reaching a length of 3–10 mm.; it has been reported from northern Massachusetts to Maine, and in northern Labrador.

Leathesia Gray, 1821

Plants forming subspherical growths, later becoming irregular, lobed or plicate; structurally obscurely filamentous, the medulla mostly pseudoparenchymatous, the close cortex of simple clavate moniliform assimilatory filaments; plurilocular gametangia uniseriate, formed on the bases of the assimilatory filaments, or on microscopic sporelings; unilocular sporangia pyriform or ovoid, in the same position.

Leathesia difformis (Linnaeus) Areschoug

Pl. 12, fig. 5; pl. 14, fig. 8

Plants epiphytic, forming large, somewhat convolute or subspherical masses, at first solid, later hollow and gas-filled, yellowish brown or olivaceous to ultimately dull yellow, reaching a diameter of 10 cm., often confluent or gregarious; spongy central medulla in young plants of loose filaments, later obsolete, the outer medulla crisp and compact, bearing the clavate assimilatory filaments which are simple beyond the basal branching, 3–5 cells long, at the top 6.5–13.0 μ diam.; reproductive structures attached at the bases of the assimilators; plurilocular gametangia obtuse-cylindrical, uniseriate, 5–10 cells and 25–35 μ long, 10–16 μ diam.; unilocular sporangia oval, 35 μ long, 25 μ diam.

North Carolina, New Jersey to Nova Scotia, Anticosti Island, Quebec, and Newfoundland; epiphytic on such coarse algae as Chondrus, or gregarious, spreading over small Corallina plants attached to rocks; a summer plant fruiting in the latter part of the season.

REFERENCES: Phyc. Bor.-Am. 130. Harvey 1852, p. 124 (as *L. tuberiformis*); Farlow 1881, p. 82.

ACROTHRICACEAE

Plants bushy, attached by a disciform holdfast, stoutly filiform, more or less branched, developing by trichothallic growth of an axial filament, which develops a primary cortex and short assimilatory filaments; sporangia lateral on the bases of the assimilators; gametangia unknown.

Acrothrix Kylin, 1907

Plants erect from a basal disk, sparingly to repeatedly branched, the branches flexuose to arcuate; structurally showing a single slender axial filament which grows at first from a terminal hair, but which may lose this and continue from an apical meristematic cell; cortex of a few layers of cells outwardly smaller, superficially bearing short assimilatory filaments; in the older portions hollow, the primary axial filament sometimes disrupted, but usually attached to the wall of the cavity; sporangia lateral at the base of the assimilatory filaments.

Acrothrix novae-angliae Taylor Pl. 13, fig. 1

Plants loosely tufted, developing 1–several main axes by wide-angled divarication near the base, and subsidiary axes laterally on these, 1.0–2.8 dm. tall, to 4.8 dm. broad, light brown; smooth but not slimy, softer in distal parts but comparatively stiff below; secondary branches radial, alternate, to 3 (–5) orders, occasionally pseudodichotomous; in diameter 0.5–1.5 mm.; main axes dissolving in long primary branches which exceed them in length; ultimate branches curved or flexuous; structurally in older parts above the base with a central cavity, an axial filament laterally displaced, a cortex of 2–3 cell layers, the inner large and colorless, the outer small and with a few chromatophores; surface loosely clothed with short assimilatory filaments usually 5–6 (–10) cells long, simple or 1–2 (–3) branched, straight or arcuate, asymmetrically moniliform above, where 6–8 (–9.5) μ diam.; colorless hairs infrequent, simple, replacing branches of the assimilatory filaments and 7.7 μ diam.; unilocular sporangia usually attached to the basal cell of the assimilative filaments, becoming subspherical, 22–31 μ diam.

Long Island and southern Massachusetts; bearing sporangia late in the spring and early summer; known to grow on stones, on Ascophyllum, on the basal parts of Zostera and certainly also on other structures, in shallow relatively protected areas.

REFERENCES: Taylor 1928b, p. 577, 1940, p. 190.

STILOPHORACEAE

Plants bushy, attached by a holdfast disk, stoutly filiform, freely dichotomously to irregularly branched, developing by the apical division of an axial group of filaments, which produce a parenchymatous cortex ultimately bearing colorless hairs and sori of unilocular sporangia or plurilocular gametangia associated with short assimilatory filaments; active plurilocular gametangia also produced on an alternate microscopic phase.

Stilophora J. Agardh, 1841

Characters of the family; filaments formed on the surface of the thallus, simple if sterile, branched if fertile, and then segregated into groups to which the reproductive organs are confined, these forming as lateral divisions of the sparingly branched fertile filaments.

Stilophora rhizodes (Ehrhart) J. Agardh

Pl. 13, fig. 6; pl. 14, fig. 6

Plants more or less erect, but entangled, from a disklike holdfast, filiform, 1–3 dm. tall, pale brown, relatively stiff and brittle, freely and dichotomously to irregularly branched without persistent main axes, the upper branches often elongate-tapering, the lower sometimes entangled; medulla of 4–5 longitudinal series of cells, or ultimately hollow, the cortex narrow, parenchymatous; branches nodulose with clumps of arcuate filaments 75–85 μ long which are slender and branched below, about 3.5 μ diam., above undivided and asymmetrically moniliform, to 9.5–13.0 μ diam., the cells with lenticular chromatophores; unilocular sporangia and uniseriate plurilocular gametangia usually on different plants, but occasionally on the same individual, arising from the bases of the peripheral filaments, the sporangia scattered, obovate or clavate, 36–56 μ long, 22–32 μ diam.; the gametangia filiform, uniseriate, with 4–9 cells, 30–50 μ long, 9.5–11.5 μ diam.

North Carolina, New Jersey to southern Massachusetts and Prince Edward Island; plants of quiet, rather warm protected bays and shallow water, loosely attached at the bases of Zostera plants, algae, or other objects, or frequently quite loose and drifting; vegetating and fruiting most luxuriantly toward the end of the summer.

REFERENCES: Algae Exsic. Am. Bor. 114; Phyc. Bor.-Am. 83. Harvey 1852, p. 112 (as *S. papillosa*); Farlow 1881, p. 90.

F. **contorta** Holden (Phyc. Bor.-Am. 679) is closely branched, entangled in dense masses, without elongated upper branches; it is not a sharply delimited form but has been reported from Connecticut.

DESMARESTIALES

Plants of filamentous construction, the branches with subapical trichothallic growth, the filaments developing uniaxial branches with often considerable cortication; plants as ordinarily described rather large, reproducing by unilocular sporangia, the zoöspores from which develop microscopic filamentous oögamous gametophytes.

DESMARESTIACEAE

Plants arising from a basal disk, filiform or compressed and with a midrib, oppositely or alternately branched; growth trichothallic at least at first, axis persistent, with an evident axial cell row and a subparenchymatous cortex, sometimes with short superficial assimilatory filaments; sporangia resulting from the conversion of superficial thallus cells, or of assimilatory filaments; gametophyte microscopic, filamentous, and either oögamous or parthenogenetic.

KEY TO GENERA

1. Uniseriate filaments persistent; sporangia seriate, in superficial tufts of filaments........................ ARTHROCLADIA, p. 152
1. Uniseriate filaments soon discarded; sporangia from superficial cells, largely immersed..................... DESMARESTIA, p. 153

Arthrocladia Duby, 1830

Plants bushy, wandlike, the axes repeatedly branched, the primary branches opposite or alternate, cylindrical, bearing whorls of simple or alternately branched, clustered filaments; structurally with a very large, thick-walled axial filament and a cortex, which shows several layers of relatively large, thin-walled cylindrical or prismatic cells within, and outwardly small cells with thick walls and brown chromatophores; unilocular sporangia seriate, moniliform, stalked, replacing lower branches of the whorled tufts of filaments.

Arthrocladia villosa (Hudson) Duby
Pl. 13, fig. 2; pl. 17, figs. 7–8

Thallus from a basal disk, to 4 dm. tall, slenderly and widely branched, brown, drying quite greenish; axis to 1 mm. thick, rather stiff, irregularly branched below, alternately or characteristically oppositely 1–3 times branched above, the intervals long; on all but the oldest parts of the plant bearing whorled tufts of simple or branched chromatophore-bearing filaments about 4 mm. long, the branching opposite below, alternate above, diameter to 21 μ below, 7 μ above in the ultimate filaments; unilocular sporangia replacing branchlets of the tufted filaments, alternately or unilaterally placed on the lower segments, to 15–20-seriate, the series 50–350 μ long, the individual sporangia 11–15 μ diam., 5.5–8.5 μ long, each producing several swarmers and developing a single lateral pore.

North Carolina, New Jersey, Rhode Island, and Massachusetts; infrequent, usually growing in moderately deep water, particularly on a stony or a shelly bottom, and drifting ashore; an annual, reaching best condition in late summer and early autumn.

REFERENCES: Algae Exsic. Am. Bor. 194; Phyc. Bor.-Am. LXXX. Harvey 1852, p. 75.

Desmarestia Lamouroux, 1813

Plants in our species bushy, erect from a small disciform holdfast, pinnately branched, the branches long, or shorter and reduced to denticulations, cylindrical or compressed, terminating in or bearing on the lateral spinules branched uniseriate brown filaments; structurally with a prominent axial cell row surrounded by a wide cortex of large and small cells; sporangia from little-modified external cortical cells, zoöspores from these unilocular sporangia germinating to form small filamentous dioecious oögamous gametophytes.

KEY TO SPECIES

1. Branching opposite; branches cylindrical, abundant, ultimately capillary . . . . D. viridis
1. Branching alternate, above markedly bilateral; branches flattened, the ultimate being spinuliform with terminal deciduous hair tufts . . . . D. aculeata

Desmarestia viridis (Müller) Lamouroux Pl. 13, fig. 3

Plants erect from small lobed holdfasts, light brown, 3–6 (–10) dm. tall, the main axis percurrent, to 3 mm. diam., abundantly oppositely 3–5 times pinnately branched, cylindrical at the base, a little compressed above, especially at the nodes; the chief branches 2–15 mm. apart, the successive divisions similar and progressively smaller to the capillary tips, or larger and smaller branches alternating on the major branches, the finer branchlets about 0.1 mm. diam.; much more delicate and attenuate in the spring, terminating in and beset with delicate branched uniseriate filaments 10–40 μ diam., which are shed by the summer, the bilateral character often obscured in the distal parts of the plant; a central axis of large cells present, but not forming a discernible costa, the cortex with large cells or air spaces separated by smaller cells; sporangia immersed at the surface of the branches, little modified from surface cells.

New Jersey to Prince Edward Island, Newfoundland, northern Labrador, Hudson Strait, and Ellesmere Island; an annual, often gregarious on rocks in moderately deep water, extending into the immediate sublittoral; reaching its climax in early summer and then disappearing.

REFERENCES: Phyc. Bor.-Am. 531, II. Harvey 1852, p. 77; Farlow 1881, p. 65; Rosenvinge 1926, p. 20; Wilce *in herb*.

This plant decays particularly quickly when confined; it should be studied and mounted as soon as possible after collecting, and never crowded with other specimens in a collecting bucket.

Desmarestia aculeata (Linnaeus) Lamouroux
Pl. 13, figs. 4–5; pl. 14, fig. 7

Plants erect from small lobed holdfasts, dark brown, somewhat cartilaginous, to 0.5 or rarely 2 meters in length, the main axis percurrent, the branching abundant, somewhat clustered, without notable distinction between main and secondary branches, 2–3 times alternate, the lower branches longer than the upper; filiform compressed, bilaterally pinnately divided, to 2 mm. broad in the young piliferous branches, but usually more slender, about 0.75 mm. diam. in the smaller branches, to oval in section, 3 mm. broad near the base, an inconspicuous immersed costa defined by an axial row of large cells extending to the tips of the branchlets, the cortex showing large and small cells intermixed; the ultimate branchlets sharply differentiated, alternate, marginal, slender-spinuliform, usually 0.5–1.0 (–2.0) mm. long, at intervals of about 4 mm.; in the spring bearing tufts of free brown uniseriate oppositely or occasionally alternately branched filaments 1–3 mm. long, in diameter to 50 μ, but the ultimate divisions more slender, about 30 μ, the cells with many lenticular chromatophores; sporangia unilocular, immersed at the surface, formed from superficial cortical cells.

New Jersey to Prince Edward Island, the lower St. Lawrence, northern Labrador, Ile Miquelon, James Bay, Hudson Bay, Hudson Strait, Baffin Island, and Ellesmere Island; often gregarious on rocks below low-tide levels, more rarely in tide pools, reaching to 10 meters depth; except sometimes in its northern range persisting over the winter, and proliferating during the following year, but in any case becoming depiliate in the summer.

REFERENCES: Phyc. Bor.-Am. 129. Harvey 1852, p. 78; Farlow 1881, p. 65; Hariot 1889, p. 155; Rosenvinge 1926, p. 20.

A very striking form has been noted from Long Island and southern Massachusetts, v. **attenuata** Taylor (1937b, p. 230), which is very long, reaching 80 cm. or more, and very slender, sparsely branched, the ultimate branchlets very delicate, to 3–7 cm. long, spinuliform determinate branchlets being practically absent; it is probably a form of deep water.

PUNCTARIALES

Plants subfilamentous or membranous-expanded, with intercalary growth and ultimately parenchymatous subdivision of the cells; plants as ordinarily described reproducing by unilocular sporangia and also with plurilocular gametangia or gametangium-like structures, but sometimes bearing functional gametangia and anisogametes on a small filamentous phase.

KEY TO FAMILIES

1. Branched plants; growth persisting longest in the small branches .... STRIARIACEAE, p. 155
1. Mostly unbranched plants, growth persisting throughout, or at least in the lower parts.................. PUNCTARIACEAE, p. 159

STRIARIACEAE

Plants filiform, sometimes tubular, variously branched; structure uniseriate to irregularly polysiphonous or parenchymatous, growth intercalary, but cell division soon ceasing below; reproductive organs formed by the direct transformation of a superficial cell or by division of it.

KEY TO GENERA

1. Reproductive organs without associated paraphyses............. 2
1. Unilocular sporangia associated with paraphyses.... STRIARIA, p. 158

2. Colorless hairs absent, sporangia sessile, gametangia in distinctive soriform clusters....................... ISTHMOPLEA, p. 156
2. Colorless hairs present, gametangia formed by division of little-modified surface cells..................... STICTYOSIPHON, p. 157

Isthmoplea Kjellman, 1877

Plants tufted, attached by basal disks, densely branching, the erect filaments freely and often oppositely branched, uniseriate with intercalary development or pluriseriate; gametangia in the continuity of the branches, formed by the division of several contiguous vegetative cells; unilocular sporangia lateral to vegetative cells, single or paired, when single generally opposite a branch, globose.

Isthmoplea sphaerophora (Carmichael) Kjellman Pl. 9, figs. 4–5

Plants tufted, 2.0–7.5 cm. tall, of repeatedly branched filaments, the spreading branches alternate or more usually opposite or whorled, below often pluriseriate, but uniseriate in the secondary and lesser branches; cylindrical segments below 40–65 μ diam., 0.5–1.5 diameters long, above about 20 μ diam., the cells with several small discoid chromatophores; plurilocular gametangia formed by subdivision of cells of the branchlets, several contiguous cells participating, 90–200 μ, probably more in length, 30–40 μ diam., below and above the gametangium the branchlets vegetative; unilocular sporangia sessile or the base immersed, usually opposite, 4-whorled, or opposite a branchlet, spherical, 40–55 μ diam.

Northern Massachusetts to Maine, and northern Labrador; rare, growing on larger algae, particularly *Ptilota serrata;* usually found fruiting in late spring.

REFERENCES: Farlow 1881, p. 74 (as *Ectocarpus sphaerophorus*); Schuh 1933d, p. 293; Kylin 1947, p. 67; Wilce *in herb.*

Stictyosiphon Kützing,[1] 1843

Plants clustered, filiform, attached by a rhizoidal holdfast, usually branched, uniseriate to parenchymatous, when solid or locally hollow below, the interior of elongate or rounded cells surrounded in some cases by a cortex of smaller cells; unilocular sporangia partly immersed at the surface, scattered or clustered; growth intercalary; colorless hairs present; plurilocular gametangia intercalary or from superficial cells, generally in scattered groups, the gametes producing filamentous or disklike plantlets which also bear intercalary or terminal plurilocular gametangia.

KEY TO SPECIES

1. Unbranched or with a few branches........... S. subsimplex, p. 158
1. Freely branching, at least above.............................. 2

2. Unilocular sporangia present; base of penetrating rhizoids; axis in considerable part monosiphonous........ S. griffithsianus, p. 157
2. Plurilocular gametangia present; the base a rhizoidal disk; the axis chiefly parenchymatous.................... S. tortilis, p. 158

Stictyosiphon griffithsianus (Le Jolis) Holmes et Batters

Plants densely tufted, 4–10 cm. tall, attached by penetrating rhizoids, the slender axes somewhat entangled, 75 μ diam. or more, freely branched; spreading branches opposite or whorled, pluriseriate, especially at the point of branching and where the sporangia are immersed; branchlets about 15–20 μ diam.; attenuate branchlet tips hairlike, but without a basal meristematic zone; unilocular sporangia in the upper branchlets, singly or in groups; plurilocular gametangia infrequent, intercalary.

Massachusetts and Maine, Prince Edward Island, and the lower St. Lawrence; an annual, epiphytic on Rhodymenia; forming unilocular sporangia in the spring.

REFERENCES: Phyc. Bor.-Am. 672. Harvey 1852, p. 138; Farlow 1881, p. 74 (both as *Ectocarpus brachiatus*) Taylor 1937c, p. 177 (as *Phloeospora brachiata*); Silva 1952, p. 302.

[1] *Stictyosiphon sorifera* (Reinke) Rosenvinge was represented in Collins' herbarium by one fragmentary and inconclusive specimen from Massachusetts, but by none from Rhode Island (Schuh 1914b, p. 152, as *Kjellmannia sorifera* Reinke). As the records are in need of confirmation, no description is included at this time.

Stictyosiphon subsimplex Holden

Plants straggling, when mature from a few millimeters to 12 cm. in length, simple or very sparingly branched, not infrequently with a few short branches at the summit; when young uniseriate, later in part pluriseriate; true hairs with basal meristematic zone present; plurilocular gametangia on slender axes, ovate-conical, lateral, the base occupying the axis, or sharing it with an opposite gametangium, or with a vegetative cell, or the base not in the axial line, and then entirely lateral, in size probably 25–45 μ long, 15–25 μ diam.; unilocular sporangia scattered or in groups, the lower portion immersed in the outer part of the axis, spherical to ovoid, or laterally somewhat compressed, 40–56 μ diam., to 56–84 μ long.

Connecticut and southern Massachusetts; epiphytic on Ruppia in salt-marsh pools, together with *Ectocarpus subcorymbosus;* ephemeral, found in late autumn.

REFERENCES: Phyc. Bor.-Am. 630. Holden 1899, p. 198.

Stictyosiphon tortilis (Ruprecht) Reinke

Plants tufted, entangled, densely crowded on thick basal disks, olive to blackish; filaments very slender, 7.5–15.0 cm. long, subsimple to usually repeatedly laterally branched, branches spreading, generally opposite, the apices uniseriate, the heavier parts incompletely hollow, the walls with large inner cells and small outer cells twice as long as broad; true hairs present, with a basal meristematic zone; chromatophores bandlike; plurilocular gametangia in sori or scattered, formed from superficial cells, becoming rounded and prominent.

Maine, Nova Scotia, northern Labrador, southeastern Hudson Bay and James Bay, Baffin Island, and Ellesmere Island.

REFERENCES: Schuh 1914a, p. 105; Wilce *in herb.*

Striaria Greville, 1828

Plants attached by a basal disk, distinctly tubular, repeatedly laterally, generally oppositely branched, delicately membranous, constructed of two layers of cells, the interior larger and round, the external smaller; chromatophores small and disklike, mostly in the surface cells, becoming somewhat elongate in the older parts; unilocular sporangia aggregated in superficial spots or in regular transverse zones with numerous unicellular paraphyses; plurilocular gametangia unknown.

Striaria attenuata (C. Agardh) Greville

Plants 1–7 dm. tall, amply branched, pale olive; axis and branches 1–5 mm. diam., the primary branches opposite or whorled, for the most part cylindrical but tapering gradually toward delicate apices and more sharply to the bases of the branches; superficial cells squarish, in irregular longitudinal rows, 19–32 μ diam.; unilocular sporangia superficial, in rounded or transverse sori banding the axis and branches at rather regular intervals of 0.25–0.50 mm., the sporangia measuring 48–60 μ diam., associated with unicellular saccate paraphyses and hairs.

Staten Island and Long Island, Rhode Island, and southern Massachusetts; producing sporangia in the spring.

REFERENCES: Phyc. Bor.-Am. 827. Harvey 1858a, p. 123; Farlow 1881, p. 90.

PUNCTARIACEAE

Plants of variable form, filiform, tubular, or phylloid, usually simple, generally stalked, from a basal disk; colorless hairs usually abundant; growth intercalary, structure parenchymatous; unilocular sporangia and plurilocular gametangia formed by the transformation of superficial cells of the erect plant, external or immersed, or on the creeping base, or on a filamentous phase of the plant.

KEY TO GENERA

1. Frond small, filamentous 2
1. Frond usually to 5 cm. tall or more 6

2. Unbranched 3
2. Branched 5

3. Axis unbranched, beset with short radiating uniseriate branchlets .... MYRIOTRICHIA, p. 160
3. Axis unbranched, pluriseriate to parenchymatous, without superficial branchlets 4

4. Uniseriate below, sparingly pluriseriate above, sometimes flattened, the gametangia notably projecting
DESMOTRICHUM, p. 164
4. Parenchymatous essentially throughout, cylindrical
LITOSIPHON, p. 162

5. Branching, if present, basal............... DESMOTRICHUM, p. 164
5. Branching sparing to abundant and scattered
RHADINOCLADIA, p. 163

6. Frond plane, leaflike.. 7
6. Frond cylindrical 9
6. Frond saccate, umbilicate after rupturing
OMPHALOPHYLLUM, p. 170

7. Thallus thick, of several cell layers with great differentiation, especially in size, between surface and medullary cells; asymmetrical at the base................... PETALONIA, p. 167
7. Thallus thinner, of a few cell layers, the outer cells smaller, but otherwise little different from the inner layers............. 8

8. Thallus delicate, filiform below, usually linear-lanceolate above; until senescent, hair-margined..... DESMOTRICHUM, p. 164
8. Thallus coarser, usually short-stalked, without notable marginal hairs .... PUNCTARIA, p. 165

9. Thallus delicate, of one cell layer............ PHAEOSACCION, p. 170
9. Thallus more substantial, of several layers..................... 10

10. Thallus when mature usually locally constricted, the sporangia and hairs scattered....................... SCYTOSIPHON, p. 168
10. Thallus unconstricted 11

11. Filiform; paraphyses and sporangia scattered... DELAMAREA, p. 168
11. Coarser, dark-flecked by groups of sporangia and hairs
ASPEROCOCCUS, p. 169

Myriotrichia Harvey, 1834

Plants filiform, the bases sometimes creeping, of uniseriate filaments attached by rhizoids, the erect axis uniseriate below, parenchymatous above, beset with short radial branchlets and usually with colorless hairs; longitudinal growth intercalary, near the base; unilocular sporangia solitary, scattered, opposite, or verticillate, attached to the creeping base, the axis, or the branchlets, globose or globose-ovoid; plurilocular gametangia lateral, solitary or aggregated into little clusters, cylindrical or subfusiform, uni- to pluriseriate; unilocular sporangia and plurilocular gametangia sometimes occurring on the same plant.

KEY TO SPECIES

1. Axis beset with simple or subsimple lateral branchlets, or sometimes nearly naked 2
1. Axis beset with forked lateral branchlets .... M. densa, p. 162
2. Axis irregular, nodose-cylindrical in aspect, enlarged by groups of very short branchlets; plurilocular gametangia in clusters M. filiformis, p. 161
2. Axis regularly clavate, densely beset with lateral branchlets, the upper ones longer; plurilocular gametangia borne singly M. clavaeformis, p. 161

Myriotrichia clavaeformis Harvey Pl. 10, figs. 4–7

Plants slenderly clavate, to 2–5 cm. tall, tufted; the primary axis simple, below naked, the segments half as long as broad, above beset with short, crowded, or even whorled uniseriate branchlets 13–25 μ diam., their length increasing toward the summit of the plant, thus giving its clavate form, each to 7–14 cells long, the cells equal or slightly longer than broad, with numerous small disklike chromatophores; plurilocular gametangia sessile, single, oval, or somewhat elongate, possibly dimorphic with contrasting larger or smaller cells and gametes; unilocular sporangia sessile, particularly in the axils of branchlets, ovate, 34–47 μ diam., 47–65 μ long.

New Jersey, Staten Island, Rhode Island, and Massachusetts; growing attached to various algae, especially Asperococcus and Scytosiphon; commonest in early summer.

REFERENCE: Farlow 1881, p. 67.

Myriotrichia filiformis Harvey

Plants filiform, to 2.5 cm. tall, flexuous, often contorted, the primary filaments simple; segments of the primary axis shorter than the diameter, uniseriate, later irregularly parenchymatously nodose, at the nodes densely bearing minute uniseriate branchlets about 2–8 cells long, or the nodes confluent; plurilocular gametangia forming dense masses surrounding the axis at intervals in clumps among the branchlets, oval or broadly conical-lanceolate, 22–56 μ long, 10–38 μ diam.; unilocular sporangia oval to spherical, 32–52 μ diam.; to 60 μ long.

New Jersey, Long Island, Rhode Island, and Massachusetts; epiphytic on such algae as Scytosiphon and Asperococcus; fruiting in the summer.

REFERENCES: Algae Exsic. Am. Bor. 167; Phyc. Bor.-Am. 734. Harvey 1858a, p. 124; Farlow 1881, p. 68 (as *M. clavaeformis* v. *filiformis*).

Myriotrichia densa Batters

Plants large, cylindrical-filiform, 1.25–3.75 cm. long, 0.2–0.25 mm. diam., the base a little attenuated; axis enveloped in short, lateral uniseriate dichotomous or unilaterally forked filaments; hairs few or absent; plurilocular gametangia solitary or 2–4 together, short-stalked, cylindrical or lanceolate, uni- or occasionally biseriate, 6–10 μ diam., 25–60 μ long; unilocular sporangia stalked, ovoid to spherical, 50 μ diam.

Maine; growing on Asperococcus and Scytosiphon; reported but once.

REFERENCE: Schuh 1933c, p. 256.

Litosiphon Harvey, 1849

Plants gregarious, from a basal disk, simple, firm, filiform, cylindrical to subclavate, at first solid but later hollow, with intercalary growth; cells with several platelike chromatophores; plurilocular gametangia nearly immersed, intercalary in the frond, formed from scattered superficial cells; unilocular sporangia scattered, rounded-elliptical; both types of reproductive organs may also be found on the basal disk and on a filamentous phase of the plant.

Litosiphon filiformis (Reinke) Batters

Plants tufted, to 10–30 mm. tall, from a basal one-layered disk; erect simple filaments at first uniseriate, later parenchymatous, 15–60 μ diam., the cells 7.5–11.5 μ diam., subrectangular, in distinct transverse and vague longitudinal rows; plurilocular gametangia and unilocular sporangia formed on the basal and the erect filaments; on the basal filaments sessile or partly immersed in the filaments, on the erect axes superficial, formed from the outer cells, singly or in groups.

Rhode Island, Maine, and Devon Island; forming an inconspicuous growth on Laminaria; spring.

REFERENCES: Phyc. Bor.-Am. 1035 (as *Pogotrichum filiforme*). Schuh 1900b, p. 206 (as *P. filiforme*).

Rhadinocladia Schuh, 1900

Plants small, often growing in loose tufts; basal layer disklike, of several cells in thickness except near the edges, originating the long branching fronds, which possess a well-defined central axis one or more cells in width; axes bearing numerous long and slender branches, with many or few branchlets and occasional hairs; plurilocular gametangia abundant, cylindrical or oblong, generally sessile and entirely superficial.

KEY TO SPECIES

1. Plants usually clustered; axis uniseriate or occasionally locally biseriate, 3–8 mm. tall; cylindrical gametangia 15–18 μ diam., crowded .... R. cylindrica
1. Plants usually solitary; axis pluriseriate except at the base, to 16 mm. tall; fusiform gametangia 20–25 μ wide, scattered R. farlowii

Rhadinocladia cylindrica Schuh

Plants 3–8 mm. tall, usually several axes arising from the same basal layer, each axis with slender divaricate branches to about the second degree; uniseriate, occasionally with binate cells; plurilocular gametangia densely clustered along the axis and the branches, their bases much compressed by crowding, very long cylindrical with rounded tips, 60–80 μ long, 15–18 μ diam.

Rhode Island and southern Massachusetts; epiphytic on Zostera; fruiting in midsummer and autumn.

REFERENCES: Phyc. Bor.-Am. 1591. Schuh 1901, p. 218.

Rhadinocladia farlowii Schuh

Plants 12–16 mm. tall, solitary from each disk, plumose, olive-brown; axis slender, percurrent, biseriate below, 40–50 μ diam., to

4-seriate in the central portion, 60–70 μ diam., bearing 30–50 flagelliform ascending branches, scattered or often opposite, 6–8 mm. long, uniseriate to rarely 2–3-seriate, hair-tipped, infrequently with a few secondary branchlets near the end; colorless hairs numerous, 12 μ diam.; plurilocular gametangia sessile, fusiform to oblong, 70–85 μ long, 20–25 μ diam.

Rhode Island and southern Massachusetts; growing on Zostera and Chorda as epiphytes; fruiting in late summer.

REFERENCE: Schuh 1900a, p. 112.

Desmotrichum Kützing, 1845

Fronds filiform or, more usually, ribbonlike, attached by rhizoids from the basal cells, or the base sometimes more ample and of creeping filaments, the erect portion in the smaller forms sometimes branched near the base, uniseriate below and cylindrical, pluriseriate to parenchymatous above, or in larger plants pluriseriate below and expanded membranous above, where of two to four cell layers; colorless hairs with basal growth scattered, or in the flat types forming a nearly continuous marginal fringe, tardily deciduous; unilocular sporangia and plurilocular gametangia on the same or separate plants; gametangia cylindrical or fusiform, with the bases either immersed or completely external and sessile, or even short-stalked; sporangia rounded-polygonal, scattered, superficial; a filamentous phase bearing gametangia also reported.

KEY TO SPECIES

1. Plant a slender filament, uniseriate below, pluriseriate by longitudinal walls above.................................. D. balticum
1. Plant a linear-lanceolate blade with a slender stalk..... D. undulatum

Desmotrichum balticum Kützing

Plant at first a simple uniseriate filament, reaching 1–10 mm. in length, with terminal hairs; later, particularly above, the axis becoming pluriseriate and cylindrical or somewhat compressed; cells with oval or, rarely, somewhat bandlike chromatophores; plurilocular gametangia conical-papillate, or even fusiform, often contiguous.

Connecticut to Maine, and James Bay; growing on Zostera and other algae; fruiting from spring to late summer.

REFERENCE: Phyc. Bor.-Am. 1080.

Desmotrichum undulatum (J. Agardh) Reinke
Pl. 15, fig. 6; pl. 16, figs. 1–2

Plants forming light brown linear-lanceolate blades with a disklike holdfast and a slender short stalk, the length usually about 3–5 cm., but reaching 2 dm., the width 2–10 mm., the base, and particularly the apex, long and acutely tapering, the margins often crisped; texture thin-membranous, not slippery; structurally of 2, rarely 4 cells in thickness and measuring 22–42 μ; surface cells little different from those within, if such are present, 9.5–22.0 μ wide, 6.5–18.0 (–20.0) μ long, sometimes piliferous, ordinarily with several small disklike chromatophores; marginal cells flattened, nearly all at first bearing a long simple hair with basal growth, 5.5–10.0 μ diam., these latter partly deciduous; plurilocular gametangia scattered, partly immersed, outwardly conical, strongly projecting, often marginal; unilocular sporangia immersed, subquadrangular in surface view, occupying the equivalent area of 1–4 neighboring surface cells, usually 13–22 μ diam., solitary or 2–4 such sporangia grouped together.

New Jersey to Nova Scotia, and Prince Edward Island; usually epiphytic on Zostera; plant in best condition during the early summer, reported to form the sporangia earlier than the gametangia.

REFERENCES: Phyc. Bor.-Am. 127. Harvey 1852, p. 115 (as *Punctaria tenuissima*); Farlow 1881, p. 64 (as *P. latifolia* v. *zosterae*).

Punctaria Greville, 1830

Thallus foliaceous or ribbonlike, tapering sharply to a short slender stalk and attached by a basal disk, beset with tufts of hairs; when mature four to seven cells thick; the surface cells not conspicuously small; unilocular sporangia scattered; plurilocular gametangia somewhat clustered, angular, immersed below, the apex prominently projecting above the surface, pluriseriate, like the sporangia formed by the metamorphosis of cortical cells; gametangia and sporangia occurring on separate plants, or on the same plant in that sequence, the swarmers developing to filamentous plantlets which may bear gametangia.

KEY TO SPECIES

1. Thallus of slender form and delicate texture, to 4 cells thick, light in color and greenish on drying.................. P. latifolia

1. Thallus broader and coarser, 4–7 cells thick, dark brown
P. plantaginea

Punctaria latifolia Greville Pl. 15, fig. 5

Plants in the form of lanceolate blades arising from minute basal disks, short-stalked, the stalks 2–5 mm. long, the blades 10–45 cm. long, 2–8 (–15) cm. broad, the base abruptly cuneate, the general contour lanceolate or oblong, flat or the margin crispate, the apex generally obtuse; texture soft and thinly membranous, subpellucid, yellowish or olive-brown, in drying becoming dull greenish, generally not adhering to paper; 2–4 cells and 50–65 (–85) μ thick, the cells in surface view 15–40 μ diam., in fairly clear rows.

New Jersey to the lower St. Lawrence; particularly common on the leaves of Zostera and on large algae; appearing during the winter and decaying by midsummer in the southern parts of its range.

REFERENCES: Algae Exsic. Am. Bor. 200; Phyc. Bor.-Am. 82. Harvey 1852, p. 116; Farlow 1881, p. 64.

F. **crispata** (Kützing) Collins (Phyc. Bor.-Am. 1388), with clearly crispate edges, has been reported from Maine, Nova Scotia, and the lower St. Lawrence.

Punctaria plantaginea (Roth) Greville Pl. 15, fig. 4; pl. 16, fig. 4

Plants in the form of broadly lanceolate blades arising from small basal disks, the stalks short, the blades with tapered bases, obovate-lanceolate, often mechanically split or truncate toward the tip, in length usually less than 2 dm., but to 6.5 dm., flat or very little undulate at the margin; texture somewhat firm, even coriaceous, 4–7 cells and 110–225 μ thick, cell membranes rather heavy, surface cells 15–40 μ diam.; plurilocular gametangia somewhat increasing the thickness of the thallus, to as much as 50 per cent, their lower parts sunken in the surface layer, oblong or obovoid, 30–48 μ long, 20–34 μ diam.; unilocular sporangia (?) nearly globose, 32–48 μ diam.

New Jersey to the lower St. Lawrence, northern Labrador, and Baffin Island; ordinarily growing on stones or coarser algae; developing vegetatively early in the year, and after fruiting drifting ashore heavily loaded with epiphytes during the early summer.

REFERENCES: Phyc. Bor.-Am. 81. Harvey 1852, p. 115; Farlow 1881, p. 64; Wilce *in herb.*

Petalonia Derbès et Solier, 1850

Frond flat, arising on a short filiform stalk from a holdfast disk; substance firm, structure internally of large cells intermixed with more slender filaments, externally of a layer of small chromatophore-bearing cells; fertile areas at first localized, later involving the whole surface of the frond, producing subcylindrical plurilocular gametangia.

Petalonia fascia (O. F. Müller) Kuntze Pl. 14, fig. 5; pl. 15, fig. 3

Plants linear-lanceolate, 0.75–4.5 dm. long, often gregarious, arising from disciform bases, with slender stalks, the bases of the blades tapering, usually asymmetrical, the margins straight or irregularly curved, the apices tapering; colorless hairs numerous over the surface; substance membranous to leathery, dark brown, particularly when fertile; thickness 145–200 μ in sterile parts and in fertile parts to 270 μ, structurally showing much larger cells within, those near the surface smaller; gametangia crowded, covering large areas of the blade, uniseriate, about 3.5–9.5 μ diam., 30–75 μ long.

Florida, North Carolina, New Jersey to the lower St. Lawrence, southeastern Hudson Bay and James Bay, Hudson Strait, Newfoundland, and Baffin Island; growing on stones and woodwork in the littoral near the high-tide line, more seldom on fuci; in active growth during the early part of the year, fruiting by early summer, disappearing toward autumn in some areas, but in others reported throughout the year.

REFERENCES: Algae Exsic. Am. Bor. 199; Phyc. Bor.-Am. 276 (both as *Phyllitis fascia*). Harvey 1852, p. 91 (as *Laminaria fascia*); Farlow 1881, p. 62; Taylor 1937b, p. 230; Wilce *in herb.*

From our area there is often secured v. **caespitosa** (J. Agardh) Taylor, which is a thinner and lighter brown form with a very markedly asymmetrical base and a width of 2.5–6.0 cm.; it comes from more northern stations, characteristic plants having been noted in New Jersey (?), Connecticut, Maine, New Brunswick, the lower St. Lawrence and northern Labrador. In addition, v. **zosterifolia** (Reinke) Taylor (Phyc. Bor.-Am. 277, 1082), which is linear,

10–20 cm. tall, 0.6–3.0 mm. wide, sometimes locally hollow, is a markedly contrasting type known from northern Massachusetts, New Brunswick, Nova Scotia, Prince Edward Island, and Newfoundland.

Delamarea Hariot, 1889

Frond filiform, cylindrical, tubular, simple, the base a disk formed by radicular fibrils, which sometimes become stoloniferous; structurally of two cell layers, the inner larger and elongate, grading to the smaller shorter chromatophore-bearing cells near the surface; paraphyses consisting of unicellular saccate enlargements of the surface cells, the large unilocular sporangia globose or ovate, scattered over the surface among the paraphyses; plurilocular gametangia conical, dioecious.

Delamarea attenuata (Kjellman) Rosenvinge

Plants coarsely filiform, to 5–8 (–30) cm. tall, usually gregarious; the base housed disciform; lower portion of the axis nearly capillary, above expanded to 1.5–2.0 mm. diam., texture firm; paraphyses widely distributed over the surface of the frond, oblong to cylindric-clavate, 60–120 μ long, 30–55 μ diam., admixed with colorless hairs; unilocular sporangia 30–60 μ long, 18–40 μ diam.; gametangia long-conical, shorter than the paraphyses.

Southern Massachusetts and Ile Miquelon.

REFERENCES: Hariot 1889, p. 156 (as *D. paradoxa*); Doty 1948, p. 262 (as *Scytosiphon attenuatus*).

Scytosiphon C. Agardh, 1811

Frond cylindrical, simple and tubular, firmly membranous, the wall of large and cylindrical inner cells, the outer rounded and angular; growth intercalary near the base; plurilocular gametangia subcylindrical, forming a broad continuous layer or localized spots, developed from the cortical cells, generally uniseriate; associated with paraphyses.

Scytosiphon lomentaria (Lyngbye) C. Agardh

Pl. 15, fig. 2; pl. 16, fig. 3

Plants gregarious, tapered-cylindrical, short-stalked from a disk-like base; length 15–70 cm., diameter 5–10 (–30) mm., characteristically locally constricted when mature, hollow; plurilocular gametangia crowded in large fruiting areas, 3.5–7.5 μ, occasionally to 9.5 μ diam., 45–65 μ long, uniseriate or frequently with oblique walls, less often biseriate with longitudinal walls, or forked; paraphyses 27–32 μ long (or more?), 8–13 μ diam., not exceeding the gametangia.

Bermuda, South Carolina, New Jersey to the lower St. Lawrence and Ile Miquelon, and northern Labrador; common on rocks in rather exposed places at and below low tide; an annual fruiting from winter to late spring, sometimes not reaching full growth until early summer.

REFERENCES: Phyc. Bor.-Am. 323. Harvey 1852, p. 98 (as *Chorda lomentaria*); Farlow 1881, p. 63; Hariot 1889, p. 156; Wilce *in herb.*

V. **complanatus** Rosenvinge (Phyc. Bor.-Am. 174), very soft and narrow, reaching 0.5–1.0 meter long but only 1.5–2 (–5) mm. diam., hollow but flattened, not constricted, without paraphyses, has been reported from Long Island and Connecticut in early spring.

Asperococcus Lamouroux, 1813

Plants simple, gregarious, stipitate from basal disks, hollow above, the wall consisting of an inner zone of a few layers of large colorless cells, decreasing in size toward the surface chromatophore-bearing layers; tufts of colorless hairs present; unilocular sporangia and plurilocular gametangia present on the same or separate individuals; sporangia sessile, with numerous subcylindrical paraphyses in small rather closely scattered superficial sori; plurilocular gametangia ovoid or ellipsoid, also in sori; swarmers developing filamentous plantlets bearing plurilocular gametangia.

Asperococcus echinatus (Mertens) Greville
Pl. 15, fig. 7; pl. 16, figs. 5–6

Plants 1–5 dm. tall, usually gregarious, arising from basal disciform holdfasts, tapering at the base, the individuals 2–5 (–25) mm. diam., straight or somewhat contorted, dark brown; texture rather rough, not slippery; superficial cells of the thallus elongate, 16–26 μ wide, 16–36 μ long, in cross section 9–22 μ thick; sori scattered, longitudinally elongate; paraphyses 3–several-celled, the terminal cells 22–32 μ diam., 16–36 μ long; unilocular sporangia ovoid to usually spherical, sessile or on one-celled stalks, 38–55 μ diam., 55–70 μ long.

Staten Island to the lower St. Lawrence, Newfoundland, and Baffin Island; generally growing attached to fuci, but also found on stones and woodwork below low tide and somewhat protected from the surf; this is a plant of the early part of the year in its southern range, fruiting in spring and largely decayed by midsummer.

REFERENCES: Phyc. Bor.-Am. 128; Phyk. Univ. 514. Harvey 1852, p. 117; Farlow 1881, p. 89; Whelden 1947, p. 174.

V. **vermicularis** Griffiths ex Harvey (Collins 1906b, p. 125), a slender form about 1 mm. diam. and to 40 cm. long, the sori in longitudinal lines on the thallus, has been recognized from Maine.

Phaeosaccion Farlow, 1882

Fronds tubulose or saccate, of one cell layer, the cells in surface view often grouped in fours; sporangia formed by division, generally by bipartition, of vegetative cells, which then discharge single zoöspores.

Phaeosaccion collinsii Farlow

Fronds gregarious, olive-brown, flaccid and subgelatinous, at first saccate, later cylindrical with open apex, to 25 cm. tall, 0.3–2.5 cm. in diam., the thallus wall 8–10 μ thick, the cells 3.8–7.0 μ diam. in surface view, with single chromatophores.

Northern Massachusetts to Maine; epiphytic on Zostera; a plant of early spring.

REFERENCES: Algae Exsic. Am. Bor. 201; Phyc. Bor.-Am. 29. Farlow 1882, p. 66; Rosenvinge 1893, p. 874.

Omphalophyllum Rosenvinge, 1893

Plants at first in the form of a saccate, stalked membrane which is early ruptured, opening out and becoming umbilicate, chiefly of one or two cell layers, without hairs; unilocular sporangia scattered, of the same form as the vegetative cells.

Omphalophyllum ulvaceum Rosenvinge

Plants short-stalked, broadly and delicately membranous, with a saccate base; stalk to 1 mm. long, solid, from a holdfast disk composed of rhizoidal outgrowths; blade to 22 cm. broad, pale olive when dry, the base funnel-shaped, but above opening out, variously lobed and lacerate with a very irregular margin; structurally of several layers of cells near the stalk but above of 1–2 layers of cells, the cells regularly disposed, subquadrate in surface view, in the thinner parts little broader than tall, with a single nucleus and several disklike chromatophores; unilocular sporangia scattered, rarely side by side, little larger than the vegetative cells.

Ile Miquelon and the west side of Ellesmere Island; dredged at 20–40 meters depth on a bottom of clay with scattered small stones.

REFERENCES: Rosenvinge 1893, p. 872, 1894, p. 106, 1926, p. 19.

DICTYOSIPHONALES

Plants as ordinarily described large and branching, terete, with initial growth from an apical cell; ultimately with diffuse cell division, becoming parenchymatous; unilocular sporangia produced at the surface, the zoöspores germinating to form microscopic filamentous gametophytes bearing isogametes.

DICTYOSIPHONACEAE

Plants slender to stout, comparatively firm, showing a definite axis derived from an apical cell, and irregularly placed branches, which may be hollow if large; delicate hairs present on the surface; often slippery but no copious mucus produced; sporangia formed from transformed superficial cells, immersed at the surface.

Dictyosiphon Greville, 1830

Plants from a small-lobed holdfast disk, erect, filiform, sparingly to copiously branched, the axis and main branches percurrent; smooth but not mucous; structurally showing a medulla of large elongated cells and a thin small-celled cortex with small disciform chromatophores, but in larger axes often partly hollow; delicate hairs visible on the surface of younger branches; the elliptical or ovate unilocular sporangia immersed at the surface, producing swarmers which give rise to filamentous plantlets which may bear plurilocular gametangia.

KEY TO SPECIES

1. Plants small, 2–8 cm. tall; subsimple, very slender.. D. eckmani, p. 172
1. Plants larger, sparingly to copiously branched.................. 2
2. Axis and branches thick, hollow, usually sparingly divided D. macounii, p. 173
2. More slender, in coarse forms hollow below only; generally freely branched 3
3. Simple, or branched to one degree.............. D. chordaria, p. 174
3. Repeatedly branched ....D. foeniculaceus, p. 172

Dictyosiphon eckmani Areschoug

Plants delicate, small, 2–5 (–8) cm. tall, 100–350 μ rarely reaching 1 mm. diam., simple or very sparingly branched above, the median portions of the axis larger than the tapered ends; the cortical cells and sporangia large, the latter about 22–34 μ in surface aspect.

Northern Massachusetts to Nova Scotia; growing on Scytosiphon in tide pools, often very densely; a plant of spring and early summer.

REFERENCES: Phyc. Bor.-Am. 533. Collins 1900d, p. 166.

Dictyosiphon foeniculaceus (Hudson) Greville

Pl. 12, fig. 4; pl. 14, fig. 2

Plants of moderate to considerable size, clustered or solitary, to 2–7 (–10) dm. tall, often bushy, the main axis slender, hardly reaching 1 mm. diam., infrequently forked, usually excurrent; long primary branches few to numerous, basal parts sometimes somewhat hollow; branching alternate to infrequently subopposite, abundant to the second or sometimes the third order, but sparing beyond; branches tapered to the tips, somewhat more slender than the axis; superficial cells small, usually in fairly distinct longitudinal series, about 12–15 μ average diam.; cells of the medulla relatively small; sporangia broadly oval, 30–55 μ diam.

New Jersey to the lower St. Lawrence, northern Labrador, Hudson Bay, James Bay, Hudson Strait, Newfoundland, Baffin Island, and Ellesmere Island, but particularly south of Cape Cod; growing on

Chordaria or other algae in exposed places; throughout the year, reaching full growth in the summer.

REFERENCES: Phyc. Bor.-Am. 673. Harvey 1852, p. 114; Farlow 1881, p. 66; Hariot 1889, p. 156; Whelden 1947, p. 116; Wilce *in herb.*

Intergrading forms between the typical and the following varieties are common. There is reported a vaguely delimited v. **americanus** Collins (Phyc. Bor.-Am. 674), with sporangia often in groups, contracted branch bases, and a softer, coarser habit and lighter color when dried, adhering closely to paper; it is found in Connecticut and Massachusetts. V. **flaccidus** (Areschoug) De Toni is of a pale color, greenish or yellowish olive, tubularly inflated, membranaceous in texture, solid only in the ultimate branchlets; the branches are more limp, the cortical cells are rather large, somewhat squarish; it is reported from the lower St. Lawrence, from Cape Breton, Nova Scotia, and Ellesmere Island. V. **hispidus** (Kjellman) Collins (Taylor 1937b, p. 226; Phyc. Bor.-Am. 677) grows to 30 cm. tall, is bushy, the axis rarely more than 1 mm. diam., sometimes hollow below, often forking, branching alternate, abundant to 4–5 orders or more; main axis and successively smaller branches beset with short, delicate lateral branchlets; it has been reported from New Jersey to Newfoundland, and in the more northern part of its range it is more common on rocky shores in late summer in rock pools or attached to Chordaria than is the type of the species. F. **hippuroides** (Lyngbye) Levring (Algae Exsic. Am. Bor. 95; Phyc. Bor.-Am. 675, 676 (as *D. hippuroides* v. *fragilis*). Farlow 1881, p. 66; Collins 1900d, p. 165; Taylor 1937c, p. 184 (all as *D. hippuroides*), consists of plants of moderate size, 10–30 (–70) cm. tall, usually gregarious; the axis and branches stout, of about the same size, sometimes partly hollow below, the tapering branches moderately abundant, of about equal length, infrequently forming branches of a second order; sporangia oval, in surface view 38–42 μ diam. This is reported from northern Massachusetts to Maine and Nova Scotia, the lower St. Lawrence, Newfoundland, and Baffin Bay; annual, on rocky shores and in low-tide pools, particularly on Chordaria; fruiting late in the summer.

Dictyosiphon macounii Farlow

Plants small, to 5–20 cm. tall, the axis simple, attenuate below, moderately to closely once laterally branched, thick, even to 6–12 mm. diam.; the branches generally simple and of about equal length, reaching 12–25 (–50) mm., and to 2–4 mm. diam., fusiform or clavate, often incurved, generally hollow, the superficial cells small, angular, about 7 μ diam.; the sporangia spherical, scattered, 38–42 μ diam.

Maine and the lower St. Lawrence; a plant of tide pools and shallow water; in good condition in midsummer.

REFERENCES: Phyc. Bor.-Am. 1843. Farlow 1889, p. 11; Collins 1900d, p. 166.

Dictyosiphon chordaria Areschoug

Plants soft, slippery, 8–20 cm. tall, simple or usually sparingly branched in one degree, the axes 0.5–1.0 mm. diam., becoming hollow below, tapering above; colorless hairs present; medulla loose, of large cells, covered by a cortex of 2–3 layers of smaller cells; unilocular sporangia formed from the inner cortical cells and immersed in the thallus, spherical or oval, 20–40 μ diam.

Rhode Island and Nova Scotia; a plant of the littoral, growing on stones and mussel shells.

REFERENCES: Collins 1904, p. 182, 1906c, p. 157; Taylor 1937c, p. 186 (all as *Gobia baltica*); Kylin 1947, p. 79.

LAMINARIALES

Plants of massive construction, parenchymatous externally but often with a filamentous medulla; growth intercalary in our genera, and in adult plants generally near the base; plants as ordinarily described reproducing by unilocular sporangia borne on general or localized portions of the surface; the zoöspores giving rise to microscopic branched filamentous oögamous gametophytes.

LAMINARIACEAE

Plants of moderate to very large size, with holdfast, stalk, and generally expanded blades of various forms, these constituting the sporophyte phase; gametophytes independent, microscopic, of branched uniseriate creeping filaments, dioecious, the antheridia small, producing a single antherozoid, the oögonia larger, producing a single egg which is fertilized and germinates while still partly enclosed in the oögonium wall.

KEY TO GENERA

1. Plant cylindrical, elongate........................ CHORDA, p. 175
1. Plant developing as a flat blade.............................. 2
2. Blade with tufts of hairs on the surface.......... SACCORHIZA, p. 176
2. Blade smooth 3
3. Blade perforate .... AGARUM, p. 185
3. Blade imperforate 4
4. Blade costate in the vegetative part................ ALARIA, p. 185
4. Blade without a midrib........................ LAMINARIA, p. 177

Chorda Stackhouse, 1797

Plants annual, subcylindrical, at first loosely filamentous within, becoming spongy, solid, hollow or septate, the outer medulla pseudoparenchymatous, of large, thick-walled cells, the cortex of cells radially arranged, bearing superficial sporangia and paraphyses.

KEY TO SPECIES

1. Free filaments rich in chromatophores covering the axis.. C. tomentosa
1. Free filaments limited to delicate colorless deciduous hairs.... C. filum

Chorda tomentosa Lyngbye

Plants elongate from a small holdfast, attaining a length of about 1 meter (reported 5–8 meters); the stipe 2–3 cm. long, slender and smooth, the upper thallus 2–4 mm. diam., its surface covered with a dense, wide, soft layer of filaments to 6–20 mm. long, 15–32 μ diam., but which taper near their apices to about 9 μ; filaments with each cell containing several lenticular chromatophores and 1.25–2.0 times as long as broad distally, but much shorter near the basal growth zone; sporangia clavate to narrowly elliptical, 75–115 μ long, 15–21 μ diam.; paraphyses cylindrical to clavate, 12–18 μ diam., slightly exceeding the sporangia in length, the protoplast subcylindrical, 4.5–7.5 μ diam., the gelatinous wall distally thick.

Long Island, Rhode Island to New Brunswick, Newfoundland, northern Labrador, southeastern Hudson Bay, Hudson Strait, and probably elsewhere in the northern part of the range; primarily an annual plant of the spring flora and of cold waters, disappearing in the southern part of its range before the advent of summer; fruiting in late spring.

REFERENCES: Phyc. Bor.-Am. 483. Hariot 1889, p. 157; Taylor 1940, p. 191; Wilce *in herb.*

Chorda filum (Linnaeus) Lamouroux Pl. 14, fig. 3; pl. 15, fig. 1

Plants long, arising singly from small irregular often gregarious disciform holdfasts, with a short slender stalk expanding into the growing region of the adult thallus, which is 3–7 mm. diam. and of quite variable length, from 0.5–5.0 (–12) meters, tapering gradually and usually partly decayed at the tip, and while young showing in the water a coating of very delicate deciduous colorless hairs; sporangia crowded on the thallus surface among obconic paraphyses 13–20 μ diam., which considerably overtop them; sporangia oblong or elliptical, 30–50 μ long by 10–15 μ diam.

New Jersey to the lower St. Lawrence, Newfoundland, northern Labrador, southeastern Hudson Bay, Baffin Bay, and the arctic; this species is very common, forming clumps on stones and shells in somewhat sheltered locations below and near the low-tide level, supported in a tolerably erect position by air in the medulla, and reaching nearly to the surface; an annual, fruiting from late summer to autumn.

REFERENCES: Phyc. Bor.-Am. 831. Harvey 1852, p. 98; Farlow 1881, p. 91; Hariot 1889, p. 157; Wilce *in herb.*

Saccorhiza De la Pylaie,[1] 1829

Plants moderately large, stalked from a primary disciform holdfast, later supplemented by whorls of simple fibers from the stalk just above the disk; the stalk at its summit expanding to a blade which is beset with tufts of hairs in cryptostomata which are evident in young plants, but later obscure; sporangia borne in patches near the base of the blade.

[1] The report by Harvey (1852, p. 91) of *Laminaria lorea fide* Despreaux from Newfoundland lacks confirmation; perhaps the plants were *Saccorhiza dermatodea.*

Saccorhiza dermatodea (De la Pylaie) J. Agardh Pl. 22, fig. 2

Plants annual, becoming moderately large; gregarious in habit; primary holdfast a conical disk of a few millimeters in diameter, later lobed and eventually obscured by adventitious subsimple outgrowths from the stalk, which cover it, the whole holdfast being rather small; stalk 15–60 cm. long, flattened; the tough blade oblanceolate from a base at first tapered, later somewhat rounded, becoming several times as long as the stalk; 30 cm. to 2 meters in length or longer, at first simple, to about 10 cm. wide, but later to 20–60 cm. wide, sometimes sparingly cleft, even down the stalk; the hair tufts in cryptostomata obvious in young blades, becoming obscure in old blades, and not evident in the lower parts; fruiting in moderately clearly defined areas about the base of the blade, the sporangia about 100 μ long by 10–14 μ diam.

Northern Massachusetts to the lower St. Lawrence, Newfoundland, northern Labrador, Hudson Strait, and Ellesmere Island; this is a species found in exposed tide pools and on surf-beaten rocks just below low-tide levels, and thence to a depth of about 50 meters.

REFERENCES: Algae Exsic. Am. Bor. 120; Phyc. Bor.-Am. 739, LXXXI. De la Pylaie 1829, p. 48; Harvey 1852, p. 92 (both as *Laminaria dermatodea*); Farlow 1881, p. 95; Hariot 1889, p. 182; Setchell 1891, p. 177; Taylor 1937c, p. 188 (both as *Phyllaria dermatodea*); Wilce *in herb*.

Collins' manuscript mentions a form of Saccorhiza which, though apparently not bearing a name, is comparable to *Laminaria agardhii* f. *angustissima,* being about 1 meter long, 1.5 cm. broad; it is reported from southern Nova Scotia.

Laminaria Lamouroux, 1813 [1]

Plants usually large at maturity, with a fibrous or disklike holdfast, distinct slender stalk, and broad flattened blade, which is smooth of surface, simple or palmately divided, plane or ruffled, and

[1] Even when the mucilage ducts are present in thin sections of Laminaria stipes and blades, it is often difficult to recognize them with a microscope. Sections are most easily made from 70 per cent alcoholic material; if these are treated briefly with very dilute aqueous methylene blue before being mounted in water for examination, the material in the ducts is usually stained very strongly.

when reproducing by sporangia is marked by thickened darker brown blotches of irregular size and form; growth basal, just above the stalk.

KEY TO SPECIES

1. Blade remaining undivided.................................... 2
1. Blade at maturity becoming palmately divided................. 7

2. Stalk normally solid throughout.............................. 3
2. Stalk hollow and enlarged in its upper part..... L. longicruris, p. 181

3. Blade relatively narrow, becoming broadly lanceolate; ranging to New England... 4
3. Blade broader, narrow to broadly oblong; mucilage ducts present in the stalk; arctic species......................... 5

4. Blade at the base long-cuneate to rounded; mucilage ducts absent .... L. agardhii, p. 179
4. Blade at the base short-cuneate to at length cordate; mucilage ducts present in the blade............... L. saccharina, p. 180

5. Holdfast branched 6
5. Holdfast disklike .... L. solidungula, p. 182

6. Stalk short; base of the blade cordate, the margins strongly ruffled .... L. groenlandica, p. 180
6. Stalk very short; base of the flat blade cuneate.. L. cuneifolia, p. 179

7. Mature plant blade subsimple or sparingly divided............. 8
7. Mature plant with the blade divided to many narrow segments... 9

8. Ducts present in the stalk.................................. 10
8. Ducts absent from the stalk, the blade not particularly dark or thick .... L. intermedia, p. 183

9. Holdfast and stalk small and very dark, nearly black; an arctic species .... L. nigripes, p. 182
9. Stalk longer, not conspicuously dark......................... 11

10. Blade cordate at the base.................... L. platymeris, p. 183
10. Blade cuneate at the base...................... L. nigripes, p. 182

11. Stalk with growth rings, plants perennial; when young and in small forms the blade but sparingly divided..... L. digitata, p. 184
11. Stalk without growth rings, long and weak; an annual L. stenophylla, p. 184

Laminaria agardhii Kjellman[1] Pl. 18, fig. 2

Plants large, stalked from a large and greatly ramified fibrous holdfast, the stalk rather short and slender, in length variable, being greatly exceeded by the blade; blade in young plants cuneate at the base, later rounded to subcordate, flat in young plants and during the winter, the succeeding summer growth thinner, ruffled and bullate, ordinarily about 10–12 times as long as broad, often reaching a width of 20–30 cm., and at times larger, lacking mucilage ducts; when fruiting producing a thick dark stripe or series of blotches of sporangia and paraphyses along the median portion of the blade.

New Jersey to southern Massachusetts and sporadically northward; the dominant kelp south of Cape Cod, although a few more northern stations, especially in Maine, but also in northern Labrador and in the arctic, have been reported; it is a plant of moderately shallow water, growing below low-tide line at 0.5–18.0 meters, and perhaps deeper, on wharves, stones, and particularly on Mytilus shells; fruiting plants are normally found in the winter, although they may occasionally be obtained early in the summer.

REFERENCES: Phyc. Bor.-Am. 1290, LXXXII. Setchell 1900, p. 147.

Two forms in addition to the typical one have been described from our area. F. **angustissima** Collins (1880, p. 117, as *L. longipes,* 1906a, p. 108; Phyc. Bor.-Am. LXXXIII) is very narrow, the stalk 5–25 cm. long, the blade flat, long-tapered at the base, 5–25 mm. broad, the length to 2.5–5.5 meters, the fruiting band about 0.6 the width of the blade; it is found on exposed coasts in Casco Bay, Maine. F. **vittata** Setchell (Phyc. Bor.-Am. 1083, 1291) is somewhat broader, the stalk 2–30 cm. long, the blade tapered at the base, 1–8 cm. wide, 30 cm. to 5 meters long, the fruiting band about half the width of the blade; it ranges from Connecticut to southern Massachusetts.

Laminaria cuneifolia J. Agardh

Plants rather small, the holdfast small and of slender strands, the terete stipe short and slender, 2–6 cm. long, with small mucilage

[1] *L. trilaminata,* mentioned by Olney and by Harvey (1852, p. 93) is a teratological form of this species; such are not uncommon. F. *zostericola* Collins (1906a, p. 108) is probably a juvenile phase, and f. *normalis* Setchell the typical form of the species.

ducts present; the simple, dark brown, rather thin blade usually cuneate at the base, sometimes obtuse or nearly transverse, 6–14 cm. broad, to 3–7 dm. long, often locally a little constricted, in section showing mucilage ducts.

Northern Labrador and nearby northern Quebec. Only known as drifted ashore, probably from deep water.

REFERENCES: Kjellman 1883, p. 228; Wilce *in herb.*

Laminaria saccharina (Linnaeus) Lamouroux

Plants moderately large, the stalk variable in length, reaching 4–5 dm., greatly exceeded by the blade, which is lanceolate to oblong, and reaches about 4.5 dm. in width, 2 meters or more in length, or about 5–10 times as long as broad; at first cuneate at the base, it very early becomes rounded to cordate; the new blade is thin in the spring, but later becomes thicker, ruffled, and bullate; the autumnal growth is again thick and plane; mucilage ducts are present in the blade.

Northern Massachusetts to the lower St. Lawrence, Labrador, Newfoundland, Hudson Strait, Baffin Island, and Cornwallis Island; it is a plant of the sublittoral, growing on rocks.

REFERENCES: Phyc. Bor.-Am. XXXII. Farlow 1881, p. 93; Setchell 1900, p. 146; Whelden 1947, p. 117.

In addition to the typical plant two forms are reported from our area. From Maine and Ile Miquelon is recognized f. **caperata** (De la Pylaie) Setchell (Farlow 1881, p. 93; Hariot 1889, p. 181, as *L. caperata*), with a stalk approaching or exceeding the length of the blade, bent above near the attachment to the blade, but solid throughout. F. **phyllitis** Le Jolis (Phyc. Bor.-Am. V, as *L. phyllitis;* Farlow 1881, p. 93), is small, plane, falcate below when very young, and though always delicate and pale it becomes somewhat thicker and less falcate with age; it is a Maine plant of tide pools in spring.

Laminaria groenlandica Rosenvinge

Plants rather small; holdfast branched, slender; stalk solid, cylindrical or, above, a little compressed, (15–) 30–75 cm. long, with round to radially elongate muciferous canals in at least the lower part, forming a conspicuous circle; blade generally linear-oblong, but ranging from this to obovate, to 1 meter long, 25–80 cm. broad; base rounded, broadly cuneate, or to cordate, the margin very deeply undulate, and with two rows of depressions (bullae) nearer the center; muciferous canals always present, sometimes only covered by a single layer of cells, sometimes deeply immersed; sporangial sori forming spots near the center toward the tip, or expanding toward the base.

Northern Labrador and Ellesmere Island. Plants of moderately deep water, dredged to 18 meters.

REFERENCES: Rosenvinge 1893, p. 847, 1926, p. 22; Wilce *in herb.*

Laminaria longicruris De la Pylaie [1] Pl. 20, figs. 1–2

Plants large, commonly 3–5 meters, and reported to 12 meters long; stalk elongated from a large, branched holdfast, commonly 1.5–2.0 meters long, and much longer in deep water, solid and slender below, becoming hollow and conspicuously inflated above and 2–3 cm. diam. or more, often bent and sharply contracted before joining the blade, which it may exceed in length; muciferous canals doubtful, variously reported; blade broad, commonly 5–7 dm., occasionally to 1.0–1.5 meters wide, but relatively short, 2–7 meters, the base broad, deeply cordate at maturity though simply rounded when very young, at which time the blades are relatively long, becoming shorter by decay of the distal parts with age; fairly thick in the median part, but quite thin and somewhat ruffled at the margins.

Connecticut and Rhode Island, a few stations at exposed sites, and found drifting off southern Massachusetts, common north of Cape Cod to the lower St. Lawrence, Ile Miquelon, Hudson Bay, Baffin Island, and Ellesmere Island; ranging from shallow to rather deep water on exposed coasts. In shallow water the inflated stipes are not particularly notable, but in deeper water, and in large drifting specimens, they extend above the water level and are a conspicuous sight off northern New England. A narrow, shallow-water

[1] The report by Bory (1822–1831, p. 188) of *L. ophiura* from Newfoundland clearly refers to this species; his report of *L. lamourouxii* (*id.*, p. 189) is not clearly referable to any particular species currently recognized. His *L. bifidans*, *L. trifidans*, and *L. delisei* (*id.*, p. 190), also from Newfoundland, are *nomina nuda*.

form has been erroneously reported as *L. faeroensis f.* (Johnson 1937, p. 103; Taylor 1937c, p. 192).

REFERENCES: Algae Exsic. Am. Bor. 117; Phyc. Bor.-Am. LXXXVI. De la Pylaie 1829, p. 41; Harvey 1852, p. 93; Agardh 1870, p. 365; Farlow 1881, p. 93; Hariot 1889, p. 182; Setchell 1900, p. 148; Collins 1902a, p. 177.

Laminaria solidungula J. Agardh

Plants fairly large, to 2–3 meters or more in length; holdfast a membranous disk, the stalk cylindrical, to 1 meter long, 1 cm. thick, the cortex in the thicker part rugose when dry; in section showing near the base a broadly elliptical medulla and a ring of usually linear mucilage canals in the interior of the cortex; blade ovate-oblong, often retaining the ragged remnants of 1–2 previous years' growth at the apex; when adult to 7 dm. long (or even to 2.5 meters), 18–38 cm. broad; laminar muciferous canals present; sporangial sori round, elliptical, or kidney-shaped, near the base of the blade.

Northern Labrador, Devon Island, and Ellesmere Island; growing in moderately deep water at 5–27 meters.

REFERENCES: Rosenvinge 1926, p. 21; Wilce *in herb.*

Laminaria nigripes J. Agardh

Plants rather small, becoming leathery and very dark with age; holdfast coarse, the branches subverticillate; stipe nearly black, 2–20 cm. long but usually quite short, terete below, flattened above, in section the cortex seen to be sharply delimited from the medulla, muciferous ducts usually evident, numerous, close to the surface; blade cuneate to cordate at the base, cleft into few to numerous segments, reaching a length of about 16 dm., in section showing numerous small to very large muciferous ducts in the outer cortex.

Northern Labrador, Killinek I., and adjacent northern Quebec. Found in extremely exposed situations.

REFERENCES: Kjellman 1883, p. 236; Rosenvinge 1926, p. 13; Wilce *in herb.*

V. **atrofulva** (J. Agardh) Rosenvinge differs from the type in having an entire or nearly entire blade cleft but a very short distance into a few broad segments; it has been found in deep protected tide pools and dredged from deep water.

Laminaria platymeris De la Pylaie Pl. 21, fig. 1

Plants of moderate size, 0.5–2.0 meters long, dark brown and leathery; the holdfast fibrous, stout, moderately small; the stalk short, 1.5–3.0 dm. long, stout, to 1 cm. diam. or even more, compressed, with small rounded mucilage ducts in a ring near the inner boundary of the narrow brown peripheral cortical zone; stalk cuneate-expanded at the top, passing abruptly into the thick broad blade, which is relatively short, the base ultimately subcordate, 30–40 cm. wide, cleft into a few (commonly 3–5) broad segments, and showing mucilage canals in section.

Northern Massachusetts to Maine, and Newfoundland; this species is reported to be an annual ordinarily found in deep water.

REFERENCES: Phyc. Bor.-Am. VI. De la Pylaie 1829, p. 52; Farlow 1881, p. 94; Hariot 1889, p. 182 (as *L. cloustoni*); Setchell 1900, p. 143.

Laminaria intermedia Foslie Pl. 19, fig. 2

Plants of moderate size, the stalk cylindrical, weak, little flattened above, with growth rings in older specimens; insertion of the blade obtuse to subcordate, the blade short, broad, divided into a few wide segments with mucilage ducts.

Connecticut to southern Massachusetts; growing at a depth of 2–10 meters.

REFERENCE: Setchell 1900, p. 145.

There are also reported three forms. F. **cucullata** Foslie, with, when fully developed, a markedly cucullate (cupped) blade on a short stalk, the blade cleft to 3–4 segments, or few more, occurs in southern New England and an isolated station in Maine. F. **longipes** Foslie has a long stalk, and the blade nearly flat, or but feebly cucullate; it has been reported from New Hampshire (Croasdale 1941, p. 214); f. **ovata** Foslie has a short stalk, a broad and entire blade; plants from Connecticut may be allocated to these two forms.

Laminaria digitata (Linnaeus) Lamouroux Pl. 18, fig. 1

Plants perennial, moderate to large in size, attached by a heavy, close fibrous holdfast, the stalk stout, clearly flattened above, to 2–3 cm. broad, reaching 6 dm., perhaps more, in length, in sections when well developed typically showing concentric growth rings, but no mucilage ducts; blade at first cuneate, later when fully developed cordate below, widely spreading, commonly to 8–12 dm., deeply cleft into numerous, narrow, flat segments (when well developed 10–30 or more), of moderately thick texture and with mucilage ducts; the rounded fruiting areas relatively small, scattered on the blade segments.

Staten Island (?), Long Island, Connecticut to southern Massachusetts, at a few exposed stations, becoming common north of Cape Cod to the lower St. Lawrence, Hudson Bay, and Ile Miquelon; commonly preferring a rocky substrate in an exposed position, growing below low-tide level; fruiting in winter.

REFERENCES: Algae Exsic. Am. Bor. 119 (as *L. flexicaulis*). Harvey 1852, p. 94; Hariot 1889, p. 181 (as *L. flexicaulis*); Setchell 1900, p. 144.

The principal variants of *L. digitata* reported for our area beside the typical plant include f. **complanata** Kjellman which has a stalk about 6 dm. long, greatly flattened above, reaching a width of 7.5 cm., or about 15 times its thickness below transition into the blade, which has been reported from northern Massachusetts to Maine, and f. **ensifolia** (Le Jolis) Foslie (Phyc. Bor.-Am. XXX), with a rather rigid stalk much shorter than the blade, somewhat curved, the blade narrow with relatively few divisions, which has been found from Long Island Sound to eastern Maine, in exposed rock pools.

Laminaria stenophylla (Kützing) J. Agardh Pl. 19, fig. 1

Plants large, reaching 15–40 dm.; stalk long, slender, and flexible; growth rings and mucilage ducts absent from the stalk; blade with a narrow base, above divided into several to many very narrow segments, becoming large, to 15–30 dm., showing mucilage ducts.

Northern Massachusetts to Maine and Newfoundland; growing exposed to heavy surf near low-water mark; fruiting in autumn and considered an annual species.

REFERENCES: Phyc. Bor.-Am. LXXXIX(?). De la Pylaie 1829, p. 55; Setchell 1900, p. 145.

The American specimens seen by the writer which had been attributed to this species were not, generally, clearly separable from *L. digitata*.

Agarum Bory, 1826

Plants with a fibrous holdfast, short slender stalk and oval to oblong, perforate blade, the center line occupied by a strong compressed midrib; growth basal, just above the stalk.

Agarum cribrosum (Mertens) Bory Pl. 22, fig. 1

Plants of moderate size; stalk usually 2–5 (–30) cm. long, cylindrical below, flattened above; blade single, cordate and irregularly crisped about the base, above plane or somewhat ruffled, to 20–30 (–60) cm. wide, 5 dm. to 1.5 meters long; the central line with a distinct heavy midrib which is flattened, to 10 mm. wide or somewhat more; lateral portions of the blade abundantly perforate, the perforations forming as small holes in the basal portion near the growth zone and increasing in size to 10–15 mm. diam.; fruiting areas irregular, darker and thicker than the vegetative portions of the blade; sporangia ellipsoidal, 35 μ long, 12 μ diam.

Northern Massachusetts to the lower St. Lawrence, Labrador, Newfoundland, James Bay, Hudson Bay, Hudson Strait, Baffin Island, Cornwallis Island, Devon Island, and Ellesmere Island; preferring deep water, but found in a dwarfed form in shallow water and exposed tide pools.

REFERENCES: Algae Exsic. Am. Bor. 112; Phyc. Bor.-Am. III (both as *A. turneri*). De la Pylaie 1829, p. 25 (as *Laminaria agarum* and *L. boryi*); Harvey 1852, p. 95 (as *A. turneri*, probably including *A. pertusum*); Farlow 1881, p. 96; Hariot 1889, p. 183; Rosenvinge 1926, p. 24 (the three as *A. turneri*).

Alaria Greville, 1830 [1]

Plants perennial, with a fibrous holdfast supporting a stalk, naked below but above expanded to a flattened rachis bearing marginal

[1] The American records of *A. membranacea* may be referable to *A. pylaii* and *A. musaefolia* to *A. esculenta* or *A. pylaii*. The margins of the midrib are not

rows of annually deciduous sporophylls, which increase in number at the upper end by meristematic activity; the rachis terminated by a thin linear blade sometimes with tufts of colorless hairs, and with a prominent flattened midrib, the blade each year replaced by new growth produced by the basal meristem.

KEY TO SPECIES

1. Midrib elevated to its margin, bordered by a distinct angle........ 2
1. Midrib less prominent, flattening out into the frond.............. 3
2. Blade long, narrow, the base acutely tapering..... A. esculenta, p. 186
2. Blade shorter and broader, the base rounded, decurrent A. musaefolia, p. 187
3. Stalk and rachis short; blade leathery.............. A. pylaii, p. 188
3. Stalk and rachis longer; blade thinner.......................... 4
4. Rachis wide, plane; sporophylls cuneate-linear, long A. grandifolia, p. 187
4. Rachis narrow; sporophylls while young broader than long; blade thin, transversely much wrinkled.... A. membranacea, p. 187

Alaria esculenta (Linnaeus) Greville Pl. 21, fig. 2

Plants of moderate size, rather narrow; stalk 1–3 dm. long, compressed, to 5–12 mm. diam., the fruiting rachis rather wide and flatter; sporophylls numerous, stalked, the bases tapered, in shape oblanceolate to spatulate, 12–25 (–50?) mm. wide, 7–25 cm. long, firm; vegetative blade at full maturity with a midrib 5–10 mm. or more in width, distinctly flattened and with evident angled margins, and a thin, ruffled, membranous border; blade reaching a width of 4–25 cm. or more and a length of 3 dm. or more, being reported to 12–30 dm.

Long Island and Rhode Island, having probably drifted from northern shores; northern Massachusetts to the lower St. Lawrence, Hudson Bay, Hudson Strait, Labrador, Newfoundland, Baffin Island, and Devon Island; often common on rocky coasts in exposed places, particularly below the tidal range.

---

usually prominent in young or small blades, so that only fully developed specimens are suitable for study. *A. delisei* (Bory) Greville, based on a single Newfoundland specimen, may be referable to a variety of *A. esculenta* (Yendo 1919, p. 109).

REFERENCES: Phyk. Univ. 463, 464; Phyc. Bor.-Am. 1236, XCIII. Harvey 1852, p. 88; Farlow 1881, p. 97; Hariot 1889, p. 194 (as *Orgyia pinnata*); Whelden 1947, p. 117.

Alaria grandifolia J. Agardh

Plants very large, the stalks elongate, 6–15 dm. long, cylindrical below, to 15 mm. or more thick, flattened above and beset with the remains of old sporophyll bases; sporophylls on the active upper rachis about 2–3 mm. apart, to 2–4 (–10) dm. long, cuneate to nearly linear at the base, above hardly to 3.5 cm. broad; the large vegetative blade ultimately subcordate at the base, ovate to long-lanceolate, 2.0–4.5 dm. broad, 1–2 (–7) meters long, the midrib appearing thinner toward its margins, the broad border much plicate, rather leathery.

Northern Massachusetts, washed ashore from deep water; reported to grow at 3.5–18 meters depth and in the arctic to fruit in late summer.

REFERENCE: Collins' manuscript.

Alaria membranacea J. Agardh

Plants moderately large, the stalks slender and elongate, the rachis also rather long; sporophylls rather distantly placed, when young broader than long, later becoming elongate-cuneate, 12–20 cm. long, 3–4 cm. broad; vegetative blade generally quite wide, to 3 dm. broad and 12–18 dm. long, the midrib not very much thickened, particularly at its margins, the lateral membranes much wrinkled and thin.

Northern Massachusetts and the lower St. Lawrence, washed ashore; reported a plant growing in 2–4 meters of water, or deeper; in arctic waters fruiting by spring.

REFERENCE: Collins' manuscript.

Alaria musaefolia (De la Pylaie) J. Agardh

Plants of moderate size, the stalks usually short and strong, 8–13 cm. in length, sometimes longer; rachis compressed, hardly flattened, bearing closely seriate sporophylls; sporophylls linear-cuneate to subfalcate, 1–4 cm. broad, 20–35 cm. long, 9–20 times as long as broad; vegetative blade linear-lanceolate from an ovate base which is very shortly decurrent, the length 9–12 dm., width 8–18 cm., the midrib strong and flattened, with prominent angled margins.

Maine, the lower St. Lawrence, and Newfoundland.

REFERENCES: De la Pylaie 1829, p. 31 (as *Laminaria musaefolia*); Yendo 1919, p. 131.

Alaria pylaii (Bory) Greville

Plants short and broad of aspect; stipes short, seldom twice the width of the blade, thicker and flattened toward the rachis; sporophylls large, broad from the first, triangularly ovate to obovate, later elongating to obovate-spatulate, to 3 cm. broad or more; vegetative blade broad, to 50 cm. across or wider, abruptly cuneate at the base, becoming ovate-cordate below, never decurrent, the blade remaining relatively short, with a little-thickened midrib which thins out marginally into the blade.

Northern Massachusetts to Maine, the lower St. Lawrence, Newfoundland, and Ellesmere Island.

REFERENCES: Phyc. Bor.-Am. 1237(?). De la Pylaie 1829, p. 29 (as *Laminaria pylaii*); Harvey 1852, p. 89; Kjellman 1883, p. 214.

FUCALES

Plants of sturdy construction, parenchymatous externally, but often with a filamentous medulla; growth from an apical cell group in the adult branches; sporophytic, the sporangia carried in superficial conceptacles with branched paraphyses; the potential mega- and microspores after meiosis continuing to divide and passing through a cytological gametophyte phase; when matured the sporangia discharging their now gametangial contents in coherent masses from the conceptacles; these disintegrate in the water, liberating gametes, the antherozoids fertilizing the eggs in a free state.

KEY TO FAMILIES

1. Axes subterete, to alate with a midrib, but not foliar; vesicles if present intercalary .... FUCACEAE, p. 189
1. Axes terete, bearing distinct foliar organs; vesicles present SARGASSACEAE, p. 196

FUCACEAE[1]

Plants with dichotomous or pinnate strap-shaped branches, in some cases with piliferous cryptostomata, often with buoyant air bladders; reproducing by eggs and antherozoids liberated in the water from morphologically heterosporous sporangial structures which undergo meiosis and subsequent mitoses in conceptacular cavities of receptacular portions of the thallus.

KEY TO GENERA

1. Branches at most compressed, but not differentiated into midrib and blade .... ASCOPHYLLUM, p. 195
1. Branches strap-shaped at least above, with a thickened midrib FUCUS, p. 189

Fucus Linnaeus, 1753

Plants erect from a disciform or irregular holdfast, generally dichotomously, but sometimes subpinnately branched, the branches strap-shaped with a more or less distinct midrib, to which by destruction of the margins of the blade the lower portions of the plant are often reduced; air-filled bladders of definite form regularly present in some species, absent in others, or exceptionally the frond in any species irregularly inflated; piliferous cryptostomata generally present on sterile portions of the blades; receptacles terminal on main or lateral branches, when young usually flat, later in some species inflated; heterosporous, becoming hermaphrodite or dioecious, the oögonia forming eight eggs.

KEY TO SPECIES

1. Margin serrate .... F. serratus, p. 194
1. Margin not serrate, though often frayed........................ 2

2. Plants small,[2] regularly under 15 cm. tall........................ 3

[1] *Halidrys siliquosa* Lyngbye is reported from Newfoundland *fide* Turner, *Cystoseira oligacantha* Kützing from Newfoundland *fide* Kützing, *Fucus distichus* Linnaeus from Newfoundland and *F. fueci* from Ile Miquelon *fide* De la Pylaie, *F. ceranoides* Linnaeus from New York *fide* J. Agardh, *Himanthalia lorea* Lyngbye from North America *fide* J. Agardh by Harvey (1852, pp. 64, 67, 70, 72); none of these appear to be currently known on our coasts.

[2] Dwarf forms of several species may be sought here, so that any specimen not agreeing with the descriptions of *F. filiformis* and *F. miclonensis* should be compared with the dwarf variants mentioned under other species, all of which show considerable ecological variation, in addition to which any specimen not

2. Plants larger, regularly exceeding 15 cm........................ 4
3. Branching wide-angled, receptacles compressed, notably warty F. miclonensis, p. 190
3. Branching narrower-angled, receptacles subcylindrical, rather long .... F. filiformis, p. 190
4. Receptacles compressed 5
4. Receptacles swollen 6
5. Receptacles 4–10 times as long as broad, tapered from the base, acute, often forked, the edges rounded........ F. edentatus, p. 191
5. Receptacles 1–2 times as long as broad, blunt, the edges acute, often ridged or winged.................... F. evanescens, p. 193
6. Receptacles spherical to oval, in the plane of the frond with a distinct line, ridge,. or wing encircling them; conceptacles bisexual .... F. spiralis, p. 191
6. Receptacles not ridged, though sometimes the margin angled when young; blades with usually paired vesicles; sexes in different plants .... F. vesiculosus, p. 192

Fucus filiformis Gmelin Pl. 23, fig. 2

Plants small, 1.0–1.5 (–2.0) dm. tall, light brown, branches dichotomously forked, the angles rather narrow, aspect flabellate, segments strap-shaped, 1.5–3.0 mm. wide, above with a narrowly margined midrib which becomes indistinct toward the tips, a little narrower but somewhat thicker below, bladders lacking, cryptostomata small and obscure; hermaphrodite, the receptacles a little broader than the blade, single or forked, narrowly cylindrical to fusiform, generally inflated, to 2.5 cm. long.

Northern Massachusetts to the lower St. Lawrence, and Newfoundland, northern Labrador, and Hudson Strait; this plant largely occurs in tide pools in the upper intertidal zone.

REFERENCES: Phyc. Bor.-Am. 233. De la Pylaie 1829, p. 97; Farlow 1881, p. 102; Hariot 1889, p. 195; Wilce *in herb.*

Fucus miclonensis De la Pylaie

Plants small, slender, the blades 3 mm. wide in the upper segments, without bladders, leathery, linear but broader in the distal

---

at full reproductive maturity may be puzzling or indeterminable. Many specimens past maturity will have the receptacles swollen out of their normal form.

segments than below, repeatedly dichotomously branched, the angles very wide and the branches spreading, somewhat distant; the blades with the membranous portion relatively wider in the upper segments; receptacles short, paired, with confluent or subseparate bases, widely spreading, compressed or little swollen, coarsely warty.

Nova Scotia, Hudson Bay, and Newfoundland, possibly with other stations to the southward and toward the arctic.

REFERENCES: De la Pylaie 1829, p. 90; Kjellman 1883, p. 206.

Fucus edentatus De la Pylaie Pl. 23, fig. 3

Plants generally of moderate size, 2–4 dm. tall, but reported to 9 dm., the holdfast a broad, low, conical disk, the blade regularly dichotomously forking, the axils generally acute, branches erect, tending to spread in a plane, the midrib distinct below, scarcely visible above though alate, but narrow, evesiculate, leathery, 5-20 mm. broad; cryptostomata rather small, often inconspicuous; toward the base the midribs long denuded, forming a firm stalk; receptacles elongate, 1–3 times divided into compressed hornlike lobes which are 5–15 mm. broad below, often tapering to acute tips, 5–10 cm. long, and are not sharply separated from the sterile portions; when dried becoming very dark brown or black.

New Jersey, Connecticut to the lower St. Lawrence, James Bay, and Newfoundland, being uncommon and reduced south of Cape Cod; growing on exposed rocks in the intertidal zone; fruiting in late winter and spring.

REFERENCES: Algae Exsic. Am. Bor. 108 (as *F. furcatus*); Phyc. Bor.-Am. XIII. De la Pylaie 1829, p. 84; Harvey 1852, p. 70; Farlow 1881, p. 102 (both as *F. furcatus*), 1889, p. 6; Hariot 1889, p. 195.

De la Pylaie recognized among his Newfoundland specimens f. **angustior** De la Pylaie (Pl. 23, fig. 1) with narrower blades and shorter, narrower receptacles than in the typical form. It has been found from Massachusetts northward.

Fucus spiralis Linnaeus Pl. 24, fig. 1

Plants of moderate size, bushy, 1.5–3.0 dm. tall, branching rather regularly and widely dichotomous, occasionally irregular, the branches of erect tendency, usually plane, occasionally a little twisted, evesiculate, cryptostomata evident; moderately wide (to about 15 mm.), strongly ribbed below, near the base denuded to the midrib; receptacles terminal, simple, or forked, obovate to long-oval and swollen, characterized by a circumscribed line or ridge in the plane of the blade, which is sometimes obscure and in spent receptacles frequently appears to be absent; hermaphrodite.

New York to the lower St. Lawrence and Newfoundland; a plant of rocks in the higher littoral zone, usually above *F. vesiculosus* and *Ascophyllum nodosum,* and often in relatively quiet water.

REFERENCES: Phyc. Bor.-Am. 234 (as *F. areschougii*), 1132, XLVI (both as *F. platycarpus*). De la Pylaie 1829, p. 78 (as *F. vesiculosus* v. *spiralis*?); Howe 1905, p. 581.

Variable, particularly in stature, as one would expect in a plant which is primarily intertidal in station. The most marked is v. **limitaneus** Montagne (Phyc. Bor.-Am. 989), reaching a height of 5 cm., with blades but 5 mm. wide and small receptacles; it occurs in Rhode Island. As a salt-marsh plant the v. **lutarius** (Kützing) Sauvageau, another narrow form, is reported (Chapman 1939, p. 27) from Massachusetts. F. *abbreviatus* Collins (Phyc. Bor.-Am. 1339) and f. *intermedius* Holden (Phyc. Bor.-Am. 1340) are simply intergrades toward the earlier described v. *limitaneus.*

Fucus vesiculosus Linnaeus Pl. 25, figs. 1–3

Plants often large, generally 3–9 dm. tall, attached by an irregular, lobed holdfast; branching usually dichotomous or a little irregular, often proliferous below, the branches above strap-shaped, about 10–15 mm. wide, with a marked midrib throughout, but below denuded of the thin margins; scattered cryptostomata and vesicles 5–10 mm. diam. usually prominent, the latter generally paired on each side of the midrib, or three together at a fork; receptacles terminal on the branches, single, paired, or forked, broadly lanceolate to obovate, usually 1.5–2.5 cm. long; sexes in different individuals, the antheridial conceptacles orange when the receptacles are opened, while the oögonial conceptacles are olive-green.

North Carolina to Ellesmere Island, and also to Hudson Bay; throughout most of our range the commonest species of the genus, and one of the commonest marine algae, typically on rocks, shells, or wharves, ranging from the middle to the lower intertidal levels, and also extending up estuaries into water of reduced salinity; fertile from autumn to spring in the southern part of its range.

REFERENCES: Algae Exsic. Am. Bor. 109; Phyc. Bor.-Am. 577. De la Pylaie 1829, p. 75; Harvey 1852, p. 71; Farlow 1881, p. 100; Hariot 1889, p. 195; Chapman 1939, p. 27; Whelden 1947, p. 118.

This appears to be the most variable of our fuci, as it is the widest-ranging and the most adaptable. Several distinctive varieties and forms are recorded from our range. F. **gracillimus** Collins (1900a, p. 14; Phyc. Bor.-Am. 578), sometimes less than 10 cm. tall, blades evesiculate, less than 5 mm. broad, most often but 1.5–2.5 mm., the midrib indistinct, receptacles slender, fusiform, is reported from tidal marshes in southern Massachusetts. V. **laterifructus** Greville (Phyc. Bor.-Am. 1085; Farlow 1881, p. 100) is much dissected in the shorter receptacular branches, which appear laterally on the lower parts of the plant, being greatly overtopped by the vegetative portions; it is reported from New Jersey, Connecticut, southern Massachusetts, and Maine. V. **sphaerocarpus** J. Agardh (Pl. 25, fig. 6; Phyc. Bor.-Am. 874) is finely dissected above, with the segments terminating in small, spherical to ovoid receptacles; it is reported from Connecticut, southern Massachusetts and Maine to the lower St. Lawrence. V. **spiralis** Farlow (1881, p. 101; Phyc. Bor.-Am. 680), perhaps the same as v. *volubilis* (Hudson) Turner but not *Fucus volubilis* Linnaeus=*Vidalia* (Pl. 25, figs. 4–5; Phyc. Bor.-Am. 680?), has closely spirally twisted blades and is of moderate size; this locally current name is applied to a growth form common near tidal ditches, although also found on rocks on the open shore, when more usually with bladders. F. **limicola** Collins (1906a, p. 109; Phyc. Bor.-Am. 1133) is of intermediate size, to 2 dm. tall, very slender, hardly twisted, usually without bladders, with small, rounded conceptacles terminating the branches; it is a salt-marsh plant, or one from muddy shores in New England, perhaps closely related to the preceding variety.

Fucus evanescens C. Agardh Pl. 23, fig. 4; pl. 24, fig. 2

Plants large, 2–6 dm. tall, the basal disks round-conical, branching widely dichotomous or alternate, the fronds markedly tending to spread in a plane, the blades wide, 1–2 cm. broad, evesiculate, the cryptostomata obvious; costate in the middle portions of the plant with broad membranous margins, the midrib disappearing above, but strong and often denuded below; receptacles flat, simple or forked, broad and short, 1.5–2.0 cm. wide, 2–4 cm. long, obtuse or slightly dentate, the apex often ridged or winged, sharply demarcate from the sterile blade.

New York to the lower St. Lawrence, Newfoundland, Hudson Bay, northern Labrador, James Bay, Hudson Strait, Baffin Island and Ellesmere Island; usually in quiet harbors, but sometimes on more exposed points, near or below low tide; fruiting in the autumn.

REFERENCES: Algae Exsic. Am. Bor. 107; Phyc. Bor.-Am. XIV. Farlow 1881, p. 101; Hariot 1889, p. 195; Whelden 1947, p. 117; Wilce *in herb.*

Material attributed to this species appears to be rather diverse in aspect. The description is appropriate to the common southern New England form illustrated. F. **cornutus** Kjellman is smaller, denuded below, above with a definite but little swollen midrib, the blades 3–4 mm. broad, receptacles fairly sharply defined, terminating most of the branches, oval to long-oval, simple or forked, swollen, with large conceptacles; it has been reported from southeastern Hudson Bay.

Fucus serratus Linnaeus Pl. 26, fig. 1

Plants moderate to large, 0.3–2.0 meters long, from a conical holdfast, irregularly dichotomously or sublaterally branched, the angles acute, the branches without bladders, 7–25 mm. broad and strap-shaped above, with a prominent midrib, evident cryptostomata and a strongly acute-serrate margin, the serrations at about 3–7 mm. intervals in mature portions, the teeth narrow, erect; below the branches often denuded to the midribs; dioecious, the receptacles compressed, oval to broadly lanceolate, 2–4 cm. long, serrate, the margins often winged.

From Chaleur Bay, New Brunswick, east and north to the end of Cape Breton, Nova Scotia, with smaller colonizations on the east side of Cape Breton Island and the southeast end of Prince Edward Island.

REFERENCES: Phyc. Bor.-Am. XCV. Harvey 1858a, p. 122; Farlow 1881, p. 101; Robinson 1903, p. 132.

Ascophyllum Stackhouse, 1809

Plants cartilaginous or fleshy, olive to yellowish, the main axis irregularly to pinnately branched with ultimate branchlets pinnately disposed; no midrib or cryptostomata present; deciduous receptacles likewise pinnately disposed on short pedicels; heterosporous and eventually dioecious, the oögonia forming four eggs.

KEY TO SPECIES

1. Plants erect, axes distinct, vesicles present; receptacles oval, short .... A. nodosum, p. 195
1. Plants entangled, axes more vague; plants of salt marshes and tidal flats 2
2. Branching irregularly pinnate, branches cylindrical or slightly compressed, slender; fruiting portions ovate to obovate A. nodosum f. scorpioides, p. 196
2. Branches fastigiate, occasionally inflated; elongate, lanceolate receptacles formed on lower parts of the plant... A. mackaii, p. 196

Ascophyllum nodosum (Linnaeus) Le Jolis Pl. 27, figs. 1–2

Plants reaching a large size, usually 3–6 dm., but occasionally to 3 meters long; erect from a disciform holdfast; main axis and principal branches compressed, coriacious, with large, single float bladders often 1.5–2.0 cm. diam. and 2–3 cm. long, at times much larger, greatly expanding them at irregular intervals; forking dichotomous or largely irregular to laterally pinnate; axis and branches pinnately beset with short simple to forked somewhat clavate compressed branchlets 1–2 cm. long, singly or in groups, which become converted into or replaced, in reproduction, by the deciduous, dioecious, oval, yellowish receptacles, which are single or in clusters of 2–5 on stalks 0.5–2.0 cm. long, the fruiting portion about 1–2 cm. long, 5–12 mm. wide.

Bermuda (adrift), New Jersey to the lower St. Lawrence, Labrador, Newfoundland, Hudson Strait, and Baffin Island; common on rocky coasts or on rocks projecting above a muddy bottom between tide levels; fruiting in winter, the receptacles deciduous in the late spring in the southern part of our range.

REFERENCES: Phyc. Bor.-Am. XV. De la Pylaie 1829, p. 110 (as *Halidrys nodosa*); Harvey 1852, p. 68 (as *F. nodosus*); Farlow

1881, p. 99; Kjellman 1883, p. 194 (as *Ozothalia nodosa*); Hariot 1889, p. 194; Wilce *in herb.*

In quiet water the plants are particularly slender, the branches subcylindrical, even to 1–2 mm. diam., and appear with reduced bladders or none at all. The extreme form may be distinguished as f. **scorpioides** (Hornemann) Reinke (Pl. 27, fig. 3; Phyc. Bor.-Am. 1039; Croasdale 1941, p. 214, fig. 1), the plants characteristically entangled, lightly attached, diffusely branching so that main axes are usually obscure, branches of all orders cylindrical or very little flattened, bladders absent or nearly so, receptacles rare, ovate or obovate, to 2 cm. long; it has been recorded from Virginia (adrift), New Jersey, southern Massachusetts, New Hampshire, Maine, Nova Scotia, and Ile St. Pierre, but no doubt is of general occurrence; it appears on mud in salt marshes along tidal ditches, generally entangled among Spartina culms or lightly attached to shells or small stones.

Ascophyllum mackaii (Turner) Holmes et Batters Pl. 26, fig. 2

Plants entangled, but habit somewhat fastigiate, holdfasts unknown, height 1.5–2.0 dm., axes dichotomously freely divided or forking irregularly, the ultimate branches slender, simple or forked, often long, curved, and at times inflated; bladders irregular, very small, often few or lacking; the large pendulous receptacles attached by slender stalks to the lower portions of the plant, occasionally to 7–8 cm. long, simple or occasionally forked, lanceolate, straight or curved.

Connecticut and southern Massachusetts; found loose on tidal flats, floating or creeping; fruiting in late winter or early spring.

REFERENCES: Phyc. Bor.-Am. 177; Phyk. Univ. 512.

SARGASSACEAE

Plants with flat costate branches or transitional stages to cylindrical branches bearing costate leaves, usually with cryptostomata, usually with buoyant air bladders; reproducing by eggs and antherozoids liberated in the water from morphologically heterosporous sporangial structures which undergo meiosis and subsequent mitoses in conceptacular cavities of receptacular portions of the thallus.

Sargassum C. Agardh, 1820

Plants when attached with rather massive lobed holdfasts, bushy, with (in our species) distinct branches and broad to filiform, entire or serrate leaf organs; lateral stalked bladders usually present; receptacles if present cylindrical, more or less forked, axillary or paniculate, heterosporous, eventually bearing eggs singly in the oögonia.

KEY TO SPECIES

1. Pelagic species, infrequently cast ashore; sterile, without holdfasts and without or rarely with cryptostomata in the leaves..... 2
1. Species when growing, attached by holdfasts; with evident cryptostomata in the leaves.................................. 3
2. Leaves narrowly linear, without cryptostomata, the marginal teeth subterete, long-tipped; the vesicles generally apiculate S. natans, p. 199
2. Leaves linear to lanceolate, often with a few cryptostomata, the marginal teeth flat, acute but not produced; the vesicles not apiculate .... S. fluitans, p. 198
3. Leaves thin, linear to lanceolate, the lowermost often forked, serrate; our only local species................ S. filipendula, p. 197
3. Leaves firmer, broad, more or less oval, entire or sparingly serrate near the obtuse tip; rarely cast ashore S. hystrix v. buxifolium, p. 198

Sargassum filipendula C. Agardh Pl. 27, figs. 4–6

Plants 3–6 (–10) dm. long, erect from a conical, spreading, lobed holdfast, the main axes smooth, usually sparsely forked, the principal branches dominant, with the branchlets attaining long pyramidal forms; branchlets bearing alternate stalked leaves 3–8 cm. long, 5–8 mm. wide, which are thin, linear-lanceolate, or on the lower portion of the plant not infrequently forked, serrate, with clearly marked midrib and numerous scattered cryptostomata, which may also appear on the stem; axillary bladders spherical at maturity, 3–5 mm. diam., stalked, the stalks about 5 mm. long, slender; receptacles simple, forked, or sparsely racemosely branching, axillary in the upper parts of the plant, the raceme 0.3–0.5 times as long as the subtending leaf, which may be considerably smaller than in the vegetative part of the plant.

From the tropics, Florida to southern Massachusetts, where not uncommon; growing attached to shells or stones in relatively quiet water, from below the lowest tides to 30 meters depth; it is probably perennial, and fruits throughout the later summer and autumn in the more northerly part of its range.

REFERENCES: Algae Exsic. Am. Bor. 101 (as *S. vulgare non* C. Agardh); Phyc. Bor.-Am. XCVII. Harvey 1852, p.57; Farlow 1881, p. 103 (both as *S. vulgare*); Simons 1906, p. 161; Taylor 1928a, p. 127.

A form designated v. **montagnei** (Bailey) Grunow (Harvey 1852, p. 58, as *S. montagnei*) has very narrowly linear leaves, frequently mucronate bladders, and receptacles 5–12 cm. long; it has been recorded from North Carolina and New Jersey to southern Massachusetts. Grunow's (1916, p. 171) f. *subcinerea* described from our area does not seem to merit recognition.

Sargassum hystrix J. Agardh, v. **buxifolium** (Chauvin) J. Agardh

Plants moderately tall, bushy, with distinct main axes closely beset with branchlets and rather firm leaves; leaves broadly lanceolate to oval, obtuse, 1.5–6.0 cm. long, 4–13 mm. broad, costa not conspicuous, margin undulate, generally entire but occasionally minutely serrate, cryptostomata small, few or none; bladders numerous, 2–8 mm. diam., on short pedicels.

Native to the Caribbean, this plant was once found washed ashore on the southern Massachusetts coast, probably having been carried north with the pelagic species in the Gulf Stream.

REFERENCES: Taylor 1928a, p. 129, 1941, p. 73.

Sargassum fluitans Børgesen

Plants widely branching, without a dominant axis, entangled, pelagic; stem smooth or sparingly spinulose, the leaves narrow to ordinarily lanceolate, 2–6 cm. long, 3–8 mm. wide, serrate, the teeth broadly flattened at the base, cryptostomata absent or few and obscure; bladders oval to subspherical, about 4–5 mm. diam., stalked, the stalks often winged, the tips not apiculate; receptacles unknown.

Southern Massachusetts, rarely washed ashore, having drifted north with the Gulf Stream and been blown ashore by gales.

REFERENCES: Taylor 1928a, p. 127, 1937c, p. 211, 1941, p. 73.

Sargassum natans (Linnaeus) J. Meyen

Plants widely branching, without a dominant axis, tangled, pelagic; stem smooth; leaves acutely linear, 2.5–5.0 (–10) cm. long, 2.0–3.5 mm. wide, serrate, the teeth slender, terete, their length to 0.5–0.75 times the width of the leaf, cryptostomata absent, midrib not prominent; bladders small, 3–5 mm. diam., borne on stalks 3–5 mm. long, apiculate or tipped with a spine or reduced leaf; receptacles unknown.

This plant, native to the Caribbean, is brought near our southern range by the Gulf Stream and not infrequently cast ashore on the exposed coast from Florida to southern Massachusetts, especially after gales from the east or southeast. It is also recorded from Nova Scotia (Lawson 1864, p. 3) and Newfoundland (Sir Joseph Banks, 1766!).

REFERENCES: Farlow 1881, p. 103 (as *S. bacciferum*); Taylor 1928a, p. 128.

RHODOPHYCEAE

BANGIOIDEAE

Plants unicellular, filamentous or membranous, the cells not connected by protoplasmic strands; cells uninucleate, with axial or parietal chromatophores and generally with pyrenoids; showing a vegetative phase which may reproduce asexually by monospores and may also show sexual reproduction; sexual organs if present consisting of spermatangia produced by subdivision of ordinary thallus cells into many small colorless units, and carpogonia of sexually modified cells which have developed a soft-walled receptive extension or trichogyne; zygote nucleus apparently undergoing meiosis in its first divisions, and often later mitosis, the fertilized carpogonium thus producing a haploid non-filamentous carposporophyte (cystocarp) of a few carpospores.

BANGIALES

Plants filamentous or membranous, simple or branched; cells usually with axial chromatophores; asexual reproduction by naked monospores; sexual reproduction if present by spermatia and carpogonia; the fertilized carpogonium developing several carpospores.

BANGIACEAE

Characters of the Subclass; axes, if pluriseriate, without differentiation of medullary and cortical tissues.

KEY TO GENERA

1. Plants of branched filaments 2
1. Plants of unbranched filaments 3
1. Plants forming membranes 4

2. Cells elongate, greenish gray .... ASTEROCYTIS, p. 201
2. Cells short, rose-red .... GONIOTRICHUM, p. 201

3. Filaments small, remaining uniseriate or nearly so, or of a few cells in section above; monosporangia lateral, formed by unequal division of vegetative cells .... ERYTHROTRICHIA, p. 202

3. Filaments larger, becoming more bulky, often of many cells in section; monosporangia formed by conversion of simple or equally divided ordinary thallus cells.............. BANGIA, p. 203
4. Plant originating as a sac, later rupturing and expanded PORPHYROPSIS, p. 205
4. Plant never saclike, developing from the start as a plane membrane 5
5. Plants minute; membrane completely attached ERYTHROPELTIS, p. 204
5. Plants macroscopic; membrane attached by a localized holdfast, but mostly free........................ PORPHYRA, p. 205

Asterocytis Gobi, 1879

Thalli bushy, filamentous, grayish green, irregularly and widely branched; cells usually irregularly uniseriately placed, ellipsoid, the walls rather gelatinous; asexual propagation by motionless spores; sexual reproduction unknown.

Asterocytis ramosa (Thwaites) Gobi

Plants to 1–10 mm. tall, in soft tufts, steel-green, the filaments 12–20 (–25) μ diam., unilaterally or subdichotomously branched; cells uniseriate, 5–8 μ diam., 8–20 μ long, oval to elongate; in reproduction the filaments generally in greater part transformed into akinetes, which become subglobose or ellipsoid when mature, about 14 μ long, with walls 2 μ thick, liberated from the filament through a lateral pore.

From the tropics to Florida, Connecticut, and Massachusetts; among other small algae as an epiphyte, particularly on Zostera; during the summer.

REFERENCES: Collins 1905, p. 230 (as *Goniotrichum ramosum*); Taylor 1928a, p. 132.

Goniotrichum Kützing, 1843

Thallus erect, filamentous, dull rose-red, pseudodichotomously branched, below attached by thickened cells, above more delicately filamentous, terete or a little irregularly thickened or flattened; cells short, often disklike, with central radiating chromatophores, the pyrenoid central and the nucleus excentric; membrane gelatinous, thick, especially laterally; reproduction by the formation of monospores, which are liberated by dissolution of the thallus membranes.

Goniotrichum alsidii (Zanardini) Howe Pl. 28, figs. 1–4

Filaments single or in rosy tufts, 0.3–6.0 mm. long and reaching a diameter of 12–20 μ, except where thickened near the base to 30–50 μ; rarely simple or subsimple, generally freely irregularly to pseudodichotomously branched; branches gradually a little attenuate to the ultimate divisions; cells uni- to pluriseriate, depressed spherical to ellipsoid or nearly cylindrical, 7–13 μ diam., 4–13 μ long, the chromatophores violet-red, tending toward green in some specimens.

From the tropics, Bermuda, North Carolina, New Jersey, Connecticut to southern Massachusetts, and Prince Edward Island; epiphytic on Zostera and coarse algae, particularly in foul, quiet water; during the late summer.

REFERENCES: Farlow 1881, p. 113; Collins 1905, p. 230 (both as *Goniotrichum elegans*); Howe 1914a, p. 75; Hoyt 1920, p. 465.

Erythrotrichia Areschoug, 1850

Thalli erect, the basal cells below dilated or the plant attached by a basal disk of a few cells; above filiform or more or less thickened and subcylindrical, seldom expanded and foliaceous; primary cells in one series, later divided by longitudinal walls; cells in various parts of the filament dividing to isolate small asexual monosporangia with dense contents; spermatia produced in the same manner from vegetative cells, minute, pale; carpogonia formed by the direct transformation of vegetative cells; cystocarps of one to few cells.

KEY TO SPECIES

1. Base unicellular, but this cell with branching lobes......... E. carnea
1. Base of short branching multicellular filaments.......... E. rhizoidea

Erythrotrichia carnea (Dillwyn) J. Agardh Pl. 28, figs. 13–15

Plants of simple terete filaments each attached by a basal cell which becomes lobed, the lobes extended to short ramified rhizoids;

0.5–2.0 (–8.0) cm. long, above 16–26 μ diam., the cells swollen, 16–32 μ long; the base gradually tapered, 9.5–13.0 μ diam., cells 15–40 μ long; usually uniseriate, occasionally with a few longitudinal walls in the thicker parts; chromatophores axial, with a central pyrenoid and radiating lobes; reproduction by monosporangia usually cut off by an oblique wall near the distal end of the cell; the spore becoming spherical on discharge, 13–15 μ diam.

From the tropics, Bermuda, Florida, North Carolina, New Jersey to Maine, Nova Scotia, and southeastern James Bay; scattered as an epiphyte on littoral or sublittoral algae in quiet waters; to be found throughout the year, but most abundant in the summer.

REFERENCES: Farlow 1881, p. 113 (as *E. ceramicola*); Hoyt 1920, p. 466; Taylor 1928a, p. 133.

Erythrotrichia rhizoidea Cleland

Plants dark red to reddish purple, upright, of terete filaments usually 1–2 (–4) mm. in length; 10 μ diam. at the base, increasing to 40–50 μ diam. above; dividing at the base into several short, irregularly downgrowing branched multicellular rhizoidal filaments which penetrate into the tissue of the host and from which often arise secondary erect filaments; cells 10–20 μ diam., dividing above in both planes into 4–8 cells; monospores reddish, 8–10 μ diam.

Southern Massachusetts; epiphytic on Porphyra; producing monospores in late summer.

REFERENCE: Collins 1918b, p. 144.

Bangia Lyngbye, 1819

Thalli erect, filiform, without branches; each strand affixed by a dilated base, wherein the original attaching cell is supplemented by intramatrical filiform extensions from the nearby cells; above more or less thickened, terete or irregularly constricted, sometimes tubular; cells with single, radially lobed chromatophores; monospores formed by the direct transformation of superficial vegetative cells, frequently in the same plants as are reproducing sexually; monoecious or dioecious, the spermatangia formed from vegetative cells by repeated division; carpogonia formed by the direct transformation of vegetative cells; cystocarps of about eight cells.

KEY TO SPECIES

1. Epiphytic, small .... B. ciliaris
1. On rocks and woodwork, becoming several centimeters in length B. fuscopurpurea

Bangia ciliaris Carmichael Pl. 28, figs. 5–9

Plants filamentous, becoming disciform at the base, the young filaments clavate, remaining simple or occasionally proliferously branched, often curved, at first uniseriate, 20–45 μ diam., later pluriseriate and thicker, long-attenuate, to about 0.5–1.0 (–2) cm. long.

South Carolina (?), New Jersey, Connecticut, Rhode Island, and northern Massachusetts and Maine; epiphytic on other algae; in summer and autumn.

REFERENCES: Phyc. Bor.-Am. 88; Phyk. Univ. 655. Collins 1905, p. 230.

Bangia fuscopurpurea (Dillwyn) Lyngbye Pl. 28, figs. 10–12

Plants filamentous, aggregated into colonies, the filaments attached at one end, floating free above, softly slippery and to 1–2 dm. long, 20 to 220 μ diam., or shorter and thicker; pale yellowish brown to brownish purple, at first smooth, later somewhat torulose, the filaments at first uniseriate, the segments equal to the diameter or to 0.25–0.33 as long, becoming quaternately and then more extensively radially divided, resulting in thick filaments internally irregularly cellular.

Bermuda, North Carolina, New Jersey to the lower St. Lawrence, and Newfoundland; abundant on exposed rocks and woodwork between tide levels; appearing in spring and early summer, in the south soon disappearing.

REFERENCES: Phyc. Bor.-Am. 87. Harvey 1858a, p. 54; Farlow 1881, p. 112.

Erythropeltis Schmitz, 1896

Thallus thin, disciform, irregularly expanded, epiphytic or ectozoic; growth marginal without erect filaments, the cells without distinct order; monosporangia formed by oblique walls from vegetative cells.

Erythropeltis discigera (Berthold) Schmitz v. **flustrae** Batters
Pl. 29, figs. 1–2

Plants forming rose-colored disks about 0.2 mm. diam., of one cell layer, the margin irregular; cells irregularly placed, polygonal, oblong to irregular, nearly isodiametric, or to 5–10 (–13) μ in length by 3–7 (–10) μ in breadth; chromatophores single in the cells, parietal; monosporangia globose, 6–7 μ diam., each isolated from the originating cell by an oblique wall.

Southern Massachusetts; ectozoic upon *Alcyonidium mytili;* early autumn.

REFERENCE: Jao 1936, p. 237.

Porphyropsis Rosenvinge, 1909

Plants at first parenchymatous, later erect, vesicular, ultimately rupturing and becoming an expanded one-layered rose-red membrane; monospores cut off by oblique walls from the vegetative cells.

Porphyropsis coccinea (J. Agardh) Rosenvinge

Plants small, gregarious, of various ages growing together; when adult 2–5 cm. long, color rosy, the individuals oval to round, inconspicuously umbilicate, subsessile, the margin at first cucullate-involute, later opening out, undulate or crisped; cells 4–7 μ diam., the single chromatophores parietal with lobed margins.

Gardiners Island, N. Y., and New Hampshire; epiphytic on Desmarestia growing below tide levels.

REFERENCES: Taylor 1940, p. 191; Croasdale 1941, p. 214.

Porphyra C. Agardh, 1824 [1]

Plants membranous, often large, each attached by a small holdfast, expanding above into a soft slippery blade of one or two cells

[1] Regarding *Conchocelis* as an alternate phase of Porphyra, see Drew (1949) and the note next following. Drew (1955) calls for caution in accepting the current accounts of the reproductive structures in Porphyra.

in thickness; cells with roughly stellate chromatophore and pyrenoid, alike except near the base where they are extended into intramatrical rhizoids to form the holdfast; asexual reproduction by monospores involving large continuous areas of the frond; sexual reproduction by spermatia produced by cell division and conversion of portions of the frond into spermatangial cells, and scattered carpogonia, which are produced from vegetative cells by formation of a short trichogyne extending to the surface of the thallus, and which produce small clusters of carpospores.

KEY TO SPECIES

1. Vegetative blade dull brown or purplish, of a single layer.......... 2
1. Vegetative blade rosy, of two layers of cells........ P. miniata, p. 207
2. Spermatangial blades with narrow whitish spots scattered over them; vegetative cells in section much higher than wide P. leucosticta, p. 206
2. Spermatangial blades with whitish or yellowish margins; vegetative cells in section subquadrate........... P. umbilicalis, p. 206

Porphyra leucosticta Thuret

Plants more or less ample, 10–15 (–70) cm. long, dull pink to reddish purple, the blades apparently sessile, the base umbilicate, the general contour oblong, becoming undulate-plicate and sometimes sparingly laciniate above; soft, the thickness 25–50 μ, monostromatic, the cells 1.5–2.0 times as high as wide; from the surface averaging 12–15 μ diam. including walls; monoecious, the spermatangia formed in elongated spots 5–10 mm. long, 1.0–1.5 mm. wide, parallel to each other near the edge of the blade.

Bermuda, North Carolina, New Jersey, Long Island, Connecticut to southern Massachusetts, and Maine; growing as an epiphyte on coarse algae in the littoral; essentially a spring plant in our area.

REFERENCES: Phyc. Bor.-Am. 376. Farlow 1881, p. 112; Hus 1902, p. 199; Taylor 1937c, p. 220 (as *P. atropurpurea*).

Porphyra umbilicalis (Linnaeus) J. Agardh[1] Pl. 30, figs. 1–3

Plants becoming large, 1–3 (–8) dm. long and 0.5–3.0 dm. wide, olivaceous to brownish purple; narrow when young but generally

[1] Drew (1949, p. 748) presents evidence that the spores of *Porphyra umbilicalis* developed into shell-penetrating filamentous plants indistinguishable from

ultimately broadly umbilicate about a rounded base, above entire, or divided into broad lobes; nearly flat to plicate; texture membranous, soft, rubbery, the thickness 30–75 μ, usually about 50 μ; monostromatic, the cells subquadrate to somewhat higher than wide, 8–25 μ diam., averaging about 18 μ, 20–30 μ high; dioecious or monoecious, when the two sexes are generally segregated on the frond, the spermatangia forming a pale deliquescing marginal band, the carpospore groups scattered in clusters over the general surface of the blade.

New Jersey to the lower St. Lawrence, Hudson Bay, and Newfoundland; growing on rocks and wharves from the intertidal zone down to extreme low-tide line; with us probably an annual, fruiting during midsummer and becoming loaded with epiphytes in the latter part of the season.

REFERENCES: Phyc. Bor.-Am. 235 (as *P. laciniata*), 1086, 1136 (as *P. laciniata* f. *umbilicalis*). Harvey 1858a, p. 53 (as *P. vulgaris*); Farlow 1881, p. 111; Hus 1902, p. 196; Collins 1903b, p. 211 (the last three as *P. laciniata*).

In our range two forms are also reported: f. **epiphytica** Collins (Collins 1903b, p. 212), blades ovate to orbicular, to 4–5 cm. long, becoming torn and irregular, found from Massachusetts to Maine, epiphytic on Ascophyllum or *Polysiphonia lanosa* in spring and early summer; and f. **linearis** (Greville) Harvey, plants small, narrowly linear, the basal disk surmounted by a short slender stalk, the base of the blade ovate to cordate, the margins plane or very slightly undulate, length to 10–40 cm., width 0.5–2.0 cm.; reported from the Gaspé and Newfoundland.

Porphyra miniata (Lyngbye) C. Agardh

Blades sessile, obovate-oblong, rounded or ovate, undulate-plicate and umbilicate below, sometimes sparingly laciniate above or perforate, 15–30 cm. diam., color rose-purple to cinnabar, texture very flaccid; at first or marginally one cell layer thick, later generally two layers in most part, the thickness typically 30–70 μ, the cells

---

Conchocelis rosea. Should the full story be worked out connecting these in the same life cycle and showing the function of the Conchocelis spores, the life histories throughout the Bangiales will have to be re-examined and the ordinal characters amended. This will also involve removal of the plants known as *Conchocelis rosea* from the Acrochaetiaceae.

subquadrate in section; fruiting marginal, the spermatangial cells intermixed with carpogonia or carpospore masses, or with vegetative cells.

Northern Massachusetts, the lower St. Lawrence, Newfoundland, northern Labrador, Hudson Strait, and the arctic; growing on stones, Corallina, or the bases of Laminaria fronds, along exposed shores; apparently fruiting at any season.

REFERENCES: Phyc. Bor.-Am. 377. Hus 1902, p. 218 (v.); Wilce *in herb*.

In addition to the typical form there may be recognized v. **amplissima** (Kjellman) Rosenvinge (Hus 1902, p. 215, as *P. amplissima*), which is larger, to 30 cm. wide by 90 cm. long, deeply plicate, the cells subquadrate to slightly vertically elongate, blades 50–80 μ thick; it has been reported from Maine and the arctic. F. **tenuissima** (Strömfelt) Foslie (Phyc. Bor.-Am. 1239; Collins 1903b, p. 212, both as *P. tenuissima*) is less folded, with a thickness in the center generally of 25–30 μ and the cells in that part of the blade 2–4 times as wide as high; it has been noted from northern Massachusetts to the Gaspé.

FLORIDEAE

Cells uni- or plurinucleate, rarely with axial chromatophores, more ordinarily with several small lateral chromatophores without pyrenoids, often conspicuously connected by a single large protoplasmic bridge; plants usually showing both diploid and haploid vegetative phases, the diploid phase sporophytic, reproducing through tetrasporangia which undergo meiosis; tetraspores formed by three parallel walls dividing the sporangium into four parts (linear or zonate type), by three intersecting walls with the spores lying parallel to each other or parallel in pairs (tetrapartite, cruciate), or by simultaneous wall formation as four pyramidal spores meeting in the center, with spherical triangular external faces (tetrahedral, tripartite); polysporangia produced by continued division from tetrasporangia, and parasporangia developed as accessory structures, also known; tetraspores germinating to develop a haploid gametophytic phase, which is sometimes also able to reproduce by monospores; sexual organs spermatangia, usually produced terminally on miniature branch systems borne on vegetative cells, and carpogonia with elongate receptive trichogynes, borne terminally on special (carpogenic) branches, sometimes associated with special cells destined to incubate the zygote nucleus (auxiliary cells, the joint structure constituting the procarp); the carpogonium or the auxiliary cell without or with extensive cell fusions producing obscure to extended branched filaments (gonimoblasts), which in greater or lesser part become converted into carposporangia and with certain associated structures constitute the diploid carposporophyte; or in other instances with auxiliary cells not located near the carpogenic branches (procarps absent), so that the zygote nucleus in the carpogonium (usually after division) is transferred by special (oöblast or sporogenous) filaments to the auxiliaries, where the formation of gonimoblasts is initiated as before; carpospores on release developing the diploid vegetative sporophytic phase.

The ordinary three-phase life cycle is termed "diplobiontic," that in which meiosis takes place in the carpogonium and the vegetative sporophyte is lacking is termed "haplobiontic" and is regularly found only in part of the Nemalionales. Recent reports suggest that some of the larger species may have microscopic filamentous phases, but these results are as yet hardly complete enough to apply in taxonomic work.

NEMALIONALES

Plants filamentous, the filaments uniseriate, creeping or erect, or the plants much larger and corticated, when developing relatively soft branches with either the central filament or the multiaxial type of structure; cells uninucleate, with axial or lateral chromatophores and with intercellular connections; asexual reproduction by monosporangia, very exceptionally by bi- or tetrasporangia; sexual reproduction by spermatia in spermatangia formed (in the larger species) from the surface vegetative cells and carpogonia with elongated trichogynes borne on few-celled carpogenic branches; the zygote itself producing the carpospores either directly or after establishing connections with nutritive cells on the carpogenic branch.

KEY TO FAMILIES

1. Plants of uniseriate, branched filaments.... ACROCHAETIACEAE, p. 210
1. Plants of more complex structure.............................. 2

2. Axes simply forking; cystocarps immersed...................... 3
2. Axes beset with small branchlets; pericarps external
BONNEMAISONIACEAE, p. 229

3. Without a continuous cortical surface or a pericarp
HELMINTHOCLADIACEAE, p. 227
3. Cortical cells approximated to form a continuous surface, the cystocarp surrounded by pericarp filaments
CHAETANGIACEAE, p. 228

ACROCHAETIACEAE

Plants small, free filaments with apical growth and more or less evident erect or prostrate axes; cell arrangement uniseriate, the cells uninucleate, with one or more chromatophores; asexual reproduction by mono-, bi-, or tetrasporangia, formed laterally or terminally; sexual reproduction by short, branched filaments producing spermatia and by carpogenic branches of one to three cells including the carpogonium, the cystocarps small, with carpospores formed in sparingly branched gonimoblast filaments.

KEY TO GENERA[1]

1. Chromatophores single or very few in each cell.................. 2
1. Chromatophores numerous, small peripheral.. RHODOCHORTON, p. 225

2. Chromatophores parietal 3
2. Chromatophores axial, stellate...................... KYLINIA, p. 211

3. Chromatophores consisting of simple or lobed plates............. 4
3. Chromatophores consisting of spiral bands..... AUDOUINELLA, p. 224

4. Sporangia without a distinctive basal cup....................... 5
4. Sporangia with a special cup-shaped base....... COLACONEMA, p. 215

5. Usually somewhat larger, macroscopic, often tufted species mostly external to their support........... ACROCHAETIUM, p. 216
5. Minute, shell-penetrating species, without free vegetative filaments .... CONCHOCELIS, p. 215

---

[1] It may be difficult, unless working with living material, to key out directly the genera and species in this family. In such cases it may be more convenient to use the key to the species in the genus Acrochaetium in the broad sense, as given in the first edition of this work, pp. 225–227, whereby in some cases it is possible to avoid using the character of the chromatophores as a distinguishing criterion.

Kylinia Rosenvinge, 1909

Filamentous, erect and tufted from a single basal cell or a cellular disk, or endophytic, solitary or gregarious; filaments branched, uniseriate, the cells each with one or occasionally more stellate chromatophores, often with pyrenoids; asexual reproduction generally by monospores, occasionally by tetraspores; sexual reproduction infrequent, the spermatangia clustered on lateral branchlets, the carpogonia sessile, lateral on erect branchlets, producing small gonimoblasts bearing terminal carposporangia.

KEY TO SPECIES

1. Base of the plant consisting of a single cell...................... 2
1. Base of the plant of more than one cell......................... 6
2. Basal cell showing a thick wall and little change in form from the original spore... 3
2. Basal cell with altered form.................................. 4
3. Cells hardly swollen, as long as broad; chromatophores central, sporangia subglobose .... K. compacta, p. 212
3. Cells moniliform above, 1–2 diameters long; chromatophores in the distal ends of the cells; sporangia ovate.. K. moniliformis, p. 211
4. Filaments unbranched, of but few cells............ K. unifila, p. 212
4. Filaments freely branched.................................. 5
5. Main filaments 11–23 μ diam.; sporangia 10–11 μ diam. K. alariae, p. 213
5. Main filaments 4–8 μ diam.; sporangia 4–9 μ diam. K. hallandica, p. 213
6. Basal disk of more than one layer of cells; branching relatively dense .... K. secundata, p. 214
6. Basal disk of one layer; branching more sparse and open K. virgatula, p. 214

Kylinia moniliformis (Rosenvinge) Kylin

Basal cell subglobose, giving rise to 2–3 (or more?) erect or ascending filaments producing a small tufted plant, 50–150 μ tall or little more, surpassed by the colorless hairs; filaments initially mostly branched from the base, the other branches rather close, somewhat irregular, composed of more or less inflated cells; diameter of filaments 7–10 μ, the cell length 1–2 diameters, the upper cells internally cask-shaped or subglobose, the walls thick, especially near the septa; chromatophores lying in the upper central part of the cell; monosporangia sessile, rarely stalked, lateral to opposite, ovate, 13.5–15.0 μ long, about 7 μ diam.

Southern Massachusetts; epiphytic on Polysiphonia in a sheltered lagoon.

There is also described from our territory v. **mesogloiae** (Jao) n. comb., (Pl. 32, figs. 15–17; Jao 1936, p. 241; Taylor 1937c, p. 228, as *Acrochaetium moniliforme* v. *mesogloiae*); plants minute, epiphytic, 40–65 μ tall; basal cell subglobose to subovoid, 8.0–9.6 μ diam., 11–17 μ long, attached to the peripheral cell of the host; from the basal cell producing mostly one, rarely 2–3 main filaments, more or less recurved; branches few, more or less dichotomous, sometimes opposite, gradually a little tapered; cells 8.0–9.5 μ diam., 6.5–9.5 μ long, cask-shaped, with a thin wall; monosporangia oval, 6.5–8.0 μ diam., 9.5–16 μ long, unilateral on the upper side of each articulation, rarely opposite; this has been described from southern Massachusetts on Sphaerotrichia.

REFERENCES: Taylor 1937c, p. 227 (as *Acrochaetium moniliforme*); Kylin 1944, p. 16; Papenfuss 1947, p. 437.

Kylinia compacta (Jao) Papenfuss Pl. 32, figs. 6–14

Plants epiphytic, small, 22–42 μ tall; basal cell subglobose or globose, remaining single, with a thick wall, 8–10 μ diam., giving rise to 2–4 (usually 3) erect main filaments more or less irregularly branched, the branchlets frequently tufted; cells somewhat cask-shaped, 5–8 μ diam., 5–10 μ long; monosporangia sessile, variously disposed, usually single, sometimes two on each articulation, ovate or subglobose, 5.0–6.5 μ diam., 6.5–7.0 μ long.

Southern Massachusetts; epiphytic on Cladophora; summer.

REFERENCES: Jao 1936, p. 241; Taylor 1937c, p. 228 (both as *Acrochaetium compactum*); Papenfuss 1947, p. 436.

Kylinia unifila (Jao) Papenfuss Pl. 32, figs. 26–32

Plants small, epiphytic, attached to the host by a single subhemispherical cell, giving rise to one unbranched 1–9-celled erect filament from the lateral face of the basal cell, rarely also to a second branch; filaments slightly arched, parallel to the surface of the host, of about equal diameter from base to apex; vegetative cells somewhat tumid, 6.5–13.0 μ diam., 13–16 μ long; hairs usually pseudolateral; monosporangia sessile, rarely on a one-celled stalk, unilateral, sometimes two on each cell, oval, 5.0–6.4 μ diam., 8.0–9.6 μ long.

Southern Massachusetts; epiphytic on Arthrocladia; late summer.

REFERENCES: Jao 1936, p. 239; Taylor 1937c, p. 228 (both as *Acrochaetium unifilum*); Papenfuss 1947, p. 437.

Kylinia hallandica (Kylin) Kylin

Plants tufted; base a single thick-walled cell, 7.5–14 μ diam., giving rise to 2–4 erect filaments 4–8 μ diam., irregularly and more or less abundantly branched above in all directions, the longer branches giving off alternate or unilateral branchlets; cells usually 2–5 diameters long; branch tips often ending in long hairs; sporangia lateral, sessile, or short-stalked, solitary or 2–3 together, 8–13 μ long, 4–9 μ diam.; spermatangia and carpogonia short-stalked, lateral.

East side of Ellesmere Island; growing on Spongomorpha; late summer.

REFERENCES: Taylor 1937c, p. 229 (as *Acrochaetium hallandicum*); Kylin 1944, p. 13; Papenfuss 1947, p. 436.

Kylinia alariae (Jónsson) Kylin

Base a single cell from which arise one or two erect filaments 0.5–1.0 mm. tall; these axes naked below, diam. 11–23 μ, the cells 2–3 diameters long, usually densely branched above, where the cells are 7–11 μ diam., the cells in sterile branches to 6 diameters long; branchlets opposite, alternate, or unilateral, when sterile often bearing hairs; when fertile the cells particularly short and clavate, each cell bearing two opposite, sessile monosporangia 17–22 μ long, 10–11 (–14) μ diam., a similar spore being terminal on the branch.

Northern Massachusetts to Maine; a common epiphyte on Alaria, growing at exposed stations.

REFERENCES: Phyc. Bor.-Am. 236 (as *Acrochaetium secundatum*). Taylor 1937c, p. 229 (as *A. alariae*); Kylin 1944, p. 13; Papenfuss 1947, p. 436.

Kylinia virgatula (Harvey) Papenfuss

Basal portion a simple disk from which arise one to many erect filaments; filaments 10–14 μ diam. below, long and straight, the cells 3–5 diameters long; branching sparing to moderately abundant, the branches long and straight, erect, 6–8 μ diam. near the end and usually terminating abruptly in a very slender hair; short 1–2 (–few)-celled branchlets abundant, scattered, opposite, or unilateral on stem and branches, mostly bearing either hairs or terminal sporangia; monosporangia terminal on the branchlets, or later lateral, sessile, ovoid, 17–20 μ long, 13–16 μ diam., occasionally replaced by tetrasporangia.

North Carolina, New Jersey to southern Massachusetts; common on various algae; fruiting in the summer.

REFERENCES: Algae Exsic. Am. Bor. 157; Phyc. Bor.-Am. 39 (both as *Chantransia virgatula*). Harvey 1853, p. 243 (as *Callithamnion virgatulum*); Farlow 1881, p. 109 (as *Trentepohlia virgatula*); Collins 1906d, p. 193; Hoyt 1920, p. 473; Taylor 1937c, p. 230 (the three as *Acrochaetium virgatulum*); Papenfuss 1947, p. 437.

In our area one may also recognize f. **luxurians** (J. Agardh) Collins (Phyc. Bor.-Am. 1393; Harvey 1853, p. 242, as *Callithamnion luxurians;* Collins 1906d, p. 194) forming, instead of isolated tufts, a uniform and continuous fringe on the edge of leaves of Zostera, the adjacent basal disks combining to a practically continuous stratum, the dimensions slightly larger than the typical form; found from Connecticut to Maine, commoner than the type, fruiting in the late summer. In specimens examined attributed to this form (and figured on Pl. 31, figs. 4–7) the monospores measured 22–35 μ long, 13–18 μ diam.

Kylinia secundata (Lyngbye) Papenfuss Pl. 31, figs. 1–3

Base a multicellular disk of 1–2 layers or more in thickness; developing several erect filaments, these simple and 7–12 μ diam. below, the cells 1.5–2.0 diameters long; above densely branched, branches mostly unilateral, seldom opposite, rather spreading, 6–8 μ diam., the cells 2–3 diameters long near the end, which often bears a hair; monosporangia generally sessile or occasionally on one-celled stalks, in unilateral or opposite series on the branches, 12–22 μ long, 10–18 μ diam.

Connecticut to Maine; found especially as an epiphyte on Porphyra, but on other algae in sheltered localities; fruiting in early summer.

REFERENCES: Phyc. Bor.-Am. 1088 (as *Chantransia secundata*). Farlow 1881, p. 109 (as *Trentepohlia virgatula* v. *secundata*?); Collins 1906d, p. 194; Taylor 1937c, p. 230 (both as *Acrochaetium secundatum*); Papenfuss 1947, p. 437.

Conchocelis Batters, 1892 [1]

Thalli penetrating into shells, minute, of branched rose-red, uniseriate filaments; cells more or less irregular, with one irregular disklike parietal chromatophore; branches here and there inflated and forming monosporangia.

Conchocelis rosea Batters Pl. 29, fig. 7

Thalli immersed, forming rounded red spots, which may run together indefinitely; primary filaments entangled into a loose layer, the cells 1.5–7.0 μ diam., 7–75 μ (or more) long, cylindrical, tortuose or irregular in form; the elongated cells with slender chromatophores, the broad ones with somewhat stellate structures; monosporangia 13–15 μ diam., formed at the surface of the host.

Southern Massachusetts to Maine and Ellesmere Island; in various types of dead shells, especially Ensis, particularly in deeper water; fruiting during the summer.

REFERENCE: Collins 1906c, p. 159.

Colaconema Batters, 1896

Plants microscopic, filamentous, epiphytic, rose-colored, the filaments irregularly branching in the wall of the host, sometimes anastomosing or laterally loosely conjoined; monosporangia sometimes terminal on main axes, sometimes on short lateral branches, formed from part of the contents of the monosporangium, the remainder persisting as a thickened cuplike base below the spore.

[1] See footnote on p. 206.

Colaconema americana Jao Pl. 29, fig. 8

Plants chiefly endophytic, the filaments much and irregularly branched, interweaving between the cortical cells of the host; branches mostly nearly at right angles, scattered or opposite; cells varying in form, 5–10, sometimes swollen to 32 μ diam., 16–55, mostly 22–38 μ long, at the cross walls but 3–5 μ diam.; from the endophytic branches arise hairlike filaments free from the host, 2 to mostly 3 cells in length, tapering toward the tip, the basal cells with ill-developed chromatophores, being 3–5 μ diam., 13–32 μ long, and the terminal cells colorless, 2.5–3.0 μ diam.; chromatophores single in the vegetative cells, slightly lobed, parietal; monosporangia sessile, globose, on the main filaments or branches, or often clustered on the short lateral branchlets, subglobose, lenticular or hemispherical and 9.5–13.0 μ diam., 6–13 μ long.

Southern Massachusetts; endophytic in *Asparagopsis hamifera;* summer.

REFERENCE: Jao 1936, p. 237.

Acrochaetium Nägeli, 1861

Filamentous, gregarious, rose-red; from a basal holdfast cell, disk, or filaments giving rise to simple or branching uniseriate immersed or erect and free filaments, which may terminate in hairs; chromatophores in the vegetative cells single, seldom divided to several, parietal, platelike or lobed, with or without pyrenoids; sporangia terminal or lateral, scattered, clustered, or in unilateral rows, usually monosporous, occasionally bi-, tetra-, or polysporous; sexual reproduction infrequent, by spermatangia borne variously but usually clustered on the lateral branchlets, and unicellular carpogonia which may be lateral or intercalary, or rarely on one-celled stalks; cystocarps formed directly from the carpogonia, the outer cells of short gonimoblast filaments producing the carposporangia.

KEY TO SPECIES

1. Base of the plant consisting of a single cell.................... 2
1. Base of the plant of more than a single cell.................... 4
2. Basal cell showing a thick wall and little change in form from the original spore; filaments generally less than 6 μ diam. A. microfilum, p. 219

2. Basal cell producing one or more rhizoidal processes, or otherwise altered in shape.. 3

3. Filaments to 8 μ diam., monosporangia about 9 μ diam., 16 μ long; basal cell occasionally supplemented by the development of a few rounded cells.................. A. dasyae, p. 217
3. Filaments to 10 μ diam., mono- or bisporangia to 13 μ diam., 23 μ long; basal cell sometimes supplemented by rhizoidal extensions .... A. intermedium, p. 218

4. Base distinctively filamentous............................... 5
4. Base disklike or massive.................................... 13

5. Free filaments absent or short and unbranched; sporangia usually terminal 6
5. Free filaments longer, often branched; sporangia usually lateral 8

6. Plant growing immersed in the jelly or the tissues of the host; cells to 4 diameters long or more.......................... 7
6. Plant superficial on the host; cells 1.0–1.5 diameters long A. radiatum, p. 223

7. Sporangia about 3 μ diam.................. A. emergens, p. 221
7. Sporangia about 5 μ diam.................... A. minimum, p. 222

8. Lateral branches elongate.................................... 10
8. Lateral branches short or absent............................ 9

9. Filaments less than 10 μ diam................ A. alcyonidii, p. 220
9. Filaments more than 10 μ diam.............. A. amphiroae, p. 223

10. Plants over 2.0 mm. tall.................................... 11
10. Plants to 0.7 mm. tall, cells 2–5 diameters long.. A. flexuosum, p. 219

11. Sporangia in series on the upper sides of the branchlets.......... 12
11. Sporangia on short, crowded branchlets near the bases of the long branches; cells 2–4 diameters long........ A. daviesii, p. 221

12. Sporangia 18–22 μ long.................... A. sagraeanum, p. 220
12. Sporangia 22–31 μ long...................... A. zosterae, p. 218

13. Cells 1–2 diameters long; sporangia scattered, 4.0–6.5 μ diam. A. attenuatum, p. 222
13. Cells 3–5 diameters or more in length; sporangia crowded near the bases of the main branches, 9–11 μ diam.... A. thuretii, p. 222

Acrochaetium dasyae Collins

Base consisting of a rounded cell, the original spore, which may remain little modified or which may put out a short descending process that sometimes branches and enlarges to a few rounded cells attached to the host; erect filaments single on each base, 6–8 μ diam., the cells 2–5 diameters long, rather sparingly branched, the branches erect, unilateral, but slightly tapering and without terminal hairs; monosporangia sessile on the upper side of the branches, solitary or in a unilateral series, one on each cell of the branch, about 16 μ long, 9 μ diam.

Southern Massachusetts; epiphytic on Dasya; fruiting in early autumn.

REFERENCES: Phyc. Bor.-Am. 1342. Collins 1906d, p. 191; Papenfuss 1945, p. 308.

Acrochaetium intermedium Jao Pl. 33, figs. 1–4

Plants 1–2 mm. tall, endophytic below, the base consisting of the original spore, which elongates downward, penetrating the host plant and giving rise to 1–3 short rhizoidal filaments embedded in the walls of the host; upper part of the basal cell 13–16 μ diam., retaining the original spore form, from each base a single erect main filament arising, ramifying by numerous long branches from near the basal cell; secondary branches regularly unilateral, cells 8–10 μ diam., (26–) 32–64 μ long, the terminal cells tapering, with a diameter of about 6.5 μ; chromatophore with a prominent pyrenoid; hairs absent; bisporangia and monosporangia cylindric-ovate to ovate, 9–13 μ diam., 19–23 μ long, unilateral, mostly sessile, rarely on one-celled stalks, 2–6 sporangia placed on the lowermost cells of the branches or branchlets, occasionally also on the upper parts of the main filaments; upper cell of the bisporangium usually smaller than the lower.

Southern Massachusetts; epiphytic on Dasya; late summer.

REFERENCES: Jao 1936, p. 242; Papenfuss 1945, p. 315.

Acrochaetium zosterae Papenfuss Pl. 33, figs. 5–7

Plants epiphytic; basal cell elongate, thick-walled, with the adjoining cell giving rise to contorted creeping filaments; solitary erect main filament arising from the modified spore cell, to 2–3 mm. tall, freely branched, the lowest branches near the base much longer than those above, branchlets numerous, unilateral or alternate; cells

6.4–9.6 μ diam., 32–70 μ long; hairs terminal, very rare; chromatophore with a pyrenoid; monosporangia mostly in series on the branchlets, sessile or on one-celled stalks, unilateral near the base of the branchlets, rarely scattered, cylindric-obovoid, 6.5–9.5 μ diam., 22–31 μ long.

Southern Massachusetts; epiphytic on Zostera and algae; late summer.

REFERENCES: Jao 1936, p. 243 (as *A. subseriatum,* but not of Børgesen); Papenfuss 1945, p. 307.

Acrochaetium microfilum Jao Pl. 32, figs. 1–5

Plants epiphytic, very small, 40–60 μ wide, 20–35 μ tall; basal cell single and globose, larger than the cells of the filament, to 6.5 μ diam., with a rather thick wall, from which are produced 3–4 free main filaments in a single vertical plane, each 2–6 cells long, the outside two more extended, straight or reflexed, the remaining filaments erect, all branching laterally or dichotomously; cells cask-shaped, 3–5 (–6) μ diam., about as long; terminal cells tapering into a hyaline hair; chromatophore filling or nearly filling the cell, with a pyrenoid; monosporangia regularly unilateral on the spreading filaments, unilateral or scattered on the erect filaments, mostly sessile, more rarely on 1–2-celled stalks, oval, 3–5 μ diam., 6.5–8.0 μ long.

Southern Massachusetts; epiphytic on *Polysiphonia fibrillosa;* late summer.

REFERENCES: Jao 1936, p. 240; Papenfuss 1947, p. 437.

Acrochaetium flexuosum Vickers

Basal portion of entangled filaments from which the erect axes form dense globular tufts to 0.7 mm. tall; erect filaments flexuous, 9–10 μ diam. near the base, the cells 2–3 diameters long, each with a chromatophore containing a pyrenoid; branching chiefly above the middle of the filament, subfastigiate, the higher branches progressively shorter, in unilateral series, erect or spreading, 6–7 μ diam., the cells about 5 diameters long; fertile branchlets unilateral, mostly 2-celled, the monosporangia mostly on the upper side, ovate-oblong, 14–16 μ long, 9–10 μ diam.

From the tropics to Bermuda, New Jersey, Rhode Island, and southern Massachusetts; growing as an epiphyte on coarse filamentous algae.

REFERENCES: Phyc. Bor.-Am. 1696. Collins 1906d, p. 192; Papenfuss 1945, p. 309.

Acrochaetium sagraeanum (Montagne) Bornet

Plants tufted or stratose, to 2–4 (–6) mm. tall; the basal layer composed of contorted and entangled filaments, from which the erect filaments arise, these about 12 μ diam. below, the cells 4–8 diameters long, the branching not very dense, the lowest branches longer, the upper gradually shorter, 7–9 μ diam., all rather erect and bearing short unilateral branchlets above the axils; one or two monosporangia borne on each articulation of the branchlets, about 18–22 μ long, 8–10 μ diam.; tetrasporangia sometimes replacing the monosporangia, 28–34 μ long, 17–27 μ diam.

From the tropics to Bermuda (f.) and Connnecticut; epiphytic on Ruppia in very sheltered stations.

REFERENCES: Phyc. Bor.-Am. 39 (as *Chantransia secundata*). Collins 1906d, p. 192; Collins and Hervey 1917, p. 97; Papenfuss 1945, p. 311.

Acrochaetium alcyonidii Jao Pl. 34, figs. 2–4

Plants endozoic below, the basal filaments much longer than the erect ones, deeply penetrating into the host, or a few superficial, all abundantly and usually dichotomously branched, the cells varying in shape and size, 6.5–13.0 diam., 13–32 μ long, often swollen in the middle; the free filaments erect, to 600 μ long, with very short branches, which are usually 1–5, sometimes to 10 cells long, unilateral or scattered, cells 6.5–9.5 μ diam., 9.5–19.0 μ long; hairs absent, but occasionally the terminal cells much tapered and containing small chromatophores with pyrenoids; monosporangia obovoid, sessile or each with a one-celled stalk, 1–2 on each articulation, terminal or unilateral, rarely irregular, in crowded groups on short branches, diameter 8.0–9.5 μ, length (13–) 16–19 μ; occasionally replaced by bisporangia.

Southern Massachusetts; partly endozoic in *Alcyonidium mytili;* fruiting in early autumn.

REFERENCES: Jao 1936, p. 245; Papenfuss 1945, p. 312.

In addition, there is described, as found with the type, v. **cylindricum** Jao (Pl. 34, fig. 5): endozoic filaments 5.0–6.5 μ diam., the cells 6.5–13.0 μ long; erect filaments to 300 μ tall, cells 4–5 μ diam., 10–19 μ long; branches short, 1–4 cells long; monosporangia cylindric-obovoid, usually terminal, rarely lateral, 4.0–6.5 μ diam., 9.5–13.0 μ long; otherwise as in the type of the species.

Acrochaetium daviesii (Dillwyn) Nägeli Pl. 31, figs. 8–10

Base of decumbent or penetrating ramifying filaments which on firm hosts may be crowded into a felted mass; bearing erect filaments to 6 mm. tall, 7–13 μ diam. near the base, usually about 8 μ diam. near the summit; branches abundant, largely unilateral, rather erect, the cells 2–4 diameters long, containing a pyrenoid; slender multicellular hairs more or less abundant; fertile branches short, 1–2 times redivided, usually attached near the bases of the longer branches, but also scattered; monosporangia sessile or short-stalked, 8–12 μ diam., 11–20 μ long, sometimes replaced by tetrasporangia which are 7–12 μ diam., 13–16 μ long.

New Jersey, Rhode Island to Maine; forming fringes along the edges of Rhodymenia, but occasionally occurring on other algae; fruiting in summer and autumn.

REFERENCES: Phyc. Bor.-Am. 880. Harvey 1853, p. 243 (as *Callithamnion daviesii*): Farlow 1881, p. 109 (as *Trentepohlia daviesii*); Papenfuss 1947, p. 435.

Acrochaetium emergens (Rosenvinge) Weber-van Bosse

Vegetative filaments spreading in the cuticular layer of the host, branches diverging at right angles; cells varying in shape, mostly swollen in the middle, 1.5–3.5 μ diam., 9.5–12.8 μ long; chromatophore in large part or completely surrounding the cell cavity, pyrenoid apparently absent; monospores terminal on a one-celled stalk and entirely free from the host, or sessile and endophytic, oval, 3.2 μ diam., 4.5–5.0 μ long.

Southern Massachusetts; endophytic on *Polysiphonia fibrillosa;* late summer.

REFERENCES: Jao 1936, p. 247; Papenfuss 1945, p. 314.

Acrochaetium attenuatum (Rosenvinge) Hamel, f. Pl. 34, fig. 1

Plants epiphytic; basal filaments fused together near the center to form a one-layered disk; erect filaments numerous, arising from the central part of the basal disk, up to 250 μ (or more) in height, ramification scattered, opposite or dichotomous, the terminal cells tapering toward the tip and forming a definite hyaline hair, the lower cells 5.0–6.5 μ diam., (6.5–) 9.5–16.0 μ long; chromatophore with a pyrenoid; monosporangia occur mostly on the upper part of the plant and usually are solitary, mostly sessile, sometimes on one-celled stalks, oval, 4.0–6.5 μ diam., 6.5–9.6 μ long.

Southern Massachusetts; epiphytic on *Polysiphonia fibrillosa;* late summer.

REFERENCES: Jao 1936, p. 244; Papenfuss 1945, p. 308.

Acrochaetium thuretii (Bornet) Collins et Hervey

Basal portion an irregular disk, 60–120 μ diam.; erect filaments 2–3 mm. tall, 7.5–12.0 μ diam. below, slightly less above, branching from the base, the branches alternate or unilateral, often ending in a hair of the same diameter as the branch; cells below 3–5, above 8–12 diameters long, with a pyrenoid; monosporangia crowded on one-celled stalks on the upper side of the branches near the base, one or two on each of the lower 1–3 cells, 14–17 μ long, 9–11 μ diam.; sexual plants monoecious, the organs borne on short branchlets near the bases of the branches; the spermatangia few, on the end and the upper segments, the carpogonium solitary on a lower segment; carpospores 18–21 μ long, 9–13 μ diam.

Massachusetts; growing on various larger algae to a few meters depth; fruiting during the latter part of the summer.

REFERENCES: Phyc. Bor.-Am. 192 (as *Chantransia corymbifera*). Collins 1906d, p. 196; Papenfuss 1947, p. 435.

Acrochaetium minimum Collins

Basal layer composed of epi- or endophytic filaments which are at first free, later more or less united, but not forming a genuine membrane, the irregular cells 2–4 μ diam., 1–4 diameters long; basal filaments supporting many short branches which extend beyond the surface of the host, 1–4-celled, in length seldom over 25 μ, less frequently to 25 cells, 2–3 μ diam., the cells 1.5–3.0 diameters long, or in the longer filaments to 8 diameters long; when fertile terminating in a pyriform monosporangium, 7 μ long, 5 μ diam., or the longer filaments occasionally also with a lateral spore which is sessile or on a unicellular stalk.

Southern Massachusetts; growing on *Desmarestia viridis;* fruiting in late summer.

REFERENCES: Phyc. Bor.-Am. 1493. Collins 1908b, p. 133; Papenfuss 1945, p. 316.

Acrochaetium radiatum Jao Pl. 32, figs. 18–25

Plants epiphytic, prostrate on the host, the patches circular or irregular, to 320 μ diam.; germinating spore dividing into two nearly equal cells, from which are produced 3–6 (rarely more) radiating filaments, which produce a disk of which the central portion becomes pseudoparenchymatous with increasingly dense and irregular branching, but the branching near the ends of the main filaments is dichotomous or unilateral and looser, their cells 5.0–6.5 μ diam., 6.5–9.5 μ long; erect branches absent, or poorly developed, only 1–3 cells long; chromatophore with a central pyrenoid; hairs terminal, mostly on the erect filaments; monosporangia borne directly on the prostrate filaments, or sometimes 1–2 terminal on erect branches, oval, (5–) 6.5–8.0 μ diam., 8.0–9.5 μ long.

Southern Massachusetts; epiphytic on *Polysiphonia fibrillosa;* late summer.

REFERENCES: Jao 1936, p. 246; Papenfuss 1945, p. 310.

Acrochaetium amphiroae (Drew) Papenfuss

Base somewhat endophytic, of filaments spreading over and penetrating between the surface layers of the host; erect filaments irregularly, somewhat unilaterally branched, cells cylindrical, 12.0–14.0 μ diam., 1.5–3.0 times as long as broad; chromatophore with a single pyrenoid; sporangia terminal and lateral, stalked, on the branchlets, sometimes clustered, ovoid, 10.0–13.0 μ diam., 16.0–19.5 μ long.

Southern Massachusetts; growing on Corallina.

REFERENCES: Drew 1928, p. 179 (as *Rhodochorton amphiroae*); Papenfuss 1945, p. 312; Doty 1948, p. 263.

Audouinella Bory, 1823

Plants small, filamentous, creeping and immersed in the substratum or erect and tufted, the filaments branched, the cells uniseriate, with one or a few spiral chromatophores lacking pyrenoids; asexual reproduction by monospores or tetraspores terminal or lateral on the branches; sexual reproduction by spermatangia clustered on the lateral branchlets and carpogonia lateral and sessile on erect branches, or terminal on one- or two-celled branchlets, or intercalary; gonimoblasts forming carposporangia terminally or in series.

KEY TO SPECIES

1. Microscopic; erect filaments short; most often found between the membranes of hydroids.................... A. membranacea
1. Large; tufted, to about 5 mm. tall, epiphytic.......... A. efflorescens

Audouinella membranacea (Magnus) Papenfuss

Pl. 31, figs. 11–12

Plants filamentous, the primary filaments creeping, uniseriate, growing between the membranes of the host and often crowded, 6–8 (–10) μ diam., the cells irregular, 1.5–8.0 diameters long; cells with a few short band-shaped chromatophores without pyrenoids; erect filaments penetrating above the surface of the host, short, but a few cells in length, simple or sparingly branched, the cells 7–8 μ diam., 1.5–2.0 diameters long; terminal cells on main axes and any branches eventually transformed into tetrasporangia, or these rarely lateral; sporangia 17–30 μ long, 12–20 μ diam.

Connecticut to Maine; most commonly growing between the chitinous layers of sertularians and probably other animals, and also reported from algae; fruiting in late spring and summer.

REFERENCES: Phyc. Bor.-Am. 99. Taylor 1937c, p. 240 (both as *Rhodochorton membranaceum*); Papenfuss 1945, p. 326.

Audouinella efflorescens (J. Agardh) Papenfuss

Plants tufted (sexual) or forming an expanded layer (asexual), the tufts to 5 (–7) mm. high, the sporangial plants to 2 mm. high; widely branched, the lower branches longer, the upper shorter, divergent, more or less unilateral; erect filaments 6–8 μ diam., the cells 6–8 diameters long; chromatophores 1–2 or more in a cell, simple or forked, spirally disposed; fruiting branches short, scattered, corymbose, the sporangia borne in series on moniliform segments, monosporangia 11–18 μ long, 5–6 μ diam., tetrasporangia 15–28 μ long, 8.0–12.5 μ diam.

South side of Ellesmere Island; growing on Chaetomorpha at 8 meters depth; fruiting in late summer.

REFERENCES: Kjellman 1883, p. 129 (as *Chantransia efflorescens*); Rosenvinge 1926, p. 28; Taylor 1937c, p. 235 (both as *Acrochaetium efflorescens*); Papenfuss 1945, p. 326.

There is a record of this plant from Marthas Vineyard, Massachusetts (Farlow 1881, p. 109, as *C. efflorescens*), growing upon Cystoclonium; it was not taken up in the Collins manuscript and must be considered doubtful.

Rhodochorton Nägeli, 1861

Plants slenderly filamentous, the lower portion consisting of ramifying decumbent filaments or cellular disks, which support erect, branching uniseriate fertile filaments; cells with numerous small chromatophores; tetrapartite sporangia terminating branches or short-stalked on special subapical branchlets; sexual apparatus imperfectly known.

KEY TO SPECIES

1. Base disklike; filaments less than 1 cm. tall........ R. penicilliforme
1. Base creeping; filaments to more than 1 cm. tall...... R. purpureum

Rhodochorton penicilliforme (Kjellman) Rosenvinge

Plants forming a disklike expansion of close, branching filaments which radiate from the center, often fusing laterally; from this disk arise the erect filaments which are 9–14 μ diam., the cells 1.5–3.0

diameters long, uninucleate, with several disklike to band-shaped chromatophores; branching when present rarely more than one degree; hairs absent; tetrasporangia terminal on short branchlets, or on stalks of 1–2 cells, obovate, 25–35 μ long, 21–27 μ diam.; male plants known, the spermatangia formed on short lateral branchlets, but carpogonial plants not reported.

Rhode Island, northern Massachusetts, Hudson Bay, Devon Island, and Ellesmere Island; external upon the membrane of tubularians, bryozoa, etc., and also known upon Chaetomorpha and Phyllophora; apparently vegetating and fruiting throughout the year.

REFERENCES: Rosenvinge 1893, p. 792 (as *R. mesocarpum* v. *penicilliforme*), 1926, p. 33; Collins 1906c, p. 160; Papenfuss 1945, p. 327.

Rhodochorton purpureum (Lightfoot) Rosenvinge

Pl. 45, figs. 1–2

Plants small, tufted or forming a turf 0.5–1.0 (–2.0) cm. tall; basal part of ramifying filaments; erect filaments sparingly branched, the branches erect-appressed, the lower longer and subfastigiate; diameter 10–15 (–20) μ below, cells usually about 1–3 (–5) diameters long, containing numerous small irregularly rounded chromatophores without pyrenoids; fruiting branches short and corymbose-crowded below the apices of main branches, alternate or opposite, simple or 1–3 times forked, the tetrapartite sporangia terminating the short branchlets and lateral on their upper segments, ellipsoid, 25–36 μ long, 14–20 μ diam.

Long Island to Nova Scotia, Hudson Strait, Baffin Island, Devon Island, and Ellesmere Island; perennial, growing on pilings, the bases of Laminaria plants, and particularly on shaded rocks near low-tide mark; vegetating continuously but fruiting in winter and early spring.

REFERENCES: Phyc. Bor.-Am. 49 (848 as *R. parasiticum*?). Harvey 1853, p. 242; Farlow 1881, p. 121 (both as *Callithamnion rothii*); Collins 1900a, p. 12 (as *R. parasiticum*); Taylor 1937c, p. 239 (as *R. rothii*); Papenfuss 1945, p. 327; Whelden 1947, p. 119 (as *R. rothii*); Wilce *in herb.*

HELMINTHOCLADIACEAE

Plants of moderate size, erect and coarsely branched, very mucous, sometimes partly calcified; structurally with an axial row of cells bearing lateral assimilators, or multiaxial with many filaments in the center developing lateral assimilative branches of the "fountain" type; monosporangia present or absent; sexual reproduction by spermatangia borne in loose clusters on the ends of the assimilatory filaments; carpogenic branches borne variously on the assimilators, usually of three cells, the terminal being the carpogonium and auxiliary cells being absent; cystocarps immersed among the assimilative filaments, without definite pericarps, the gonimoblasts closely branched and the outer cells producing the carpospores.

Nemalion Targioni-Tozzetti, 1818

Fronds terete or a little compressed, soft and slippery, simple or furcate-branched; structure multiaxial, the filaments of the axis with scattered anastomoses in a pale medullary cord, giving off dichotomous corymbose fascicles of peripheral chromatophore-bearing filaments; spermatangia hyaline, near the apices of peripheral filaments; carpogenic branches terminal as divisions of the peripheral filaments, immersed, usually three-celled; cystocarps compact, the end cells of the gonimoblasts producing the carpospores without special investment.

Nemalion multifidum (Weber et Mohr) J. Agardh Pl. 35, fig. 6

Plants dull reddish brown, elastic, arising from a small disk to 5–20 (–30) cm. tall, subsimple or 1–5 times subdichotomously branched, the branches to 1–2 mm. diam., expanded below the rounded axils, otherwise cylindrical, upwardly tapering; outer cells of the assimilators oval to rounded, 12–16 μ diam., 13–23 μ long, the chromatophores somewhat radiate-lobed, with a pyrenoid; slender colorless unicellular hairs numerous; monoecious or apparently dioecious; spermatangia numerous, colorless, formed on the terminal forkings of assimilators, usually 1–4 to a cell, opposite or terminal; carpospore masses not visible to the unaided eye, the spores spherical, 12–14 μ diam.

Long Island, to Nova Scotia; growing in slippery colonies on rocks in the intertidal zone, particularly in exposed places; in good fruiting condition in midsummer, later overgrown with epiphytes and decaying; at least the erect plants annual.

REFERENCES: Phyc. Bor.-Am. 193. Harvey 1853, p. 134; Farlow 1881, p. 117; Chester 1896, p. 340; Wolfe 1904, p. 607; Cleland 1919, p. 323.

CHAETANGIACEAE

Plants of moderate size, erect and bushy of habit, soft in texture; sometimes partly calcified; structurally multiaxial, the filaments in the center developing lateral branches of the "fountain" type, the outer cells of which may be joined into a continuous epidermis; monosporangia or tetrasporangia absent from most genera, but sometimes present; spermatangia scattered over the surface of the plant as little cells in small groups or in conceptacular sacs; carpogenic branches three-celled, borne on inner forks of the lateral filaments, the cell below the carpogonium capable of originating nutritive cells; cystocarps immersed, the gonimoblasts developed within a pericarp of slender crowded filaments, eventually discharging through a pore at the surface.

Scinaia Bivona, 1822

Plants bushy, repeatedly dichotomously branched, the texture firm-gelatinous, a more or less indistinct axial strand being present in the branches; structurally showing three layers, the strand in the center giving off lax dichotomous filaments containing most of the chromatophores and constituting the assimilatory zone, these ramifications ending in large cells without conspicuous chromatophores which are closely joined laterally in a smooth epidermis; monosporangia spherical, scattered, formed between the epidermal cells; the spermatangia spherical, in small sori, where two or three are pushed to the surface between the epidermal cells; carpogenic branches developed at forkings of the medullary filaments, consisting of three cells, the outer being the carpogonium, the second developing before fertilization four lateral nutritive cells, and the third or lowest developing the rudiments of the pericarp filaments; the terminal cells of the gonimoblasts producing the carpospores; cystocarps lying below the peripheral layer, enclosed in a dense filamentous pericarp which eventually opens to the surface by a pore.

Scinaia furcellata (Turner) Bivona Pl. 53, fig. 1

Plants forming soft, pinkish-red, hemispherical clumps 4–8 cm. tall, the axis single below, abundantly dichotomously branched, rather closely fastigiate; the cylindrical branches acutely spreading, sometimes a little contracted at their bases or elsewhere and where constricted often proliferous, a little greater in diameter in the upper parts of the plant, usually about 1–2 mm. diam., the apices pointed; the axial strand very obscure, in dried specimens sometimes faintly visible in the middle parts of the plants; the surface cells rounded, 16–24 μ diam.; monoecious; carpospore masses scattered, immersed, often numerous, just visible to the naked eye, in pericarps 150–250 μ diam.

Long Island to southern Massachusetts; infrequent, a plant of rather deep water, secured either by dredging or as washed ashore by heavy storms; forming carpospores in late summer.

REFERENCES: Phyc. Bor.-Am. 194. Harvey 1853, p. 136; Farlow 1881, p. 117; Setchell 1914, p. 79.

BONNEMAISONIACEAE

Plants of moderate size, slenderly branched, with an evident axis and extensive branch systems; growth apical, an axial row of cells being surrounded by short, compact, branched cell rows constituting a pseudoparenchymatous cortex with a continuous surface; sporangia unknown; spermatangia in dense masses covering small lateral branchlets; carpogenic branches lateral on the corticating cell rows, forming after fertilization a cystocarp with a definite superficial pericarp, which is chiefly developed from adjacent tissue and which surrounds the gonimoblasts and associated nutritive cells.

Asparagopsis Montagne, 1840

Plants bushy, above repeatedly slenderly branched to brush-like tufts, the alternate branches beset with short branchlets; the stem and larger branches corticated, becoming subtubular; spermatangia covering enlarged branchlets; each carpogenic branch on the fifth to seventh cell of a corticating cell series, on which three pericentral cells are formed, the oldest functioning as the supporting cell of the three-celled carpogenic branch; gonimoblasts eventually borne on a large fusion cell; pericarps subpyriform, becoming pedicellate by transformation of a branchlet.

Asparagopsis hamifera (Hariot) Okamura [1]

Pl. 35, fig. 5; pl. 40, fig. 1; pl. 44, fig. 4; pl. 52, fig. 3

Plants bushy, deep rose-red or red-purple, attached by rhizoidal clasping branches; erect axes 5–10 cm. tall, about 1–2 mm. diam.; branching repeatedly alternate, vaguely bilateral, distinctly pyramidal, but the habit often becoming confused by entangling and concrescence; ultimate branchlets generally alternately somewhat bilaterally disposed, tapering or slightly spindle-shaped, the bases contracted, about 1–3 mm. long, 0.1–0.2 mm. thick, with small colorless refractive glands scattered among the chromatophore-bearing surface cells; occasional branchlets enlarged and hooked, about 5 mm. long, scattered, but mostly near the tips of the branches; pericarps short-stalked, urceolate; sporangia and spermatangia unknown in our area.

Long Island and southern Massachusetts; in the sublittoral and somewhat deeper water, often attached to coarse algae by the hooked tips, or washed ashore in tangled masses; forming infertile pericarps during midsummer.

REFERENCES: Lewis and Taylor 1928, p. 197, 1933, p. 150; Taylor 1940, p. 192, 1941, p. 73.

GELIDIALES

Plants with slender, wiry, subsimple or redivided branches; from an apical cell or less often a small group of cells ultimately developing a multiaxial type of structure, and corticated; asexual reproduction by tetraspores formed at or below the thallus surface; sexual reproduction by spermatia in spermatangia formed from the surface cells and carpogonia on carpogenic branches loosely associated with chains of nutritive cells; the carpogonium producing filamentous gonimoblasts, on which carposporangia are borne among the nutritive cells.

[1] See footnote on p. 291.

GELIDIACEAE

Plants small to moderately large, tough and wiry, branched; asexual plants bearing tetrasporangia, the sporangia generally embedded in the cortex of localized areas or branchlets of the plant; spermatangia borne on the surface of considerable areas or only on branchlets of the male sexual plants; carpogenic branches borne on inner filaments, of the female plants, associated with chains of nutritive cells in the same areas; the carposporophyte ultimately produced by elongated gonimoblast filaments which run out from the fertilized carpogonium through the surrounding tissue, producing the carposporangia laterally, but not making definite connections with the nutritive cells.

Gelidium Lamouroux, 1813

Plants of moderate to small size, cylindrical or flattened, with the axis erect and usually laterally branched; structurally developing from an apical cell, the axial row of cells giving rise to pericentral cell rows which are later paralleled by many other secondary medullary filaments, the whole surrounded by short, compact, radial-branched cell series constituting a firm assimilatory cortex; slender thick-walled filaments (rhizines) generally present in the axis between the larger cells, localized in the inner cortex or somewhat dispersed, but not localized in the medulla; tetrapartite sporangia present, formed on independent plants in the cortex of restricted areas of the branches; spermatangia forming a more or less extensive coating on local areas of the branches; three-celled carpogenic branches developed on the pericentral cell rows; from these rows there are also developed chains of nutritive cells; after fertilization the gonimoblast filaments spread among the nutritive cells and form a bilocular cystocarp with a median septum embedded in the thallus, the carposporangia terminal on the gonimoblasts, the discharge pores not particularly elevated.

Gelidium crinale (Turner) Lamouroux
Pl. 35, figs. 1–3; pl. 40, fig. 3; pl. 41, fig. 5

Plants forming expanded tufts 2–5 cm. tall; dull purple to yellowish brown, cartilaginous; primary branches rhizomatous, spreading over the substratum, about 0.5 mm. diam., firmly attached by holdfasts; above forming erect subterete branches 1.5–7.5 cm. tall which are irregularly alternate below, irregularly to distinctly pinnate above; superficial cells of the thallus 5–16 μ diam., 6–16 μ deep; in tetrasporic plants the branchlets more pinnately disposed and flattened, often spatulate; tetrasporangia 10–22 μ diam., 13–25 μ long; sexual plants dioecious; cystocarps in somewhat spindle-shaped enlargements of the branchlets, usually solitary, occasionally paired.

Bermuda, Florida to northern Massachusetts and possibly to Maine; also from the Gulf of St. Lawrence; growing on rocks between the tide marks, particularly on precipitous rock and cliff faces, but also along marshy shores; fruiting during the latter part of the summer.

REFERENCES: Phyc. Bor.-Am. 195. Harvey 1853, p. 116 (as *G. corneum p. p.*); Farlow 1881, p. 158.

CRYPTONEMIALES

Plants showing various shapes from filiform to fleshy-membranous or rocklike; corticated, with either the multiaxial or the central filament types of structure; asexual reproduction by tetraspores formed in sporangia at the thallus surface, or in sunken pits or conceptacles; sexual reproduction by spermatangia borne on surface cells or the lining of conceptacles and by carpogonia on carpogenic branches sunken in the cortex or in conceptacles, the carpogenic branches associated with typical auxiliary cells formed before fertilization, usually on special accessory axes which are neighboring to the carpogenic axes or more commonly remote; carpogonium after fertilization ordinarily producing oöblast filaments which transmit the diploid nuclei to the auxiliary cells, from which the cystocarps are produced.

KEY TO FAMILIES

1. Plants encrusted with lime.................. CORALLINACEAE, p. 241
1. Plants free from lime.. 2

2. Plants reduced or parasitic epiphytes..... CHOREOCOLACACEAE, p. 258
2. Plants crustaceous .... SQUAMARIACEAE, p. 236
2. Plants erect; bushy or foliaceous............................ 3

3. Branches cylindrical, texture rather tough; reproductive structures in nemathecial areas............ RHIZOPHYLLIDACEAE, p. 235

3. Branches tapering or foliaceous; reproductive structures not sharply localized 4

4. Branches nearly filiform, subgelatinous; carpogenic branch of 3 cells, the auxiliary in a separate axis on the same supporting cell .... GLOIOSIPHONIACEAE, p. 254
4. Branches coarser, firmer or foliaceous......................... 5

5. Carpogenic branch of 3 cells, auxiliary the supporting cell of the carpogenic branch .... KALLYMENIACEAE, p. 256
5. Carpogenic branch of 5 cells, auxiliary cells in separate unrelated axes .... DUMONTIACEAE, p. 233

DUMONTIACEAE

Plants of medium size, more or less branched, or plane and entire, soft; the original axial filament and apical growth soon obscured, the plants then appearing to have a pseudoparenchymatous cortex and a filamentous, often hollow medulla; sporangia tetrapartite; spermatangia superficial on the male plants, widely distributed; carpogenic branches scattered, of five cells, the carpogonium fusing with an intermediate cell in the branch from which oöblast filaments go out to auxiliary cells in separate axes, which in turn give rise to the scattered cystocarps.

KEY TO GENERA

1. Thallus subcylindrical, branched; sporangia scattered DUMONTIA, p. 233
1. Thallus flat, somewhat cleft; sporangia in localized patches DILSEA, p. 234

Dumontia Lamouroux, 1813

Plants erect from a radial disk, cylindrical or compressed, ultimately tubular, widely branched, the branches subsimple; showing a pseudoparenchymatous cortex about a spongy longitudinally filamentous, often hollow medulla; sporangia immersed in the peripheral layer, scattered, round, tetrapartite; spermatangia forming a more or less continuous superficial layer; carpogenic branches of five cells, curved; auxiliary axis resembling the carpogenic branch, of four cells; fertilized carpogonium fusing with the third (rarely the second) cell as a nutritive pseudoauxiliary cell, from which in turn two or three oöblast filaments are produced that carry the diploid nuclei to the second or the third cell of auxiliary axes; all the cells in the gonimoblasts forming carpospores, each cystocarp lying in the cortex without investment by any pericarp.

Dumontia incrassata (O. F. Müller) Lamouroux
Pl. 35, fig. 7; pl. 41, fig. 6

Plants generally tufted, more or less gregarious, from a scutate disk, 1–6 dm. tall, dull reddish, when old or exposed to the light becoming dull yellowish; axis subsimple or irregularly alternately 1 (–2) times branched, the branches long, flagelliform, like the axis usually twisted, at first tapering toward both ends, but the distal portions usually decaying on the larger branches, so that they appear truncate; solid near the base and in young branches, but in larger and older branches generally hollow, sometimes compressed, sometimes inflated, generally to 2–5 mm. diam., reaching 20–26 mm. diam. in large plants; tetrasporangia borne on the inner forks of the cortical filaments, tetrapartite, 45–60 μ diam., 60–80 μ long, formed throughout the plant, progressively maturing from the top, the shedding associated with decay of the thallus.

Long Island, Rhode Island to Nova Scotia, and James Bay; in shallow water and tide pools, not laid bare by the tides, usually on small stones and shells; in best condition in the spring, the tetrasporic reproduction being on the wane in southern Massachusetts by the first of July.

REFERENCES: Phyc. Bor.-Am. 2149 (as *D. filiformis*). Dunn 1916, p. 271, 1917, p. 425; Setchell 1923, p. 33; Howe 1927, p. 25B; Lewis and Taylor 1928, p. 195; Whelden 1928, p. 121 (the last four as *D. filiformis*).

Dilsea Stackhouse, 1809

Plant forming short-stalked blades, cuneate below, above broadening, simple or sparingly cleft, flat and somewhat fleshy; structurally showing a medulla of densely intermeshed filaments surrounded by a cortex of subradially disposed filaments, the inner part of large cells rather distinctly delimited from the outer of smaller cells; sporangiferous plants bearing colorless hairs; sporangia in localized areas on the plant, borne between the filaments of the inner part of the cortex, tetrapartite; carpogenic branches and auxiliary cell branches relatively long, associated with many short-celled vermiform sterile filaments; cystocarps in very great numbers in the fertile portions, immersed in the medullary network, relatively small and not projecting, the spores released by the decay of the outer tissues.

Dilsea integra (Kjellman) Rosenvinge

Plant foliaceous, the stalk short, the blade generally lanceolate but sometimes subreniform or linear-lanceolate, with an entire margin, or occasionally broadly lobed; length of large fronds to 32 cm., breadth to 10 cm.; color light wine-red to deep red-brown; substance membranous to fleshy; cystocarps scattered toward the tips, generally in small groups.

Labrador, southeastern Hudson Bay, and Hudson Strait; dredged in 18 meters; fruiting in late summer.

REFERENCES: Kjellman 1883, p. 152 (as *Sarcophyllis arctica*); Howe 1927, p. 25B; Wilce *in herb.*

RHIZOPHYLLIDACEAE

Plants erect, bushy, dichotomously branching and tough; structurally with a medulla of many longitudinal filaments and an assimilative cortex of branches turning outward; tetrapartite sporangia between the cortical cells of the sporangial plants, scattered or in nemathecia; spermatangia in nemathecia, lateral on short, crowded, superficial filaments; carpogenic branches associated with quite similar auxiliary axes, the carpogonium fusing with an intermediate cell in the carpogenic branch, from which oöblast filaments go out to auxiliary cells in separate axes, which give rise to the crowded cystocarps in swollen nemathecia.

Polyides C. Agardh, 1823

Plants bushy, the coarse divisions dichotomously ramified; the axis showing a medulla of elongated colorless filaments surrounded by an inner cortex of short broad cells and an outer cortex of smaller cells in moniliform radiating filaments; tetrapartite sporangia between the cells of the outer cortical layer; spermatangia lateral on short filaments formed in patches on the surface of the male plants;

carpogenic branches of five to seven short cells, auxiliary axes and sterile filaments aggregated in swollen nemathecia; gonimoblasts formed from an enlarged projection of the auxiliary, the end cells forming carpospores.

Polyides caprinus (Gunnerus) Papenfuss

Pl. 35, fig. 8; pl. 40, fig. 2

Plants bushy, color dull red, drying blackish; texture firm, elastic, drying to a horny consistency; arising from a relatively large basal disk, which bears one to usually several or even many erect axes, these to 8–21 cm. tall, 1–2 mm. thick, little-tapered, subsimple below, above 6–8 times dichotomous, fastigiate in habit, the angles of forking rather narrow, the apices of the branches subacute to distinctly obtuse; tetrasporic branches locally thickened.

Long Island to the lower St. Lawrence, Prince Edward Island, Labrador, and Hudson Strait; primarily a plant of deep water.

REFERENCES: Phyc. Bor.-Am. 850. Harvey 1853, p. 128; Farlow 1881, p. 160; Taylor 1937c, p. 251 (all as *P. rotundus*); Wilce *in herb.*

SQUAMARIACEAE

Plants spreading, crustaceous; structurally usually with a basal layer of radiating branched filaments from which an upper layer of compact erect filaments arises, the whole sometimes a little encrusted with lime; growth peripheral from the ends of the filaments of the basal layer; sporangia tetrapartite, scattered between the erect filaments in nemathecial groups or in crater-like conceptacles; spermatangia in tufts on the ends of erect, paraphysal filaments; carpogenic branches scattered or in nemathecia, short, lateral on paraphysal filaments; after fertilization the carpogonium fuses with an intermediate cell in the carpogenic branch, from which oöblast filaments go out to auxiliaries formed laterally at the bases of other paraphyses; cystocarps small, scattered.

KEY TO GENERA

1. Tetrasporangia not in conceptacles 2
1. Tetrasporangia in sunken conceptacles .... HILDENBRANDIA, p. 240

2. Erect filaments closely united; rhizoids present. PEYSSONNELIA, p. 240
2. Erect filaments not closely united; rhizoids absent 3

3. Paraphyses associated with the sporangia...... RHODODERMIS, p. 237
3. Paraphyses not present........................ CRUORIOPSIS, p. 238

Rhododermis Crouan, 1852

Plants forming horizontally expanded membranous growths; composed of a basal layer of flabellate filaments without rhizoids, not sharply distinguished from the erect filaments which arise from it; reproductive organs grouped on the upper surface, associated with somewhat curved paraphyses; sporangia tetrapartite; sexual organs unknown.

KEY TO SPECIES

1. Plants on Zostera leaves; tetrasporangia sessile, paraphyses of but few cells................................ R. georgii, p. 237
1. Plants on coarse algae, stones, shells, etc.; tetrasporangia stalked, paraphyses of a considerable number of cells........... 2
2. Plant very thin; in section cells broader than tall.. R. elegans, p. 238
2. Plant thicker; cells taller than broad........... R. parasitica, p. 237

Rhododermis georgii (Batters) Collins

Plants small, forming dark red patches 0.1–1.0 mm. diam., becoming increasingly pulvinate with age, often growing together; cortex of a few layers of cells; tetrasporangia in convex sori, with somewhat curved and club-shaped paraphyses 3–4 cells long, of cells 3–4 μ diam., 6–9 μ long; tetrasporangia sessile, 20–36 μ long, 14–18 μ diam.

Long Island, Rhode Island to Maine; epiphytic upon the margins of Zostera leaves; from March onward, forming tetraspores in the latter part of the summer.

REFERENCES: Phyc. Bor.-Am. 1299. Collins 1906c, p. 160.

Rhododermis parasitica Batters

Plants forming blackish-red round or irregular crusts 0.3–4.5 cm. diam., to 0.1–0.2 mm. thick; with vertical filaments of 12–30 cells, diam. 7–8 μ, the cells in section taller than broad; tetrasporangia about 28–37 μ long, 12–21 μ diam., among paraphyses which are simple, curved, about 5 cells and 32–60 μ long, 4–5 μ diam.

Northern Massachusetts; growing on the stipe and the blade of Laminaria.

REFERENCE: Collins 1906c, p. 160.

Rhododermis elegans Crouan

Plants disklike, color bright red, thin, the margins irregular; of radiating filaments about 5.5 μ diam. with cells as long as broad, or one-half longer, in section lower than broad; this basal layer bearing erect cell rows to form a compact tissue 4–20 cells thick or more, the rows 7–9 μ diam.; sori in the form of roundish spots, the sporangia tetrapartite, 24–33 μ long, 16–24 μ broad, among 4–6-celled, slightly curved paraphysal filaments; spermatangia superficial, aggregated, 10–11 μ long, 4 μ diam.

Northern Massachusetts, Maine, Devon Island, and Ellesmere Island; on the shells of live crabs and on pebbles.

REFERENCES: Phyc. Bor.-Am. 1248. Collins 1906c, p. 160; Rosenvinge 1926, p. 36.

Cruoriopsis Dufour, 1864

Plants crustaceous, rather flat, attached by the whole lower surface without rhizoids; basal layer of radiating rows of cells bearing the erect filaments, which are more or less united by jelly; sexual and asexual reproduction may occur on the same plant; tetrapartite sporangia scattered or aggregated; spermatangia forming small tufts borne on the tips of the erect filaments; carpogenic branches and auxiliary cells on separate special filaments; gonimoblasts small, forming two to four carposporangia in a row, the more or less numerous cystocarps about a single fertilized carpogonium forming more or less evident immersed sori scattered over the plant.

KEY TO SPECIES

1. Plant to about 20 μ thick; tetrasporangia over 15 μ diam. .... C. ensis
1. Plant to 100 μ thick; tetrasporangia less than 15 μ diam. 2
2. Erect filaments 5–8 cells long .... C. hyperborea
2. Erect filaments 20–30 cells long .... C. gracilis

Cruoriopsis ensis Jao Pl. 29, figs. 3–6

Plants forming an irregularly expanded rose-pink crust, to 3 mm. diam., 13–19 μ thick; radiating basal filaments mostly dichotomously branched, closely appressed to each other, the cells (3.2–) 6–11 μ diam., 6–16 μ long; during the fruiting stages there are formed erect filaments in groups upon the basal layer, 2–4 cells long, their cells 6.5–9.5 μ diam., 9.5–13 μ long, each cell with several small parietal disk-shaped chromatophores; tetrasporangia produced among the erect filaments, on one-celled stalks, ovate, 16–25 μ diam., 25–32 μ long, irregularly tetrapartite.

Southern Massachusetts; encrusting shells of dead *Ensis directus*, dredged at a depth of 7–8 meters; summer.

REFERENCE: Jao 1936, p. 247.

Cruoriopsis hyperborea Rosenvinge

Plants crustose, 1–2 cm. diam., intensely red, 80–100 μ thick, the basal layer of radiating filaments, remaining as one layer, the cells 4.5–5.5 μ broad, 8.0–10.5 μ high, 2–3 times as long as broad; erect filaments arising from the basal layer, consisting of 5–8 cells, reaching 7.5–11.0 μ diam., the cells about twice as long as broad, with a single chromatophore; sporangia tetrapartite, terminal on the erect filaments or lateral near the top, usually sessile, obovate to short-oblong, 15–23 μ long, 11–14 μ diam.

Ellesmere Island; growing on stones, barnacles, and Lithothamnium.

REFERENCE: Rosenvinge 1926, p. 35.

Cruoriopsis gracilis (Kuckuck) Batters

Plants crustose, the red crusts to 1.5 cm. diam.; basal layer 1–2 cells thick, bearing erect filaments which are 3.0–5.5 μ diam., or to 7.0 μ at the base, and are 20–30 cells long, the cells shorter than broad below, but to one-third longer above and terminally often rounded-conical; oval sporangia usually sessile, lateral, 15–22 μ long, 6–14 μ diam.

Maine; fruiting in the summer.

REFERENCES: Phyc. Bor.-Am. 1650. Collins 1911a, p. 281 (as *Plagiospora gracilis*).

Peyssonnelia Decaisne, 1841

Plants crustose, little to considerably calcified, attached directly or by more or less scattered rhizoids, the margin of the plant often somewhat free; basal layer of radiating, usually dichotomously branched filaments parallel to the substratum, upper layers formed of filaments borne upon these, simple or forked once or twice, becoming erect and closely laterally united; reproductive organs in nemathecia of unconsolidated filaments or paraphyses on the surface of the thallus; tetrapartite sporangia associated with paraphyses; plants often monoecious; spermatangia lateral on paraphysal filaments; carpogenic branches lateral on the bases of paraphyses, four- or five-celled; after fertilization fusion occurs with the second cell of the carpogenic branch as a nutritive cell from which oöblast filaments arise; auxiliary axes also on the bases of the paraphyses, the auxiliary cell usually the second cell of the axis; after fusion with oöblast filaments the auxiliaries give rise to scattered cystocarps consisting of a few large carposporangia.

Peyssonnelia rosenvingii Schmitz

Plants forming brownish-purple, thin, rounded crusts, the margins somewhat undulate, strongly adherent to the substratum below by abundant rhizoids; becoming 0.5 mm. thick or more, and cracking when dried, on the lower face encrusted with lime except about the narrow margin, the upper surface minutely radially striate; the lower layer monostromatic, the cells 3–4 diameters long; ascending filaments sparingly dichotomously branched, more slender above than below, 20–30 (–38) μ diam. below, the cells longer than broad, but above the cells shorter; cystocarpic nemathecia hardly elevated above the general surface.

Northern Massachusetts and Maine; forming crusts attached to lithothamnia, shells, etc.

REFERENCES: Farlow 1881, p. 115 (as *P. dubyi*); Rosenvinge 1893, p. 782.

Hildenbrandia Nardo, 1834

Plants crustose, widely spreading, the lower side strongly adherent, composed of a horizontal filamentous layer from which arise erect cell rows to form the upper layer of the plant, the whole strongly united to a firm though thin uncalcified crust; tetrasporangia irregular in division, borne in sunken conceptacles which discharge by a wide pore.

Hildenbrandia prototypus Nardo Pl. 36, figs. 9–10

Plants forming a thin orange-red to brownish-red, ultimately dark purplish-red crust, widely expanded, 0.2–0.5 mm. thick; composed of a basal layer of cells in rows, supporting vertical, crowded, infrequently branched rows of cells 4.0–6.5 μ diam.; conceptacles scattered over the whole surface of the plant, about 100 μ diam., with a relatively large opening; tetrasporangia attached to the bottom and walls of the conceptacle, 16–30 μ long, 9–14 μ diam., irregularly divided into 4 cells, usually obliquely; paraphyses absent, but old sporangial walls persisting and evident in the conceptacles.

Florida, New Jersey to Nova Scotia, and Baffin Island; growing on rocks and shells between the tide limits and extending a little lower; fruiting in summer.

REFERENCES: Phyc. Bor.-Am. LXXV. Farlow 1881, p. 116 (as *H. rosea*); Whelden 1947, p. 118.

CORALLINACEAE

Plants with a thin basal layer which may constitute the whole thallus, or which may develop into a massive calcareous crust or a system of rigid branches, or from which may arise erect, slender-branched, jointed axes, throughout showing the multiaxial type of structure with chromatophores in the peripheral cells, and the whole calcified except where there are intervening flexible joints; reproductive organs in conceptacles with a definite perithecium-like wall, sunken in the crust, or terminal on lateral enlarged branches; tetrasporangia transversely divided (zonate), often associated with sterile paraphysal filaments; spermatangia on short filaments crowded in conceptacles; carpogenic branches usually three-celled, in the central part of the cystocarpic conceptacle; after fertilization union of the carpogonium with the lower cell of the carpogenic branch occurs and then by oöblast filaments with auxiliary cells at the bottoms of other branches, after which general fusions occur, so that the carposporangia ultimately arise marginally from a large fusion cell.

KEY TO GENERA

1. Plant bushy; branches with flexible uncalcified joints
   CORALLINA, p. 253
1. Plant crustose or with unarticulated branches.................. 2

2. Tetrasporangia in sori with more than one pore................. 3
2. Tetrasporangia in individual conceptacles, each discharging through a single pore.................................... 5

3. Thallus thin, of 1–5 cell layers.................. MELOBESIA, p. 249
3. Thallus thicker, often massive................................ 4

4. Conceptacles superficial or subimmersed, the top flat or convex
   LITHOTHAMNIUM, p. 244
4. Conceptacles immersed, the top concave.... PHYMATOLITHON, p. 242

5. Vegetative thallus of two zones; conceptacles clearly partly immersed; tetrasporangia surrounding paraphyses
   LITHOPHYLLUM, p. 250
5. Vegetative thallus one cell thick except about the conceptacles; conceptacles superficial or very slightly immersed, with no paraphyses at maturity...................... FOSLIELLA, p. 252

Phymatolithon Foslie, 1898

Plants forming substantial crusts; a basal layer (hypothallus) sometimes distinct from an upper one (perithallus) of erect cell rows; the sporangial conceptacles crowded, immersed, the roof depressed, traversed by numerous pores, the sporangia bi- or tetrasporic; cystocarpic conceptacles at first immersed, later elevated and cup-shaped, discharging through a central pore and eventually decorticated.

KEY TO SPECIES

1. Pores generally less than 30................................ 2
1. Pores generally more numerous............................... 3

2. Conceptacles of sporangia usually less than 200 μ diam., ultimately overgrown and becoming immersed... P. compactum, p. 243
2. Conceptacles usually larger, not becoming overgrown
   P. evanescens, p. 243

3. Sporangia bisporic; surface of plant smooth to somewhat glossy, the crust less than 1 mm. thick............ P. laevigatum, p. 244
3. Sporangia tetrasporic; the surface of the plant uneven, the crust somewhat thicker .... P. polymorphum, p. 243

Phymatolithon evanescens (Foslie) Foslie

Plants forming irregular crusts, moderately closely adherent to the substratum but ultimately easily removed, rather thicker at the center than at the margins, the surface somewhat uneven, glaucescent, the margin slightly crenate; cells of the vegetative tissue in the upper layers quadrate to generally rectangular, 8–10 μ by 6–8 μ; sporangial sori with a somewhat depressed roof, the margin distinctly angled, 200–250 μ diam., perforated by 15–20 large, crowded, angular pores; sporangia bisporic, 130–200 μ long, 40–55 μ diam.

Northern Massachusetts to Maine; growing on small stones.

REFERENCE: Foslie 1929, pl. 41 (as *Clathromorphum evanescens*).

Phymatolithon compactum (Kjellman) Foslie

Plants forming crusts 2–5 mm. thick, pale wine-purple, yellowish, or whitish, generally circular, in the young sterile state smooth with radiating striae concentrically disposed, at first strongly adherent, ultimately by superposition of additional growths forming deposits 2.0–2.5 cm. thick and becoming separated from the substratum; the hypothallus feebly developed, usually of but one layer, the cells about 6–8 μ long, 4 μ diam.; the perithallus of cells 6–8 (–11?) μ long, 4 (6–10?) μ diam.; sporangial sori immersed, globose to subglobose, 150–200 (–350) μ diam., the roof over the sorus depressed, perforated by 10–20 (–30) moderately large pores; the sorus eventually becoming overgrown; bisporic sporangia 120–160 μ long, 50–80 μ diam.

Northern Massachusetts to Maine, Nova Scotia, Newfoundland, southeastern Hudson Bay, and Ellesmere Island; growing upon stones to 40 meters depth.

REFERENCES: Phyc. Bor.-Am. LXII (as *Lithothamnium circumscriptum*), LXIV. Kjellman 1883, p. 101; Foslie 1929, pl. 41 (as *Clathromorphum compactum*); Rosenvinge 1926, p. 37; Howe 1927, p. 26B.

Phymatolithon polymorphum (Linnaeus) Foslie

Plants forming heavy rocklike crusts, of considerable breadth, layers 2–6 mm. thick; variously and vaguely lobed and elevated; hypothallus of dichotomous rows of cells in several layers (the cells about 12–22 μ long, 7–11 μ diam.), from which widely curving rows run upward to give rise to the perithallus, with the cells about 6–8 μ diam., 7–12 μ long, and other lesser series go downward toward the substratum; sporangial conceptacles immersed, concave, frequently with a distinct border, 150–300 μ diam., the roof depressed, with 30–70 pores; tetrasporangia 80–110 μ long or longer, 25–45 μ diam.; cystocarpic conceptacles 250–400 μ diam., densely scattered, when young sometimes convex; spermatangial conceptacles 100–120 μ diam.

Rhode Island, New Brunswick, also the arctic to about 75° N. L. at Beechey Island and other stations; in depths of 2–9 meters.

REFERENCES: Phyc. Bor.-Am. LXIX (as *Lithothamnium polymorphum* f. *papillatum*). Farlow 1881, p. 182; Kjellman 1883, p. 103; Foslie 1929, pl. 39.

Phymatolithon laevigatum (Foslie) Foslie

Plants forming strongly adherent crusts 0.3–0.8 mm. thick, the surface very smooth, almost glossy, violaceous to purplish or yellowish; margin crenate-lobed and zonate; hypothallus not strongly developed, cells 4.5–7.0 μ diam., 13–21 μ long; cells of the perithallus about 5–9 μ long by 4–7 μ diam.; sporangial conceptacles in the marginal parts of the plant, somewhat crowded, depressed, the margin hardly elevated, the diameter 150–200 μ, the roof penetrated by 40–55 pores; sporangia bisporic, 120–150 μ long, 35–60 μ diam.

Rhode Island, Maine, Nova Scotia, Newfoundland, and southeastern Hudson Bay; growing on rocks and stones in the lower littoral and down to about 18 meters, in somewhat more protected situations than *P. polymorphum*.

REFERENCES: Foslie 1929, pl. 39; Howe 1927, p. 26B.

Lithothamnium Philippi, 1837

Plants forming firm calcareous crusts, or branching from a crust-like base; of two cell layers, the basal spreading, the upper of erect cell rows with the cells in transverse superposed zones; sporangial conceptacles soriform, superficial, or somewhat immersed, at first isolated, later the septa of adjoining units in part disappearing leaving a common cavity with the roof traversed by several pores; sporangia bisporic or tetrasporic, sometimes both types intermixed;

cystocarpic conceptacles superficial or slightly immersed, conical or subconical, the fugacious tip short and thin, with an apical pore; carposporangia arising from the peripheral portion of the fusion cell, the central parts of which for a time support withered remnants of the trichogynes.

KEY TO SPECIES[1]

1. Thallus crustose, without definite upgrowths 2
1. Thallus crustose or free, when adult with more or less of erect simple or branched outgrowths 4

2. Crust smooth or almost smooth; sporangial conceptacles large..... 3
2. Crust frequently irregularly squamulose or margins imbricate; the conceptacles of sporangia smaller, 200–350 μ diam., with 25–45 pores .... L. lenormandi, p. 245

3. Conceptacles clearly depressed and margined by an elevated border, with 40–60 pores .... L. foecundum, p. 246
3. Conceptacles superficial or a little raised, with 80–120 pores L. laeve, p. 246

4. Thallus at first attached, eventually becoming free.............. 5
4. Thallus free, only exceptionally growing attached................ 6

5. Branches irregularly or dichotomously forked.. L. colliculosum, p. 247
5. Branches little forked.................... L. glaciale, p. 247

6. Branching abundant, slender, often crowded... L. norvegicum, p. 248
6. Branching sparse to crowded, divisions more coarse.............. 7

7. Branches coarse, somewhat crowded and coalescing below, freer above; sporangia bisporic, 40–80 μ diam.; 50–80 pores in the roofs of the conceptacles.................... L. tophiforme, p. 249
7. Branches crowded and frequently coalescing throughout; sporangia tetrasporic, 35–50 μ diam.; with about 50 pores in the roofs of the conceptacles.................... L. ungeri, p. 249

Lithothamnium lenormandi (Areschoug) Foslie

Plants forming strongly adherent crusts 2 cm. or much more in diam., 100–600 μ thick, dull purplish rose to chalky in color, the margin of rounded lobes, the surface zonate; adjacent plants confluent; surface at first smooth but dull, later squamulose; the hypo-

[1] The type of *Lithothamnium labradorense* Heydrich is considered by Foslie (1905, p. 31) to have had a tropical origin, and in any case to have been in an indeterminable condition.

thallus of usually 7 or 8 close layers of cells, these 15–22 (–30) μ long, 3–6 μ diam., the perithallus of rather smaller cells, 6–13 μ long, 4–6 μ diam.; sporangial conceptacles rather crowded, depressed-hemispherical or disk-shaped, sometimes somewhat sunken, 200–350 μ diam., the roof penetrated by 25–45 pores, ultimately decorticate, leaving a large cavity, and often overgrown; sporangia occasionally bisporic, generally tetrasporic, 60–110 μ long, 20–48 μ diam.; cystocarpic conceptacles 320–350 μ diam., hemispherical to subconical; spermatangial conceptacles similar, 150–350 μ diam.

New Jersey, Connecticut, southern Massachusetts, New Hampshire, Maine, Nova Scotia, the lower St. Lawrence, and perhaps to the arctic; from the tide pools and the lower sublittoral to 14–30 meters, in either sheltered or exposed situations.

REFERENCES: Phyc. Bor.-Am. LXVII. Farlow 1881, p. 181 (as *Melobesia lenormandi*); Kjellman 1883, p. 103; Foslie 1929, p. 51, pl. 3.

Lithothamnium foecundum Kjellman

Plants crustose, the young individuals at first strongly adherent, later free, singly about 2 mm. thick, but often superposed to form rather thick crusts; surface smooth and somewhat glossy when young, pale pinkish, margins undulate-lobate; sporangial conceptacles densely crowded over the whole surface except a marginal band, 300–600 μ diam., immersed below, the roof perforated by 40–60 pores, later disappearing and the cavity bordered by a perceptible ridge, eventually overgrown and buried in the thallus; the conceptacles infrequent, convex, subhemispherical or conical, 300–450 μ diam.; spermatangial conceptacles 200–250 μ diam.; plants often with conceptacles of diverse types on the same plant.

Maine, Devon Island, and Ellesmere Island; in water to 100 meters depth.

REFERENCES: Kjellman 1883, p. 99; Foslie 1929, pl. 2; Rosenvinge 1926, p. 37.

Lithothamnium laeve (Strömfelt) Foslie

Plants forming firmly attached thin crusts, roundish, the surface remaining smooth, scarcely zonate, or when thick indistinctly stratified; hypothallus of few layers of cells, the lowermost cells 15–35 μ long, 6–11 μ diam.; upper thallus portion of squarish cells about 7 μ diam., or vertically elongate, up to 12 μ long, 11 μ diam.; sporangial conceptacles depressed-hemispherical or disk-shaped, superficial or nearly so, 0.4–1.0 mm. diam., often crowded, the roof perforated by 80–120 pores; sporangia bisporic, 126–129 μ long, 67–72 μ diam., or more generally tetrasporic, 200–600 μ long, 100–200 μ diam.; cystocarpic conceptacles conical, 600–800 μ diam.; spermatangial conceptacles conical, about 350 μ diam.

Connecticut, northern Massachusetts to Maine, Newfoundland, southeastern Hudson Bay, Devon Island, and Ellesmere Island; from the lower littoral, but especially from deeper water to 35 meters; reproducing from late spring to early autumn.

REFERENCES: Phyc. Bor.-Am. LXVI. Foslie 1929, pl. 3; Collins 1905, p. 235; Rosenvinge 1926, p. 36; Howe 1927, p. 26B.

Lithothamnium colliculosum Foslie

Plants at first crustaceous, strongly adherent, somewhat rose-colored, 0.5–1.5 mm. thick, the crusts solitary or crowded, when the edges become pushed up by the contact; eventually producing wart-like upgrowths which extend to short, subcylindrical branches about 4 mm. in length; with age the basal crust disappearing, the outgrowths repeatedly branching, the upper portion increasing in extent and becoming free, so that masses 12 cm. diam. and 5 cm. thick may result; cells of the vegetative thallus quadrate or subrectangular, 8–12 μ by 5–8 μ diam.; sporangial conceptacles not prominent, convex, often crowded in the upper part of the branches, 300–400 (250–500) μ diam.; the sporangia bisporic, 140–220 μ long, 60–100 μ diam.; cystocarpic conceptacles 300–600 μ diam.; spermatangial conceptacles 200–300 μ diam.

Northern Massachusetts to Maine, with some doubt.

REFERENCES: Phyc. Bor.-Am. LXIII. Foslie 1929, pl. 21.

Lithothamnium glaciale Kjellman

Plants originating as a crust 2–3 mm. thick, frequently surrounding the supporting object, adherent to the substratum, the surface somewhat pinkish or reddish, the margin with rounded lobes and with submarginal ridges; developing conical to subcylindrical branches which remain short and distinct, 7–8 mm. long, 5 mm. diam. near the base; sporangial conceptacles on the basal layer and especially on the branches, 250–350 (–500) μ diam., obscure to slightly projecting, the roof with 30–60 (–70) crowded, angular pores 7–10 μ diam., ultimately often disintegrating, the cavities often also overgrown; sporangia bisporic, 80–180 μ long, 20–80 μ diam.; cystocarpic conceptacles about as large as those of sporangia, conical; spermatangial(?) conceptacles probably about 200 μ diam.

Massachusetts to Maine, Nova Scotia, Anticosti Island, Newfoundland, Labrador, southeastern Hudson Bay, and Ellesmere Island; from the sublittoral down to 55 meters on open coasts; conceptacles produced from spring to early winter.

REFERENCES: Phyc. Bor.-Am. LXV. Farlow 1881, p. 182 (as *L. fasciculatum*, probably); Kjellman 1883, p. 93; Foslie 1929, pls. 23, 24; Rosenvinge 1926, p. 37; Howe 1927, p. 26B.

Lithothamnium norvegicum (Areschoug) Kjellman

Plants bushy, early becoming free on the bottom; branches cylindrical, a little tapered at the apices, except below hardly more than 2 mm. diam., irregularly disposed and forking, habit crowded to fasciculate, uneven, often coalescent, 2–3 cm. long; cells of the upper portions 8–12 μ long, 5 μ diam.; sporangial conceptacles 300–400 μ diam., on somewhat thickened branch apices, the roof perforated by about 50 canals, ultimately dissolving; the sporangia tetrasporic, 90–130 (–140) μ long, 25–40 (–55) μ diam.; cystocarpic conceptacles conical, 350–400 μ diam., with a single canal which is about 20 μ diam.

Maine and perhaps to the arctic; in sheltered places, at times forming considerable accumulations at a depth of 2.5–20.0 meters; generally sterile, but found reproducing during winter, spring, and summer.

REFERENCES: Phyc. Bor.-Am. LXVIII. Kjellman 1883, p. 93.

F. **pusilla** Foslie (1929, pl. 26), with a diameter of 1–3 cm., the branches partly crowded, partly sparingly divided and more or less diverging, is considered the most representative form of the species and has been reported from Maine.

Lithothamnium tophiforme Unger

Plants soon detached, spherical to subspherical, reaching 8 cm. diam., rose-purple; branches radiating from a solid center, dichotomously subdivided, free or coalesced in older plants, cylindrical or compressed, smooth, elongate or curved, with rounded apices; cells of the upper part of the branches subquadrangular, about 5–8 μ diam., 11–17 μ long; sporangial sori usually crowded, somewhat convex, superficial, 300–500 μ diam., the roof with 50–80 pores; sporangia bisporic, 90–160 μ long, (20–) 40–80 μ diam.; cystocarpic conceptacles about 400–600 μ diam., somewhat more elevated-conical; spermatangial conceptacles 200–300 μ diam.

Maine and Nova Scotia; forming small banks in moderately deep water.

REFERENCE: Foslie 1929, pl. 20.

Lithothamnium ungeri Kjellman

Plants soon free, forming rounded masses with the branches radiating from the center, reaching about 2 dm. diam. and 1 dm. thick; branches erect-spreading, short, often curved, little-tapered, the ends rounded or somewhat enlarged, approaching a fastigiate habit, 1.5–2.0 mm. diam., often with lateral projections or side branches; the whole finally crowded and somewhat coalescent; sporangial conceptacles immersed in the roughened upper part of the branches, frequently crowded, convex to somewhat prominent, 300–500 μ diam., the sporangia tetrasporic, 100–150 μ long, 35–50 μ diam., cystocarpic conceptacles conical, about as large as the sporangial form.

Maine and the arctic.

REFERENCES: Phyc. Bor.-Am. LXX. Kjellman 1883, p. 91; Foslie 1929, pl. 27.

Melobesia Lamouroux, 1812

Plants crustose, completely attached to the support, composed of but a single layer of cells in the vegetative parts, each cell cutting off obliquely a small superficial associated cell, but in the neighborhood of the conceptacles of four or five layers of cells; tetrasporangia in conceptacles with several pores; spermatangia lateral on two-celled filaments distributed over the bottom of the conceptacle cavities; cystocarpic conceptacles showing groups of erect filaments arising from the basal layer, those in the center developing into three-celled carpogenic branches, those next outside developing into incomplete sterile branches, and the peripheral ones contributing to the wall of the conceptacle; after fertilization the carpospores are produced marginally from the central fusion cell in each conceptacle.

Melobesia membranacea (Esper) Lamouroux

Plants at first a thin membrane, subrugose, later becoming calcareous, sometimes overlapping, reddish to purple or white; thin, of one layer of cells except about the conceptacles, where 4–5 layers exist; cells of the monostromatic part of the thallus when seen from above 9–20 μ long by 5–9 μ broad, and in section 5–18 μ deep; conceptacles generally hemispherical; tetrasporangial conceptacles about 110–140 (–200) μ (or by Foslie, 160–350 μ) diam., scattered superficially over the crust, sometimes confluent, wartlike, with 8–27 distinct pores; the apertures of the sexual conceptacles contracted by filiform cells projecting from the surrounding walls.

Florida, Long Island to northern Massachusetts; growing as an epiphyte on various plants; appearing from March onward.

REFERENCES: Phyc. Bor.-Am. LIX. Foslie 1929, p. 49 (as *Lithothamnium membranaceum*).

Lithophyllum Philippi, 1837

Plants forming thin but opaque, bistratose, calcified crusts; conceptacles partly immersed, hemispheric-conical, with an apical pore; the tetrasporangia with a short stalk arising from the basal tissue peripherally between the persistent paraphyses; the carpospores produced around the periphery of the fusion cell, the central parts of the latter supporting remnants of the trichogynes.

KEY TO SPECIES

1. Sporangia dividing to 2 cells only; thallus somewhat thicker, of 8 cell layers or more 2
1. Sporangia dividing into 4 cells; thallus of 2–3 cell layers, rarely to 8 layers, and 350 μ thick .... L. pustulatum

2. Growing on Corallina; sporangia to 88 μ long, 32 μ diam. L. corallinae
2. Growing on Chondrus and fuci; sporangia to 140 μ long, 60 μ diam. .... L. macrocarpum

Lithophyllum corallinae (Crouan) Heydrich

Plants forming round, gray-lilac to rufescent crusts, oval or irregularly rounded, 1–5 mm. diam., convex, the entire lower face fastened to the support, or the margin sometimes free; thickness 80–400 μ; the sporangial conceptacles deeply immersed, outwardly wartlike, 200 (–350) μ diam.; sporangia bisporic, 50–88 μ long, 18–32 μ diam.; cystocarpic conceptacles of about the same size, the cells about the canal papilliform-elongate; spermatangial conceptacles not projecting, small.

Rhode Island to Maine; growing on plants of Corallina.

REFERENCES: Phyc. Bor.-Am. LVIII (as *Melobesia corallinae*). Collins 1888b, p. 87, 1901a, p. 134 (as *M. corallinae*).

Lithophyllum pustulatum (Lamouroux) Foslie

Plants at first flat, suborbicular, completely attached by the under face, becoming thick-convex pulvinate, or ultimately often confluent, and superimposed; 2–10 mm. diam., 2–3 cells thick, or rarely to 8 cells, and 350 μ thick; conceptacles conspicuous, scattered over the thallus, 300–500 μ diam., the tetrasporangia 80–130 μ long, 30–70 μ diam.

North Carolina, New Jersey to New Hampshire; growing on coarse algae, particularly fuci, but sometimes on Zostera; from midwinter, fruiting in the latter part of the summer.

REFERENCES: Phyc. Bor.-Am. 300 (as *Melobesia pustulata*). Farlow 1881, p. 181 (as *M. pustulata*).

Lithophyllum macrocarpum (Rosanoff) Foslie

Plants at first flat, suborbicular, completely attached by the lower face, becoming thick-convex, 4.0–7.5 mm. diam., the disks sometimes overlapping and confluent, the edges even free; at the margin of one primary basal layer and a small-celled cortical layer, together 25–42 μ thick, but nearer the center of 2–3 primary layers and the thickness 100 (–600) μ, cells of the basal layer vertically elongated, twice as long as broad; conceptacles of cystocarps and sporangia large, prominent, hemispheric-conical, 300–600 μ diam., the sporangia bisporic, 60–140 μ long, 30–60 μ diam.

Rhode Island and Massachusetts; epiphytic on various algae, particularly Ascophyllum and Chondrus.

REFERENCE: Farlow 1881, p. 181.

Fosliella Howe, 1920

Plants forming thin, rather lightly calcified crusts, the lower face adherent to the substratum; structurally showing one to few cell layers, the basal consisting of a radially disposed closely united filament system; conceptacles of sporangia superficial or slightly immersed, conical or hemispheric-conical, with a single apical pore; tetrasporangia with a short foot, arising from the lining tissue of the conceptacle, and associated with evanescent trabecular filaments; cystocarpic conceptacles smaller than those of sporangia but otherwise similar to them.

KEY TO SPECIES

1. Colorless swollen cells at intervals in the vegetative thallus. F. farinosa
1. No such cells present.................................. F. lejolisii

Fosliella farinosa (Lamouroux) Howe

Plants forming whitish fragile crusts on the support, usually to 2–5 mm. diam. or more, often crowded; vegetative cells in surface view variable in size, sometimes about 12–20 μ long by 7–10 μ broad, sometimes 18–30 μ long by 15–18 μ broad, often bearing superficial cells cut off on the upper side near the distal ends of the cells of the primary layer; the colorless swollen cells ("heterocysts") which are distinctive of this species each terminating a cell row, 22–40 μ long, 12–30 μ broad, commonly prolonged to a hair on the upper surface; sporangial and cystocarpic conceptacles 140–250 μ diam., those of spermatangia 60–80 μ; tetrasporangia (40–) 50–90 μ long by (20–) 30–50 μ diam.

From the tropics, Florida, North Carolina, New Jersey, Rhode Island to northern Massachusetts; growing on the leaves of marine plants, particularly Zostera, in relatively shallow water; fruiting during the summer.

REFERENCES: Phyc. Bor.-Am. 200. Farlow 1881, p. 180 (both as *Melobesia farinosa*).

Fosliella lejolisii (Rosanoff) Howe Pl. 36, figs. 6–8

Plants in the form of delicate dull pink or white crusts often closely crowded on the host, about 0.5–2.0 mm. diam., 15–30 μ thick, the cells as seen from the surface 6–10 μ long, 6–7 μ wide, or larger if bearing a fork and resembling heterocysts; in section the disk near the margin one cell thick, or with small, superficial cells cut off on the upper side near the distal end of each cell of the primary layer, but toward the central part each cell of the filament is seen to be divided in the plane of the substratum so that the disk is 2–4 cell layers in thickness, of which the lowest is of cells 2–3 μ high, the second layer of cells 17–20 μ high, and the superficial layers like that at the base; sporangial and cystocarpic conceptacles convex or subhemispherical, frequently crowded, or almost confluent, 150–250 (–300) μ diam., the apical pore apparently without filiform cells; tetrasporangia 50–80 μ long, 30–50 μ diam.; conceptacles of spermatangia 75–100 μ diam.

Florida, New Jersey to Nova Scotia, Prince Edward Island, and perhaps to the arctic; particularly common on Zostera leaves, but also growing on other plants; from midwinter onward, fruiting during the summer.

REFERENCES: Phyc. Bor.-Am. 149. Farlow 1881, p. 180; Kjellman 1883, p. 105 (all as *Melobesia lejolisii*).

Corallina Linnaeus, 1758

Plants with calcified, often confluent crustose basal disks spreading on the substratum; these bases giving rise to an indefinite number of erect axes which are terete to compressed, generally branching in a plane, the branching usually oppositely pinnate; articulated, the joints cylindrical to flattened, the short, flexible nodes ecorticate, with only thick-walled longitudinal cells; conceptacles formed by the conversion of lateral or terminal pinnules, or occasionally lateral, or occupying branchlets of a single segment lateral to intercalary segments, naked or bearing hornlike projections, the apex with a pore.

Corallina officinalis Linnaeus Pl. 36, figs. 1–5

Plants tufted, the initial stages pinnately much-branched, the main branches in turn to tripinnate, the ultimate pinnae to some extent with cylindrical pinnules; segments in general more cylindrical in the lower parts of the plant, more flattened in the distal parts, to flat-cuneate; conceptacles ovate-subspherical, without horns, except in the spermatangial conceptacles.

Bermuda, Long Island to the lower St. Lawrence, and Newfoundland; growing in tide pools and on rock faces from the lower intertidal zone to fairly deep water; through the year.

REFERENCES: Phyc. Bor.-Am. 349. Harvey 1853, p. 83; Farlow 1881, p. 179; Hariot 1889, p. 196.

This species varies very much under different ecological conditions, in particular producing dwarf, matted forms in the rough water of the intertidal zone. It is doubtful to what extent these forms are sufficiently discontinuous to make it advantageous to give them segregated descriptions and names. V. **profunda** Farlow (Phyc. Bor.-Am. 1550; Farlow 1881, p. 179) has slenderly elongate branches, sparingly divided, nearly cylindrical throughout, the terminal conceptacles often with horns; it is reported from deep water off the northern Massachusetts and Maine coasts. V. **spathulifera** Kützing (Phyc. Bor.-Am. 350) is a low form about 4 cm. tall with short, crowded blades, the upper branches much flattened, often laterally united into broad fan-shaped expansions; this is found in dense masses in water shallow at low tide, particularly in northern Massachusetts and Maine.

GLOIOSIPHONIACEAE

Plants soft and bushy, structurally composed of a central filament with compact whorls of assimilatory filaments; sporangia tetrapartite; carpogenic branches of three cells, carried on a supporting cell common to the associated auxiliary axis of seven or eight cells, the fifth being the functional auxiliary; cystocarps attached to the bases of the assimilatory filaments, without a pericarp.

Gloiosiphonia Carmichael, 1833

Plants bushy, with simple or sparingly branched main axes which bear many repeatedly divided lateral branches; structurally the axial filament may become invested by rhizoidal downgrowths from the cortical assimilatory whorls, and eventually disappear as the plant becomes hollow; sporangia tetrapartite, borne on the outer forks of the assimilators; monoecious, the spermatangia borne on the assimilators at the surface; procarps present, each carpogenic branch of three cells, the subcarpogonial enlarged, borne upon a supporting cell attached to the base of a cortical assimilator; branched auxiliary axes attached to the same support, seven or eight cells long, the fifth functioning as the auxiliary; after fertilization forming cystocarps without pericarps near the bases of the assimilatory filaments, all the cells of the gonimoblasts forming carposporangia.

Gloiosiphonia capillaris (Hudson) Carmichael Pl. 37, figs. 3–4

Plants bushy, dark red, nearly gelatinous, usually with several pale axes from a common base, these 5–30 cm. tall, 1–5 mm. diam., usually rather naked below, irregularly cylindrical or tapering to both ends, axis and main branches ultimately hollow; habit virgate, repeatedly laterally divided, lateral branches irregularly disposed, red, divaricate, to 2–6 cm., the ultimate divisions about 0.1 mm. diam., 1–3 mm. long, tapering to both ends; surface cells of the branchlets (end cells of assimilatory filaments) 8–13 μ diam.; chromatophores 1 (–3), platelike, chiefly on the outer faces of the cells of the assimilatory filaments; delicate colorless hairs numerous; tetrasporangia infrequent; cystocarps commonly found.

Connecticut to Maine, Nova Scotia, the lower St. Lawrence, and Newfoundland; growing on stones and shells in the littoral just below low-tide levels, though sometimes in tide pools; an annual, fruiting in the late spring and then disappearing.

REFERENCES: Phyc. Bor.-Am. 849, 1700. Harvey 1853, p. 202; Farlow 1881, p. 141.

KALLYMENIACEAE[1]

Plants foliaceous or branched, erect and rather soft; when mature with a filamentous or subparenchymatous medulla and thin, small-celled, assimilative cortex; sporangia scattered or in the branch tips, irregular to tetrapartite; procarps consisting of a supporting auxiliary cell which bears a three-celled carpogenic branch and a sterile cell, or a number of carpogenic branches; gonimoblasts formed from the auxiliary cells or the basal cells of the carpogenic branches which then have wide connections with the auxiliaries; cystocarps invested and penetrated by a nutritive tissue of filaments developed from the cells near the procarp.

KEY TO GENERA

1. Plant freely and narrowly branched more or less in a plane
   EUTHORA, p. 256
1. Plant simple or broadly lobed................. KALLYMENIA, p. 257

Euthora J. Agardh, 1847

Plants of a bushy habit, the branches flat, subpinnate, repeatedly divided, broadest somewhat above the base, tapering to the slender and short apical segments; at maturity showing internally large, roundish or oblong cells with slender branching filaments intermixed, the cortex of smaller cells, usually in one or two layers; sporangia immersed in the thickened tips of branches, irregularly divided, principally rather obliquely tetrapartite; cystocarps marginal, swollen; the procarps developed on outer cells of the cortex, originating as two-celled axes, in each the lower forming on its inner side a carpogenic branch usually of three cells and functioning as a combined supporting and auxiliary cell, the whole becoming overgrown and embedded.

Euthora cristata (Linnaeus ex Turner) J. Agardh Pl. 60, figs. 2–3

Plants bushy, 3–5 (or more) cm. tall, the axes below compressed, dark, above freely subflabellately branched, rose-pink; branches in the central part of the blades to 1–3 mm. diam., the branching re-

[1] *Calophyllis laciniata* Kützing, Delaware (Harvey 1853, p. 171), is a doubtful record.

peatedly subpinnate, or in part subdichotomous, the segments successively smaller, closer, the angles rounded, spreading, terminal segments linear; tetraspores in the thickened tips of the segments; cystocarps small, marginal, sometimes mostly on the broader branches.

New Jersey, Long Island, Rhode Island, and southern Massachusetts, rare and only at exposed stations, becoming a more common form north of Cape Cod, to Nova Scotia, Prince Edward Island, Newfoundland, northern Labrador, Hudson Bay and Strait, Baffin Island, and Ellesmere Island; in the southern part of its range found in deep water, and usually attached to the bases of laminarias and other objects, but in the northern part it may be found in the deeper tide pools; fruiting in the summer.

REFERENCES: Phyc. Bor.-Am. 40, 744. Harvey 1853, p. 150; Farlow 1881, p. 153; Rosenvinge 1926, p. 32; Wilce *in herb.*

With transitions to the ordinary form there is reported v. **angustata** Kjellman (approached by Phyc. Bor.-Am. 744b), which has a much narrower frond, becoming 15 cm. tall, the segments more distant, not crowded even at the tips, ultimately almost capillary; it has been found at various stations within the range of the type, with extreme examples at the Gaspé.

Kallymenia J. Agardh, 1842

Plants forming expanded blades, without veins, subsimple or broadly lobed; structurally with a medulla of branched, interlaced, and anastomosing filaments, which support externally an inner cortex of large polygonal cells and, beyond these, smaller rounded chromatophore-bearing cells to the surface; sporangia scattered in the cortical layer, tetrapartite; procarps consisting of a large supporting cell which functions as an auxiliary, bearing a number of three-celled carpogenic branches, all of which may become functional; gonimoblast filaments arising from the first cells of the carpogenic branches, which have been in open communication with the auxiliary cell; cystocarps immersed or projecting; carpospores formed in large masses, invested by rhizoidal outgrowths from the neighboring cells.

KEY TO SPECIES

1. Plants small, the blades 4–8 cm. long, and to 1.5 times as broad
K. reniformis
1. Plants larger, the blades to 20–40 cm. wide and about as long
K. schmitzii

Kallymenia reniformis (Turner) J. Agardh

Plants foliaceous, each borne on a short stalk, attached by a disk-like holdfast, the bright red blade broadly ovate to reniform, sometimes cleft into a few obovate or cuneate irregular segments, the margin often proliferating, the length 4–8 cm., the width 1.0–1.5 times as great; cystocarps small, scattered somewhat thickly at the surface of the blade.

Northern Massachusetts; probably growing in deep water, having been found cast ashore; spring.

REFERENCE: Collins 1900b, p. 49.

Kallymenia schmitzii De Toni

Plants in the form of rounded soft-fleshy red to purplish blades; base a disklike holdfast bearing a short stalk, the base of the blade cuneate, becoming cordate, above asymmetrical, undulate-plicate, to 20–40 cm. wide, 130–190 (–280) μ thick, the margin sinuate to broadly crenate, the lobes somewhat overlapping; clavate refractive cells often present among the medullary filaments.

Northern Labrador and Ellesmere Island; growing in 10–25 meters of water, attached to stones.

REFERENCES: Rosenvinge 1926, p. 30; Wilce *in herb.*

CHOREOCOLACACEAE

Plants minute, partly or when colorless completely parasitic; consisting of a penetrating basal portion and an emergent part, usually cushion-shaped or with short, stout branches; reproductive organs in the outer part; sporangia generally tetrapartite; spermatangia produced in series from the surface, or on the lining of conceptacles; carpogenic branches of three or four cells, the supporting cell acting as an auxiliary or cutting one off after fertilization; cystocarps showing a pericarp and an ostiole.

KEY TO GENERA

1. Emergent portion with short, stout branches.... CERATOCOLAX, p. 259
1. Emergent portion cushion-shaped............................... 2

2. Growing on Polysiphonia.................... CHOREOCOLAX, p. 259
2. Growing on Rhodomela....................... HARVEYELLA, p. 260

Ceratocolax Rosenvinge, 1898

Thallus endophytic at the base, the part within the host relatively small, sharply delimited; free above with cylindrical branches forming a hemispherical tuft; branches with a medullary layer of rounded-angular cells, the outer layers smaller, and with a cortical layer of cells in short radiating series; nemathecia globose, set on a branch or at a fork, the sporangia tetrapartite; carpogenic branches three- or four-celled, borne on a large (auxiliary?) cell, the first cell of the branch bearing a lateral sterile cell; cystocarps unreported.

Ceratocolax hartzii Rosenvinge

Plants forming small red branched fleshy outgrowths on the host; branches with a spread of 2.5–5.0 mm., short and densely divided, generally 0.5 mm. thick, or to 1.0 mm. at the base; sporangial nemathecia mostly on swollen ends of the branches, to 0.5 mm. diam., the tetrasporangia 6–7-seriate, developed from the inner cells of the radial cortical rows of the nemathecium, tetrapartite, 9–26 μ long, 8–16 μ diam.; sexual plants with branches much divided, the ends not swollen.

South side of Ellesmere Island; growing on *Phyllophora interrupta* at about 25 meters depth.

REFERENCE: Rosenvinge 1926, p. 29.

Choreocolax Reinsch, 1875

Plants parasitic, forming minute whitish cushions attached by branching rhizoids; structurally composed of radiating cell series, with the external larger; tetrasporangia immersed at the surface of the cortex, tetrapartite; dioecious, the spermatangia congested into thin cushions on the surface layer; carpogenic branches of four cells, an auxiliary forming after fertilization from the supporting cell of the carpogenic branch; cystocarps enveloped in a pericarp and discharging by an apical pore.

Choreocolax polysiphoniae Reinsch [1]

Plants in the form of whitish-brown lumps, hemispherical to globose, usually irregular and somewhat lobed, reaching a maximum diameter of about 4 mm., but usually much smaller; basal endophytic part of cells 3.5–11.0 μ diam., exterior cells 28–34 μ by 2.5–4.5 μ; sporangia generally tetrapartite, rarely tripartite, about 40 (–80) μ long by (15–20–) 28 μ diam.

Connecticut to Maine and Nova Scotia; parasitic upon fronds of *Polysiphonia lanosa;* forming tetraspores (and perhaps sexual organs) during spring and early summer.

REFERENCES: Phyc. Bor.-Am. 286. Richards 1891, p. 46.

Harveyella Reinke et Schmitz, 1889

Plants parasitic, forming somewhat convex lumps on the host attached by penetrating rhizoidal filaments; sporangia in the outer layer of the emergent part, tetrapartite; dioecious, the spermatia forming a thin coating on the surface; carpogenic branches of four cells, an auxiliary formed after fertilization from the supporting cell of the carpogenic branch; cystocarps immersed, occupying most of the outer part of the emergent portion of the plant.

Harveyella mirabilis (Reinsch) Schmitz et Reinke

Parasitic, the basal portion of penetrating filaments, the upper part causing a small lump on the host; composed of compacted filaments, the cells 27–40 μ long by 9–11 μ diam.; tetrasporangia obovoid, 25–45 μ long, 17–20 μ diam., irregularly tetrapartite; carpospores ellipsoid, 17 μ long, 13 μ diam.

Rhode Island to Maine; parasitic on Rhodomela; through the year.

REFERENCE: Phyc. Bor.-Am. 1847.

[1] Reinsch (1875) described in addition *C. rabenhorstii* on Phycodrys, *C. americanus* on Rhodomela, and *C. tumidus* on Ceramium; little is known of them, and with nothing beyond his inadequate characterization no proper disposition of them can be made.

GIGARTINALES

Plants showing various forms from filiform to fleshy-membranous or crustose; corticated, with either the multiaxial or central filament type of structure; asexual reproduction by tetraspores formed in sporangia scattered over the plant just below the surface, or in restricted areas on branches; sexual reproduction by spermatangia borne on surface cells in more or less restricted areas and by carpogenic branches sunken in the cortex; typical auxiliary cells when present more or less remote from the carpogonia, established before fertilization, consisting of enlarged intercalary cells of the cortex filaments; carpogonium after fertilization producing oöblast filaments which transmit the zygote nuclei to the auxiliaries, from which the carpospore-bearing gonimoblasts are produced.

KEY TO FAMILIES

1. Plants encrusting .... CRUORIACEAE, p. 262
1. Plants erect, foliaceous or bushy 2

2. Texture softly fleshy or nearly gelatinous. NEMASTOMATACEAE, p. 264
2. Texture firmer 3

3. Branches essentially terete 4
3. Axes and main branches flat or compressed 7

4. Branch tips often hooked; axes near the tips showing a central filament .... HYPNEACEAE, p. 271
4. Branch tips always erect 5

5. Tetrasporangia zonate; inner medulla of longitudinal filaments.... 6
5. Tetrasporangia tetrapartite; medulla parenchymatous throughout .... GRACILARIACEAE, p. 272

6. Lateral branches terete, similar to the axis; structurally developed from several initial axial filaments. SOLIERIACEAE, p. 266
6. Lateral branches clearly differentiated from the axes; structurally with one initial axial filament (soon obscured) RHODOPHYLLIDACEAE, p. 267

7. Tetrasporangia zonate .... RHODOPHYLLIDACEAE, p. 267
7. Tetrasporangia tetrapartite 8

8. In section, outer cortical cells radially seriate 9
8. Outer cortical cells not radially seriate...... GRACILARIACEAE, p. 272

9. Branches compressed to subfoliar 10
9. Branches terete 11

10. Axes terete below, above usually compressed to flabellar; sporangia in superficial nemathecia.... PHYLLOPHORACEAE, p. 274
10. Axes generally flattened, above flabellar to subfoliar; in our species rarely terete nearly throughout; sporangia in immersed sori .... GIGARTINACEAE, p. 280

11. Branches firm-fleshy, dichotomously divided; the axis in section clearly filamentous............. FURCELLARIACEAE, p. 270
11. Branches very slender and wiry, dichotomous to irregular and often with many adventitious branchlets; axis in section not obviously filamentous .... PHYLLOPHORACEAE, p. 274

CRUORIACEAE

Plants forming expanded crusts with marginal growth, adhering by the under surface, above forming erect, crowded filaments lightly united; zonate or irregularly tetrapartite sporangia known; carpogenic branches short, the oöblast filaments developing from the carpogonium to make vegetative unions with various assimilatory filaments, and ultimately to give rise directly to the gonimoblasts, of which all cells are converted to carposporangia.

KEY TO GENERA

1. Sporangia zonate, lateral on the erect filaments...... CRUORIA, p. 262
1. Sporangia irregularly tetrapartite, intercalary..... PETROCELIS, p. 263

Cruoria Fries, 1835

Plants forming expanded thick and fleshy crusts adherent by the lower face to the substratum, without or with but rudimentary rhizoids; basal layer of cells arranged in flabellate-radiating series, bearing the erect-ascending filaments which are densely set and involved in a soft jelly; sporangia scattered among the erect filaments, lateral, zonate; spermatangia lateral, tufted, on the upper ends of the erect filaments; carpogenic branches two- or three-celled, lateral on the erect filaments; cystocarps small, producing five to ten carpospores.

Cruoria arctica Schmitz

Plants in the form of crusts to 0.7 mm. thick, the margin irregularly rounded to lobate, thin toward the edges and reddish brown when dry, somewhat thicker toward the center and very dark red to blackish brown; strongly adherent or locally free from the substratum, without rhizoids; texture firm-fleshy; basal zone of one layer of cells, or sometimes somewhat distromatic, the cells 6–8 times as long as broad; an erect filament on each cell of the basal layer, once subdichotomously branched below, 10–12 μ diam., the cells as long as broad, or in thick crusts to 3–4 diameters long; between the erect filaments elongate gland cells occur; sporangia are scattered in the thicker crusts, laterally affixed to erect filaments, cylindrical, zonate, 53–80 μ long, 16–27 μ diam.

Ellesmere Island; growing upon barnacle shells.

REFERENCES: Rosenvinge 1893, p. 784, 1926, p. 35.

Petrocelis J. Agardh, 1852

Plants crustose, horizontally expanding on the substratum; attached by a radially filamentous basal layer which does not develop rhizoids and which bears erect filaments loosely associated in a general gelatinous matrix; sporangia intercalary in the erect filaments, or terminal, single or seriate, irregularly tetrapartite; spermatangia borne on the ends of the erect filaments; carpogenic branches two-celled, rarely three- or four-celled, lateral on the erect filaments; nutritive cells formed from somewhat enlarged cells of other erect branches; carpospores formed in small spindle-shaped masses, scattered in the cortical layer.

KEY TO SPECIES

1. Erect filaments of the thallus 7–9 μ diam. . . . . P. polygyna
1. Erect filaments 3–4 μ diam. . . . . P. middendorfii

Petrocelis polygyna (Kjellman) Schmitz

Plants forming dark red crusts, the lower layer monostromatic, the erect filaments 7–9 μ diam., cells 1.5–2.0 diameters long, the outermost cells rounded; cystocarps numerous in fertile crusts, spindle-shaped, nearly 100 μ long and the cells uniseriate at the ends, the carposporangia 9–12 μ diam.

Devon Island.

REFERENCES: Rosenvinge 1926, p. 34; Søren Lund, *in litt.*

Petrocelis middendorfii (Ruprecht) Kjellman

Plants forming rather thick brownish- to purplish-red patches about 1 cm. diam., gelatinous, slippery, marked by concentric lines; tough and strongly adherent to the substratum; basal layer well developed, the erect filaments occasionally branched, laterally coherent for their lower half or more, diameter 3–4 μ, cells of the erect filaments about as long as broad or a little longer in the lower portion, in the upper middle portion 2–4 times as long as broad, and near the cuticle again shorter, enveloped by a very soft jelly except at the surface of the plant, where the tips of the filaments are held together by a rather firm cuticle; tetrasporangia solitary, developed in the middle of the erect filaments, obovoid to nearly spherical, 9.0–9.5 μ diam.

Long Island (?), northern Massachusetts to Maine, reported as north to the Arctic Ocean; on rocks in deep tide pools and below low water on exposed shores; vegetating throughout the year, fruiting in autumn and winter.

REFERENCES: Phyc. Bor.-Am. 899 (not *P. cruenta*), 1548. Farlow 1881, p. 115 (as *P. cruenta non* J. Ag.); Kjellman 1883, p. 140; Collins 1908c, p. 158.

NEMASTOMATACEAE

Foliaceous or branching plants, rather soft, with longitudinally filamentous medulla and compact radially filamentous cortex; sporangia tetrapartite or lacking; spermatangia developed on the outermost cells of the cortex; carpogenic branches of three to seven cells developed on the cortical filaments; auxiliary cells scattered, intercalary, developed in inner segments of corticating filaments; the gonimoblasts completely maturing into carposporangia, the cystocarps without a pericarp.

KEY TO GENERA

1. Thallus branched, compressed to subcylindrical...... PLATOMA, p. 264
1. Thallus foliaceous, more or less lobed........... TURNERELLA, p. 265

Platoma (Schousboe) Schmitz, 1889

Plants subcylindrical to flat, simple or irregularly branched, often marginally proliferating, rather gelatinous, the medulla comparatively thick, the primary interlaced filaments in turn invested by rhizoidal downgrowths; cortex of radiating branched fascicles of filaments forming the surface of the plant, the outermost often beset with small gland cells; sporangia tetrapartite, scattered in the outer cortex; carpogenic branches three-celled, borne in inner segments of the cortical filaments; auxiliary cells on different inner cortical filaments; cystocarps scattered, immersed.

Platoma bairdii (Farlow) Kuckuck Pl. 43, fig. 7

Plants erect from a basal layer, compressed or nearly cylindrical, irregularly and sparingly forking, the upper divisions often tapering, somewhat intestiniform, to about 2 cm. wide and 14 cm. tall; gelatinous, color rose to carmine; basal layer of closely placed vertical filaments, which may bear tetrapartite sporangia; erect plant with a soft, filamentous medulla, the cortical filaments of larger cells within, smaller cask-shaped cells without, containing 1–3 disk- or band-shaped chromatophores and a nucleus; the inner part loose, the outer more compact, often bearing colorless hairs; tetrasporangia borne near the surface, ovoid, 17–20 μ diam., 11–12 μ long, spermatangia unknown; cystocarps large, 65–75 μ diam.

Southern Massachusetts; apparently a plant of the lower littoral, growing on stones; very rare in this country.

REFERENCES: Farlow 1875, p. 372, 1881, p. 142 (as *Nemastoma bairdii*); Taylor 1941, p. 72.

Turnerella Schmitz, 1896

Thallus foliaceous, flat, undivided or irregularly lobed, gelatinous to membranous; subfilamentous in structure, the medulla loose, the cortex compact, often with large gland cells; mature tetrasporangia not reported; monoecious, the spermatangia developed on the surface cells near the growing tips; carpogenic branches of five to seven cells, attached to one of the inner cells of the cortex; auxiliary cells consisting of somewhat enlarged and dense inner cortical cells; cystocarps immersed near the surface, hardly projecting, discharging through a pore but without a definite pericarp.

Turnerella pennyi (Harvey) Schmitz

Plants pale red-brown, to 20 cm. (or more?) in diameter; orbicular, reniform to somewhat elongate, subsimple to slightly lobed, or margin vaguely dentate; thickness 225 μ or somewhat more; cortex fairly thick, the outermost layer of cells about 12–15 μ high, 4–6 μ diam., pyriform refractive gland cells 15–30 (–60) μ diam. projecting between them from below, the subcortical cells 10–26 μ diam., the medulla loose; cystocarps immersed near the surface of the upper part of the blade.

Labrador, eastern Hudson Bay, north and south sides of Somerset Island, Baffin Island, Cornwallis Island, and Ellesmere Island; dredged from 20 to 36 meters on rocks.

REFERENCES: Harvey 1853, p. 172; Kjellman 1883, p. 162; Rosenvinge 1893, p. 815, 1926, p. 30.

SOLIERIACEAE

Plants plane or bushy, subsimple or branched, in structure the medulla clearly filamentous, the cortex obscurely so, appearing subparenchymatous with large cells within; tetrasporangia scattered at the surface, zonate; carpogenic branches of three or four cells borne on the inner cortex; auxiliaries more or less evident, scattered, borne similarly; cystocarp showing a central mass of sterile tissue, often connected by strands with a filamentous sheath, the carpospores discharged through a pore.

Agardhiella Schmitz, 1896

Plants branching, the branches terete; the medulla filamentous, diffuse, the cortical tissue outwardly compact, small-celled, often with delicate unicellular hairs, the inner cells larger, usually becoming multinucleate, developing downgrowing rhizoidal filaments which contribute to the medulla; tetrasporangia zonate, scattered, immersed, developed from the surface cells; spermatangia formed in patches on young branches, each cortical cell carrying three to five spermatangial mother cells, on each of which two or three spermatangia are produced; carpogenic branches lateral on inner cortical cells, three-celled, occasionally four-celled, the trichogyne extending to the surface; auxiliary cells intercalary in the cortical branching series, asymmetrically enlarged; after fertilization the auxiliaries produce in the inner cortex large cystocarps showing a sterile central tissue from the surface of which the carposporangia are produced, the whole invested by a partly filamentous tissue which is connected with the central mass by strands, and discharged through a pore.

Agardhiella tenera (J. Agardh) Schmitz

Pl. 38, fig. 4; pl. 40, fig. 7; pl. 41, fig. 2; pl. 59, fig. 9

Plants bushy, to about 3 dm. tall, from deep rose-red to pinkish, translucent, firm-fleshy, the attachment at first disklike, later fibrous; branches 1–3 (–20) mm. diam., terete throughout, 1–3 times alternately or somewhat unilaterally branched; branchlets with constricted base and elongate-acuminate apex; tetrasporangia long-oval, zonate, often germinating and producing plantlets *in situ;* cystocarps scattered, though evident to the unaided eye yet little projecting.

From the tropics to Florida, North Carolina to New Hampshire, though only in specially favorable stations north of Cape Cod; a plant of quiet shallow water, descending to 2–4 meters, attached to stones and shells; developing in the spring, fruiting in the summer, occasionally persisting into the autumn.

REFERENCES: Algae Exsic. Am. Bor. 143 (as *Rhabdonia tenera*); Phyc. Bor.-Am. 138, 539, 2143. Harvey 1853, p. 121 (as *Solieria chordalis p. p.*); Farlow 1881, p. 159 (as *Rhabdonia tenera*); Osterhout 1896, p. 403; Lewis 1912, p. 239; Hoyt 1920, p. 479; Taylor 1928a, p. 147.

In addition to the coarse, sturdily bushy form with rather sparingly divided flagellate branches, one often finds a decidedly slender plant (Pl. 38, fig. 5). It branches much more freely, the ultimate branches may be almost capillary, and sometimes it is rather notably entangled. It is found in very quiet pools, often when a little brackish, and is dredged from water deeper than that ordinarily populated by this species. This must be sharply observed to distinguish it from forms of Gracilaria in similar situations, but the difference in color, the structure in section, and the much less prominent cystocarps serve readily to characterize it.

RHODOPHYLLIDACEAE

Plants plane or bushy, subsimple or branched; when mature the medulla either diffusely filamentous or apparently parenchymatous, the cortex of larger cells within, smaller without; procarps present, the carpogenic branches of three cells, each with an auxiliary borne on the same supporting cell; cystocarp developed from the enlarged auxiliary within a pericarp formed by division of the neighboring cortex cells.

KEY TO GENERA

1. Plant bushy, with a strong main axis and shorter cylindrical lateral branchlets .... CYSTOCLONIUM, p. 268
1. Plant subdichotomous, the branches flat...... RHODOPHYLLIS, p. 269

Cystoclonium Kützing, 1843

Plants bushy, with coarsely fibrous holdfasts and prominent main axes, widely branched; the compact cortex of radial cell series bearing delicate colorless hairs, fairly compact and pseudoparenchymatous within, the medulla filamentous; zonate sporangia developed from surface cells, becoming immersed, the branchlets bearing them slightly thickened; spermatangia in irregular confluent groups on the upper branchlets, each cell of the assimilative layer developing four to six spermatangium mother cells carrying two or three spermatangia; procarps present, the carpogenic branch developed on the inner side of a superficial cell, curved, of three cells in length, the lowest developing laterally a sterile cell; auxiliary cell borne on the same supporting cell close to the carpogonium, becoming the lower segment of a cortex filament as the whole becomes immersed by extension of the cortex; cystocarps in markedly swollen branchlets, the gonimoblasts relatively loosely branched, without a filamentous pericarp or pore, but the adjacent cortical tissue thickened.

Cystoclonium purpureum (Hudson) Batters [1] Pl. 37, figs. 5–7

Plants each at first attached by a small disk, this later replaced by a coarsely fibrous holdfast; bushy, 1–5 dm. tall, the lower part of the axis sometimes naked, subsimple or sometimes sparingly forked, 1–3 (–8) mm. diam., above with numerous lateral branches 3–5 times alternately divided, often close-subcorymbiform; branches of all orders terete, the lesser tapering to both ends, 0.25–0.50 mm. diam., about 2–10 (–25) mm. long; texture firm-fleshy; color dark

[1] *Cystoclonium purpureum* f. *stellatum* Collins (1906a, p. 111; Phyc. Bor.-Am. 1140) with radiating burrlike growths on the branches is probably a pathological state.

purplish red or brownish, later bleached, yellowish or greenish, the cystocarps often remaining bright red; zonate sporangia at the surface of branchlets little distinguished from sterile ones; cystocarps single or seriate, causing nodose swellings in the branchlets.

From New Jersey to the lower St. Lawrence, and Newfoundland; attached to rocks and shells, growing in the lower littoral; annual, or perennating from the basal attachment, maturing early in the summer and then rather abundantly washed ashore.

REFERENCES: Phyc. Bor.-Am. 690. Harvey 1853, p. 170; Farlow 1881, p. 147 (all as *C. purpurascens*).

V. **cirrhosum** Harvey (Pl. 37, fig. 8; Phyc. Bor.-Am. 489; Harvey 1853, p. 170; Farlow 1881, p. 147) is distinguished by its less-tufted ultimate branches, more general gradation from the more slender main axis, more tangled habit, and essentially by the attenuate, arcuate to spiraled tendril-like shape of many of the ultimate branchlets; it is known from New Jersey, Rhode Island, and southern Massachusetts.

Rhodophyllis Kützing, 1847

Plants membranous, the dichotomous divisions abundantly proliferous at the margins; the medulla of branched filaments simulating veinlets, sometimes associated with large angular cells, the primary cortex of one layer of large cells on each face; tetrasporangia scattered, zonate, superficially immersed in the blade or in marginal proliferations; plants monoecious; spermatangia scattered over the surface; procarp present, the three-celled carpogenic branches borne on a primary cortical cell series consisting of three cells, of which the lowest is the supporting cell and the second the auxiliary which lies close to the carpogonium, the gonimoblasts producing carposporangia terminally; the prominent cystocarp wall formed by thickening of the overlying cortex.

Rhodophyllis dichotoma (Lepeschkin) Gobi Pl. 60, fig. 1

Plants arising from a branching holdfast, erect and somewhat bushy or flabellate, to 5–8 (–20) cm. tall, dark rosy or reddish purple; branches flat, 7–10 mm. wide, or sometimes much wider (to 6 cm.), irregularly a few times dichotomously forked, the divisions with proliferously ciliate margins, the proliferations subulate to linear-lanceolate bearing numerous tetrasporangia; cystocarps densely aggregated, often confluent, about the bases of similar projections.

Northern Massachusetts to Nova Scotia, Newfoundland, Labrador, Hudson Strait, Baffin Island, and the west side of Ellesmere Island; growing in deep water on shells or stones; fruiting in summer, and also reported in autumn and winter.

REFERENCES: Algae Exsic. Am. Bor. 146 (as *R. veprecula*); Phyc. Bor.-Am. 691. Harvey 1853, p. 105 (as *Calliblepharis ciliata*); Farlow, 1881, p. 152 (as *R. veprecula*); Collins 1902a, p. 176; Wilce *in herb.*

F. **setacea** Kjellman (Harvey 1853, p. 105; Farlow 1881, p. 152, both as v. *cirrhata*), with very slender blades which are even more conspicuously ciliate-proliferous, has been recorded from Nova Scotia and is, northward, the more abundant form.

FURCELLARIACEAE

Plants bushy, the branches variously disposed, terete to compressed, or foliaceous; structurally showing a filamentous medulla without any central axis, the cortex compact, of larger cells within, smaller toward the surface; sporangia formed in the cortex; spermatangia superficial; carpogonial branches formed on inner cortical cells, composed of three or four cells; auxiliary cells conspicuous, formed on other inner cortical cells before fertilization; gonimoblasts formed between the cortex and the medulla, most of the cells being matured as carposporangia, cystocarps without pericarps or discharge pores.

Furcellaria Lamouroux, 1813

Thalli firm, the holdfasts of coarse rhizoidal strands, above terete, freely dichotomously branched, the branches with an axis of longitudinal and a cortex of radial filaments; sporangia tetrapartite, in the cortex of the branches; spermatangia formed on small pale rose-colored inflated terminal branchlets, usually two spermatangia being produced by each surface cell; carpogonial branches of three or four cells, simple or forking, developed two or three together on the innermost cells of the cortex; auxiliary cells also formed in the inner cortex, cystocarps without pericarps lying between the cortex and the medulla.

Furcellaria fastigiata (Hudson) Lamouroux

Plants erect from a fibrous base, bushy, to 5–20 cm. tall, dark brown and firm, dichotomously much branched, the branches erect, 1–2 mm. diam., tetrasporangia and cystocarps formed in the uppermost branches.

Nova Scotia and Newfoundland; a perennial growing well below low tide level on stones; fruiting in the winter.

REFERENCES: Harvey 1853, p. 195; Bell and MacFarlane 1933a, p. 270.

HYPNEACEAE

Plants bushy, laterally branched, branches terete, often with spinulose branchlets; structurally showing a persistent central filament developed from an apical cell, and a cortex which matures into a pseudoparenchymatous structure with large cells within, small cells without; tetrasporangia more or less localized in somewhat swollen branchlets, formed at the surface, zonate; procarps present, the carpogenic branches of three cells on a supporting cell that forms additional corticating series, an inner cell of which forms an auxiliary near the carpogonium; cystocarps swollen, showing a group of filamentous cells anchoring the central mass to the pericarp, the carpospores formed terminally on close-branched gonimoblasts.

Hypnea Lamouroux, 1813

Plants bushy, branches slender, terete; habit virgate or spreading; tetrasporangia terminal on the corticating cell series, becoming immersed, zonate, grouped in the swollen parts of special short, lateral branchlets which are often fusiform; cystocarps enclosed in a hemispherical poreless pericarp.

Hypnea musciformis (Wulfen) Lamouroux Pl. 37, fig. 2

Plants bushy, their texture somewhat fragile, fleshy; color dull purplish, gray-violet, or bleached; the bases disklike, ill-defined;

erect branches to 10–45 cm. tall, about 1–2 mm. diam., with relatively few leading branches, these dividing several times, beset with numerous short spur branchlets 2–20 mm. long; tips of the main and principal lateral branchlets often elongate, naked, typically swollen and crozier-hooked; tetrasporic branchlets somewhat siliquose or spindle-shaped and rostrate, with the sporangia numerous, girdling the wider part; cystocarpic branchlets divaricate, often with spurlike divisions.

From the tropics, Bermuda, Florida to southern Massachusetts; in the northern extent of its range on stones and shells in exceptionally warm and protected coves; fruiting in summer.

REFERENCES: Algae Exsic. Am. Bor. 144; Phyc. Bor.-Am. 196. Harvey 1853, p. 123; Farlow 1881, p. 156; Hoyt 1920, p. 485; Taylor 1928a, p. 156.

GRACILARIACEAE

Plants branched, to even bushy, the branches slender to coarse, terete to strap-shaped, firm and often cartilaginous; axes developing from apical cells, forming a parenchymatous medulla and a narrow small-celled assimilative cortex which may bear delicate colorless hairs; sporangia tetrapartite, scattered, just below the surface; spermatangia scattered, cut off from surface cells; carpogenic branches of two cells, the carpogonium after fertilization producing a large fusion cell which gives rise to the gonimoblasts; cystocarp with a sterile basal placenta and a thick projecting pericarp opening by a pore.

Gracilaria Greville, 1830

Plants usually bushy, from a small discoid base, terete or flattened, fleshy to cartilaginous, dichotomously, irregularly, or proliferously branched; sporangia formed just below the surface of the plants, tetrapartite; spermatangia cut off from the lining cells of vase-shaped crypts; cystocarps hemispherical, with a large cellular basal placenta tissue, which may be connected by filaments with the wall, being within a prominent superficial pericarp composed of several layers, the outer cell rows radiating; discharging through a pore.

KEY TO SPECIES

1. Branches cylindrical, even at the forkings; plants of involved wide-spreading habit, color purplish to dilute grayish or greenish translucent .... G. verrucosa [1]
1. Branches typically broadly flattened, especially at the forkings, or if subterete, the habit subfastigiate............... G. foliifera

Gracilaria verrucosa (Hudson) Papenfuss Pl. 38, fig. 1

Plants bushy, with age often becoming free; texture firm-fleshy, color dull purplish red to purplish, grayish or greenish translucent; branches 0.5–2.0 mm. diam., repeatedly dividing, alternately or occasionally nearly dichotomously branched, with numerous lateral proliferations, terete throughout, tapering to the ultimate branchlets; tetrasporangia numerous, scattered over the branchlets, oval, from the surface 22–30 μ diam., in section 30–33 μ long; cystocarps very prominent, hemispherical, often numerous.

From the tropics, Florida to southern Massachusetts, and Prince Edward Island; in shallow, warm, quiet bays, often adrift on the bottom in the latter part of the summer.

REFERENCES: Phyc. Bor.-Am. 1041. Harvey 1853, p. 108; Collins 1903c, p. 231; Hoyt 1920, p. 483; Taylor 1928a, p. 152, 1937c, p. 293; Bell and MacFarlane 1933a, p. 270 (all as *G. confervoides*).

In the New England form of this species the bushy habit is crisply entangled, not elongate and flagellar, as often seen southward.

Gracilaria foliifera (Forsskål) Børgesen
Pl. 38, figs. 2–3; pl. 41, fig. 1; pl. 59, fig. 7

Plants sparingly bushy, 10–30 cm. tall, dull purple or faded; the lower part relatively slender, above more coarse, thick, terete to compressed or even expanded, when 2–15 mm. or somewhat more in width, sublinear or laciniate, the margin often proliferous; branching of one to several degrees, usually in the plane of the blade, ditrichotomous or alternate; tetrasporangia 20–35 μ diam., 30–45 μ long,

[1] Sometimes easily confused on habit alone with certain forms of Agardhiella found in warm, quiet and brackish coves; color, structure, and reproductive organs all serve to distinguish these two unrelated forms.

formed in the upper mature branches just below the surface; pericarps projecting strongly on faces or margins of the blades.

From the tropics, Florida to New Hampshire; on shells and stones in quiet bays and salt-marsh ditches, in water very shallow at low tide; maturing from late summer to autumn.

REFERENCES: Harvey 1853, p. 107; Farlow 1881, p. 164 (both as *G. multipartita*); Hoyt 1920, p. 484 (as *G. lacinulata*); Croasdale 1941, p. 214.

V. **angustissima** (Harvey) Taylor (Phyc. Bor.-Am. 240, 634; Farlow 1881, p. 164, all as *G. multipartita* v. *angustissima;* Taylor 1937b, p. 230) is recognized by its subterete branches, the flattening evident only below the forkings, and usually unaccompanied by much proliferation, the habit being open and fastigiate; it has been secured from North Carolina to southern Massachusetts in quiet bays. In addition, there has been noted in partly landlocked ponds a very tangled, bushy plant, highly proliferous, with much flattening of the lower branching; this is to be interpreted as a form of the present species rather than of *G. verrucosa,* which it rather resembles; it is darker and less adherent on drying, and has been secured in southern Massachusetts.

PHYLLOPHORACEAE

Plant bushy, dichotomously branched, the branches cylindrical to membranous, firm in texture; growth from a marginal or apical meristem producing a pseudoparenchymatous medulla and a close cellular cortex; sporangia tetrapartite, seriate, crowded in nemathecial sori; spermatangia developed on outgrowths from superficial cells; procarps present, the carpogenic branch of three cells carried on a supporting cell which serves as the auxiliary; gonimoblasts first growing inward, becoming interwoven with thin sterile filaments producing irregular masses of carposporangia without a definite sheath; sporangial or gametangial reproduction greatly modified in some species.

KEY TO GENERA

1. Branches at least in the distal part strap-shaped and complanate, to foliaceous .... PHYLLOPHORA, p. 277
1. Branches narrow, filiform to narrowly flattened 2

2. Plants small, to 5 cm. tall, moderately firm; in some the branches flattened .... GYMNOGONGRUS, p. 276
2. Plants larger, to 10 cm. tall, blackish and wiry; altogether cylindrical .... AHNFELTIA, p. 275

Ahnfeltia Fries, 1835

Plants bushy; the cylindrical branches developing from several independent axial initials, which produce a medulla of slender longitudinal filaments, and an outer cortex of radial moniliform laterally coherent filaments; reproductive structures apparently reduced to monosporangia grouped in small cushion-like nemathecia on the branches.

Ahnfeltia plicata (Hudson) Fries [1] Pl. 37, fig. 1; pl. 40, fig. 6

Plant forming bushy tufts, dark violet or purplish black, stiff or wiry in texture, arising from a thin diffuse disk and reaching a height of 6–20 cm., dichotomous or unilateral to irregularly proliferously branched, to 10 forkings or more, 1–4 cm. apart, spreading or erect, the axils rounded, closer above, often laterally proliferous, the branches slender, little tapered, to 0.5–0.9 mm. diam.; nemathecia forming small oblong swellings on the branches, constituted of radiating rows of cells, the outer in localized spots forming monosporangia.

New Jersey to the lower St. Lawrence, Newfoundland, northern Labrador, James Bay, Hudson Bay and Strait, Baffin Island, and Devon Island; perennial, growing on stones in the lowest tide pools, in rock crevices, and on the protected faces of boulders and cliffs; fruiting in winter.

REFERENCES: Phyc. Bor.-Am. 743. Harvey 1853, p. 168; Farlow 1881, p. 147; Kjellman 1883, p. 166; Hariot 1889, p. 196; Wilce *in herb.*

F. **furcellata** Collins (1911b, p. 187; Phyc. Bor.-Am. 1645) is simply a type in which the upper branching is regularly and equally flabellately dichotomous; it is reported from southern Massachu-

[1] The long-disputed *Sterrocolax decipiens* Schmitz (Phyc. Bor.-Am. 1595) appears to be a monosporic stage of Ahnfeltia. Setchell (1905, p. 136) states that *Sphaerococcus torreyi* C. Agardh is really this plant. Much about the description of *Cordylecladia? huntii* (Harvey 1853, p. 155), except the flattened character, suggests that it also may be referable here.

setts; v. *fastigiata* Farlow (1881, p. 147) differs little from the typical plant and is hardly worth designation. There is a form in partly landlocked pools which is very sparingly divided, but the branches are thickly beset with squarrose radiating branchlets about 3–6 mm. long; the writer has noted it in southern Massachusetts.

Gymnogongrus Martius, 1833

Plants small, bushy, repeatedly forking; from moderately firm to horny, the branches cylindrical to flat, sometimes proliferated laterally; structurally showing a medulla of angular rounded cells, the cortex of firmly coherent radial rows of several small cells; sporangia borne in the outer layer of small swellings or nemathecia on the branches, often imperfectly tetrapartite.

KEY TO SPECIES

1. Branches cylindrical, except below the forkings and near the branch tips .... G. griffithsiae
1. Branches flat, except at the extreme base, to 2 mm. or more in width .... G. norvegicus

Gymnogongrus griffithsiae (Turner) Martius [1]

Plants in pulvinate tufts, from a basal disk, to 2–5 cm. tall, dark purplish; repeatedly branching, fastigiate-dichotomous, sometimes polychotomous; branches slender-filiform, little tapering, to about 0.3–0.7 mm. diam., the tips pointed or a little flattened; nemathecia scattered, at first unilateral and pulvinate on the lower side of the branches, later more extensive, about 1 mm. diam.; sporangia generally imperfectly tetrapartite.

From the tropics, Florida, North Carolina, rare in Rhode Island and southern Massachusetts; growing on rocks just below low-water mark; fruiting in early summer.

REFERENCES: Phyc. Bor.-Am. 239b. Collins 1901a, p. 133; Hoyt 1920, p. 477.

[1] The structures called *Actinococcus aggregatus* Schmitz (Phyc. Bor.-Am. 786) seem to be the nemathecia of this Gymnogongrus.

Gymnogongrus norvegicus (Gunner) J. Agardh Pl. 35, fig. 9

Plants tufted, the tufts dense, rarely more than 5 cm. tall, reddish purple, firmly membranous; terete near the base only, above repeatedly dichotomously divided and in the upper segments sometimes polychotomous, the branches all flat, 2–5 mm. wide, the apices generally obtuse or emarginate; tetraspores in nemathecia formed on the surface of the branches, to about 2 mm. diam.; cystocarps immersed, about 1 mm. diam., numerous in the terminal segments.

Northern Massachusetts to New Brunswick; in deep tide pools or cast ashore.

REFERENCES: Phyc. Bor.-Am. 381. Harvey 1853, p. 166; Farlow 1881, p. 146.

Phyllophora Greville, 1830

Plants with a more or less evident stalk below, above branched and expanded into flat, often proliferous blades; structurally with a medulla of large angular cells covered by a cortex of small chromatophore-bearing cells, often in vertical rows without; sporangia tetrapartite, in chains, closely aggregated into nemathecia borne on the surface of the blades of their margins; dioecious, the spermatangia on special small leaflets, where the surface cells produce branch systems of small, rather slender cells which bear the spermatangia on the ends; carpospore-bearing blades short, thick, lateral and sessile or pedicellate on the stalklike parts of main blades, the procarps numerous between the inner cortical cells, consisting of a supporting auxiliary, a three-celled carpogenic branch and sterile cell series; the cystocarp developed from slender gonimoblasts which ultimately develop close clusters of carpospore-forming filaments, or, by suppression of the carpospores and the independent tetrasporophyte, produce tetrasporangia directly.

KEY TO SPECIES

1. Plants large 2
1. Plant very small, less than 2 cm. tall, simple or once-forked, the cystocarps on marginal projections.......... P. traillii, p. 279

2. Blade on an indistinct stalk, flattened from near the base and gradually expanded; the apparent nemathecia spherical, stalked, on the upper margin of the blade 3
2. Blades on long, branched, cylindrical stalks, which expand sharply at the ends into forked or laciniate membranes; nemathecia broadly expanded on the face of the blade P. membranifolia, p. 278
3. Branching usually limited to the flabellate ends of the primary blades, but when successive the secondary blades with stalk-like bases .... P. brodiaei, p. 278
3. Branching progressive, the divisions generally not stipitate; the apices broadly rounded.................... P. interrupta, p. 279

Phyllophora membranifolia (Goodenough et Woodward) J. Agardh Pl. 39, fig. 1

Plants bushy, to 10–15 (–50) cm. tall, each from a small disk, the stalk and lower branches terete, the blades broadly and sharply expanding from the branches with a wedge-shaped base and a sparingly or freely forked or laciniate distal end, the whole spreading 3–5 cm., the divisions 1–5 mm. broad, dull purplish, texture firm-membranous; sporangial nemathecia forming large, dark, little-elevated spots near the bases of the blades, the cortical cells in rows, and the tetrasporangia seriate; cystocarps about 2 mm. diam., stalked, attached to the lower parts of the branches, but rarely to the blade; spermatangia on small bright blades about 0.5 mm. wide, 2 mm. long, attached to the edge.

New Jersey to the lower St. Lawrence, and Baffin Island; growing on stones in very low tide pools and in deeper water to several meters; through the year, fruiting in autumn and winter.

REFERENCES: Phyc. Bor.-Am. 379. Harvey 1853, p. 165; Farlow 1881, p. 145.

Phyllophora brodiaei (Turner) J. Agardh [1] Pl. 39, fig. 2

Plants from small disks, to about 1 dm. tall, the stalks solitary or few together, compressed or flattened, expanding into flat wedge-shaped blades to 3 cm. wide, dull red-purple and of membranous

[1] *Actinococcus subcutaneus* (Phyc. Bor.-Am. 429) consists of the nemathecial tetrasporiferous phase of *P. brodiaei*.

texture; blades simple or dichotomously 1–2 times forked; carpogenic branches formed on crisped upper margins of the blades; after fertilization these develop the short-stalked dark red nemathecial spheres, about 1–2 (–3.5) mm. diam., near the surface of which tetrapartite sporangia are borne in radial series.

New Jersey to the lower St. Lawrence, and Newfoundland; perennial, growing most actively between winter and early summer on rocks and shells in somewhat deep water well below tidal range; forming nemathecia in the summer.

REFERENCES: Phyc. Bor.-Am. 428. Harvey 1853, p. 164; Farlow 1881, p. 145.

There may be distinguished v. **catenata** (C. Agardh) Collins, slender, with rather long narrow attenuate blades which bear new stalked blades at their summits in repeated succession, reaching a length of over 2 dm.; this has been found in Rhode Island and southern Massachusetts, probably washed up from deep water.

Phyllophora interrupta (Greville) J. Agardh

Plants bushy, to 15 cm. tall, deep red, firmly membranous, briefly stipitate below, above expanded, abundantly di- to polychotomously branched, the branches at their bases often somewhat narrowed but not stipitate, the segments simple or undulate-constricted, strap-shaped or, if closely forked, nearly triangular, 5–20 mm. broad, often progressively broader from the lower divisions, the tips very broad and rounded or retuse, 10–25 mm. wide; nemathecial plants smaller, darker, with segments smaller and more cuneate than in sterile individuals.

Recorded from northern Massachusetts, but known with more certainty from Nova Scotia, Labrador, southeastern Hudson Bay, Hudson Strait, Devon Island, and Ellesmere Island.

REFERENCES: Rosenvinge 1926, p. 28, as *P. brodiaei* f. *interrupta;* Kjellman 1883, p. 164; Howe 1927, p. 22B; Wilce *in herb.*

Phyllophora traillii Holmes

Plants small, to 1–2 cm. tall, with a minute stalk which is cylindrical and may fork; blade ovate, lanceolate or oblong, wedge-shaped at the base, entire or simply forked; margin often beset with small subcylindrical blades which bear the minute immersed cystocarps.

Connecticut to Maine; on vertical rocks and in tide pools, near low-water mark, generally overhung with Fucus; probably fruiting in winter; when sterile difficult of determination.

REFERENCE: Phyc. Bor.-Am. 689(?).

GIGARTINACEAE

Plants plane or bushy, subsimple or branched, developing with an axis of several filamentous initials, the medulla ultimately pseudoparenchymatous, the cortex obscurely filamentous, the small outer cells in rows and containing chromatophores; sporangia in sori in the blade, developed from forking cell systems among the medullary filaments, tetrapartite; spermatangia formed in sori on the surface; procarps present, the three-celled carpogenic branch carried on a supporting auxiliary which after fertilization forms slender-branching gonimoblast filaments which ramify into the medulla and produce a diffuse cystocarp.

KEY TO GENERA

1. Blades relatively little proliferous, the nemathecia somewhat swollen .... CHONDRUS, p. 280
1. Blades much proliferated on the face, the proliferations usually nodular to subcylindrical, in part containing the cystocarps GIGARTINA p. 281

Chondrus Stackhouse, 1797

Plants bushy, the stalk slender and compressed below, the branches above flattened, linear to broadly cuneate, repeatedly dichotomous, firmly membranous to subcartilaginous; the medulla vaguely filamentous with lateral radial filaments forking dichotomously to produce a small-celled cortex, the outermost cells chromatophore-bearing; sporangia tetrapartite, numerous, in obvious sori formed below the superficial layer; spermatangia in superficial sori on young branches; the three-celled carpogenic branches each borne upon a large supporting auxiliary cell in the inner cortex; after fertilization branching gonimoblast filaments of slender cells grow out into the medulla, and eventually from these forking series of shorter cells arise which form the carposporangia in a fashion very similar to the origin of the tetrasporangia.

Chondrus crispus Stackhouse Pl. 39, figs. 3–6; pl. 40, fig. 4

Plants with several blades from a disklike holdfast, generally 8–15 cm. tall, forming loose to often dense clumps, dark red-purple, the more or less slender compressed stalk expanded sharply into the wedge-shaped base, which is divided dichotomously, less often somewhat irregularly, into a flabellate blade; segments linear-compressed to narrow bandlike, or broad-membranous, usually somewhat closely divided and crisped in the ultimate segments, width of segments 2–15 mm., but much wider across the forkings; apices of branches rounded; margins entire, or sparingly proliferous; tetrasporangia in irregular, solitary or scattered, little swollen, dark-red sori in the terminal segments; cystocarps in similar sori about 2 mm. diam., often unilaterally swollen.

New Jersey to Nova Scotia, Prince Edward Island, and Newfoundland; growing throughout the year on rocks, shells, or woodwork, in tide pools and the intertidal belt, descending to moderate depths, often exceedingly common, particularly on exposed stations; where shaded developing normal coloration, but where exposed to the light bleached and greenish, yet actively growing; through the year, fruiting with carpospores during the summer and with tetraspores during the later autumn.

REFERENCES: Phyc. Bor.-Am. 488, 785. Harvey 1853, p. 181; Farlow 1881, p. 148; Hariot 1889, p. 195; Papenfuss 1950, p. 191.

This plant is exceedingly variable in aspect; in general, the writer has met narrower forms in the southern part of its range, and in the more northern collections wide forms as well.

Gigartina Stackhouse, 1809

Plants foliar to bushy, firm-fleshy, dark reddish purple, simple to abundantly subdichotomously branched, the branches compressed to bladelike; developing a filamentous medulla surrounded by a cortex of radiating branched rows of successively smaller cells with the outermost containing the chromatophores; tetrapartite sporangia in immersed, rather spreading sori of indefinite form, the sporangia developed from cells of the inner cortex; spermatangia in groups developed terminally on close-branched filaments from superficial cells; cystocarps generally crowded in special fertile nodules or branchlets; procarps present, developed on the inner cortex, the supporting auxiliary cell bearing a three-celled carpogenic branch; after fertilization the carpogonium and the auxiliary cell fuse and the latter inwardly and loosely develops laterally branched gonimoblast filaments which consist of larger nutritive cells supporting more distal divisions of smaller cells which form the carpospores, the mass surrounded eventually by a limiting layer of the medulla; cystocarps unilaterally projecting, the fleshy pericarp eventually rupturing.

Gigartina stellata (Stackhouse) Batters Pl. 35, fig. 4

Plants tufted, rather small, to 5 (–15) cm. tall, dark purplish brown, arising from a basal disk, the axis closely branching, the branches compressed to flat and expanded, 2–4 (–10) mm. wide, sometimes somewhat channeled, forking dichotomously or somewhat irregularly, the angles wide, the upper divisions closer than below; faces and margins of the branches becoming papillose, the papillae rounded to ligulate, infrequently developing into branching blades like the main branch system; cystocarps at the summits of the papillae, immersed.

Rhode Island, northern Massachusetts to Nova Scotia and Newfoundland; on rocks in the lower tide pools and on exposed rocky shores near low-water mark; perennial, fruiting in the winter.

REFERENCES: Phyc. Bor.-Am. 838. Harvey 1853, p. 175; Farlow 1881, p. 148; Hariot 1889, p. 196 (all as *Gigartina mamillosa*).

RHODYMENIALES

Plants showing various shapes from filiform to fleshy-membranous, sometimes hollow; corticated, with a modified multiaxial type of structure, commonly appearing parenchymatous; asexual reproduction by tetraspores in sporangial sori or scattered over the plant just below the surface; sexual reproduction by spermatangia borne on surface cells in more or less restricted areas and by carpogonia in procarps sunken in the cortex; auxiliary cells established by segmentation indirectly from the cell supporting the carpogenic branch; cystocarps enveloped by a pericarp.

KEY TO FAMILIES

1. Foliaceous, or, if bushy, with a distinct main axis
RHODYMENIACEAE, p. 283
1. Bushy; without a distinct main axis.......... CHAMPIACEAE, p. 286

RHODYMENIACEAE [1]

Plants plane or bushy, nearly simple or somewhat freely divided, the divisions flat and subdichotomous or subcylindrical and radially branched, solid or hollow, soft to tough-membranous; developing from an apical meristem, the innermost cells large, forming a parenchymatous medulla, the surface cells small, often in short radial series and containing the chromatophores; sporangia sometimes in sori, tetrapartite, formed between these superficial cells; carpogenic branches three-celled, the supporting cells each also cutting off an auxiliary mother cell; gonimoblasts extensively branched, most of the cells forming rather small carpospores, the whole eventually enveloped by a loose pericarp.

KEY TO GENERA

1. Plant laterally branched, branches narrow, thick. HALOSACCION, p. 283
1. Plant in the form of broad, thin, dichotomously cleft blades
RHODYMENIA, p. 284

Halosaccion Kützing, 1843 [2]

Plants erect, the holdfasts small, stalks slender and short, the main portions subcylindrical to obovate, simple, sparingly or proliferously branched, hollow; structurally with an inner part of two or three layers of large colorless cells and an outer cortical layer of small chromatophore-bearing cells; sporangia borne below the surface, tetrapartite; spermatangia formed over large areas of the cortex, the surface cells dividing into smaller units and then again parallel to

[1] *Plocamium coccineum* Lyngbye, northern Massachusetts (Farlow 1881, p. 151), is a doubtful record.

[2] The plant from Maine recorded by Collins (1896a, p. 6) as *H. scopula* Strömfelt is probably a Gymnogongrus.

the surface, so that the contiguous spermatangia are carried on elongated stalklike cells.

Halosaccion ramentaceum (Linnaeus) J. Agardh

Plants gregarious, purplish, reddish, or faded; firm-fleshy in substance; short stalks arising from the basal holdfasts, supporting tubular cylindrical or compressed axes 10–40 cm. long, reaching a diameter of 5–15 mm., tapering to base and apex, branched freely from the base, or more sparingly above, or proliferously; tetraspores numerous, large.

Northern Massachusetts to the lower St. Lawrence, Newfoundland, northern Labrador, Hudson Strait, Baffin Island, Devon Island, and Ellesmere Island; perennial in shallow water, such as tide pools, on stones and shells, generally near low-water mark, in exposed as well as sheltered places.

REFERENCES: Algae Exsic. Am. Bor. 78; Phyc. Bor.-Am. 400. Harvey 1853, p. 194; Farlow 1881, p. 143; Kjellman 1883, p. 153; Hariot 1889, p. 195; Rosenvinge 1926, p. 32; Whelden 1928, p. 124; Wilce *in herb.*

This plant is exceedingly variable in habit. It is possible to recognize the main tendencies under the following forms. V. **gladiatum** Eaton (1873, p. 347; Farlow 1881, p. 143) has relatively few branches, and they are long, subsimple, somewhat incurved and inflated like the main axis. F. **ramosum** Kjellman (Phyc. Bor.-Am. 400a, 1344) has a more or less slender (2–4 mm. diam.) cartilaginous main axis, sparingly forked, beset with very many softer branchlets 1.0–1.5 (–4) cm. long, which diverge at right angles or nearly so. F. **subsimplex** Kjellman (Phyc. Bor.-Am. 1343) is also firm, almost cartilaginous, and slender in the main axis; this is simple or sparingly branched.

Rhodymenia Greville, 1830

Plants forming large membranous fronds, simple to dichotomously or palmately divided, often proliferous from the margins, in some cases the divisions narrow, strap-shaped; structurally of two layers, the internal of oblong colorless cells, the cortical of small cells with chromatophores, becoming of increased thickness with cortical cells in anticlinal series in fertile portions; sporangia formed between these series of superficial cells, in sori, tetrapartite; procarps present, the carpogenic branch consisting of three cells borne on a large multinucleate supporting cell, on which there is also borne a two-celled axis, the lower cell of which cuts off the auxiliary so that it lies beside the carpogonium; cystocarp with a somewhat swollen loose pericarp opening by a distinct pore.

Rhodymenia palmata (Linnaeus) Greville

Pl. 41, fig. 7; pl. 42, fig. 3

Plants large, from small disklike holdfasts, solitary or a few together, simple below or branching from the base, the stalk inconspicuous, the blade gradually expanding, above dichotomously or palmately divided into broad segments, the total length to 5 dm., width to 16 cm. above the forks, but usually to about 7 cm.; frequently the blade subsimple, abundantly ligulate-proliferous from the margin, the proliferations often forking and large; color purplish red, texture leathery-membranous, the thickness 170–260 μ, increasing to 350 μ in the sporangial areas; reproduction only known by tetrasporangia in scattered sori appearing as somewhat darker spots on the blade, associated with an increase in the thickness of the cortical layer; sporangia 30–50 μ diam., 50–70 μ long lying at the surface of the blade.

New Jersey to the lower St. Lawrence, Newfoundland, northern Labrador, James Bay, Hudson Bay and Strait, Baffin Island, and Ellesmere Island; through the year, growing from the intertidal zone down into rather deep water, and in the southern part of its range principally restricted to deep water. Here also it appears to fruit in the winter.

REFERENCES: Phyc. Bor.-Am. 591. Harvey 1853, p. 148; Farlow 1881, p. 150; Kjellman 1883, p. 147; Hariot 1889, p. 196; Wilce *in herb.*

Though quite variable, this little affects the general habit. In f. **angustifolia** Kjellman (Phyc. Bor.-Am. 1398; Kjellman 1883, p. 148) blades are long-linear, the width 1–2 cm. In v. **sarniensis** (Mertens) Greville (Farlow 1881, p. 150) the lower part is expanded and the upper parts are repeatedly filiform-dissected, the segments about 2–5 mm. diam., and the plants are small, about 10–15 cm. tall; both varieties occur within the range of the type.

CHAMPIACEAE

Plants usually bushy, the branches cylindrical or compressed, delicately membranous to quite soft; from an apical meristem developing superficial small assimilatory cells, an inner cortex of large cells and a medullary cavity traversed by longitudinal filaments, the filaments bearing lateral secretory cells; tetrahedral sporangia formed from cortical cells and lying just below the surface; spermatangia borne on groups of surface cells of the small male plants; carpogenic branches three- or four-celled, each borne on an inner cortical cell, the auxiliary secondarily derived from the supporting cell; after fertilization the gonimoblasts and in turn the large carposporangia are formed from a large fusion cell, the whole covered by a prominent ostiolate pericarp.

KEY TO GENERA

1. Plant septate throughout........................ CHAMPIA, p. 288
1. Plant not septate, bases of branches often contracted and closed LOMENTARIA, p. 286

Lomentaria Lyngbye, 1819

Plants tufted, repeatedly irregularly or unilaterally branched, the slender branches tapering near each end, terete or compressed, if hollow usually closed at the bases of the branches; tetrahedral sporangia somewhat grouped; spermatangia formed in patches on the surface of the plant, the mother cells each bearing two or three spermatangia; procarp developed on a primary cortical cell, at maturity consisting of a supporting cell bearing a carpogenic branch of three cells and one or two auxiliaries; after fertilization the active auxiliary cell uniting with other structures to form a fusion cell which in turn bears gonimoblasts, the basal portions of these remaining sterile but the upper and larger part forming carposporangia enclosed within a projecting ostiolate pericarp.

KEY TO SPECIES

1. Branches alternate, elongate, cylindrical-tapered........ L. baileyana
1. Branches opposite, flat, about 2–10 times as long as broad L. orcadensis

Lomentaria baileyana (Harvey) Farlow

Pl. 35, fig. 10; pl. 41, fig. 4; pl. 43, fig. 6

Plants tufted, often widely expanded, 3–7 (–20) cm. tall, densely branching; dull purplish red to pink, usually bleached below, drying much brighter; substance soft; branching irregularly alternate, the hollow branches terete, tapering near each end, curved, the ultimate segments 1–2 cm. long, often somewhat unilateral, 0.5–1.5 mm. diam.; surface of the thallus showing both larger and smaller cells; tetrasporangia 30–55 μ diam., scattered under the cortical layer in the branches; cystocarps usually scarce, sessile, external on the branches, somewhat produced at the ostiolate tip.

From the tropics to Florida, the Carolinas, New Jersey to northern Massachusetts; on stones and shells in sheltered shallow water, down to about 7–9 meters; fruiting in the summer, probably annual.

REFERENCES: Algae Exsic. Am. Bor. 75 (as *Lomentaria baileyana*); Phyc. Bor.-Am. 886, 1399 (as *L. uncinata non* Meneghini). Harvey 1853, p. 185 (as *Chylocladia baileyana*); Farlow 1876, p. 698; 1881, p. 154; Hoyt 1920, p. 492 (all as *L. uncinata*).

This plant is usually very constant in form; however, a few variants are worth mentioning. V. **filiformis** (Harvey) Farlow (1881, p. 155; Harvey 1853, p. 185, as *Chylocladia baileyana* v. *filiformis*) has very slender filamentous branches which are little or not attenuate at the base; it is reported to occur with the type, and probably in the warmer coves. A very distinctive variety, which may be called v. **siliquiformis** Taylor (1937b, p. 231), has the usually unilateral, sometimes bilateral or irregular ultimate branches 2–3 mm. broad immediately above the sharply contracted base; they taper thence evenly to the subacute apices and are 1–2 cm. long; the rather small plants are unusually brightly colored. This plant was reported from New Jersey and is occasional in the tide pools of southern Massachusetts. V. **valida** (Harvey) Collins (Harvey 1853, p. 185, as *C. baileyana* v. *valida;* Taylor 1937b, p. 226) is described as a robust form with generally unilateral branching, short and stout ultimate branches which taper equally to the ends. It seems to occur in South Carolina and with the type in New Jersey and southern New England.

Lomentaria orcadensis (Harvey) Collins

Pl. 40, fig. 8; pl. 41, fig. 3; pl. 43, fig. 5

Plants tufted, small, bright rose-colored, soft, 2–5 cm. tall, briefly stalked; the flat branches opposite, 1–2 times pinnate, lanceolate or linear-lanceolate, abruptly constricted at the base, the apex obtuse, 1.5–3.0 (–4.5) mm. wide, 1.–3 cm. long; the surface of the thallus showing larger cells to 75 μ diam., 95 μ long, and superposed smaller cells to 12–16 μ diam.; medulla of somewhat reticulate filaments, the cells 6–12 μ diam., 30–110 μ long or more; tetrasporangia 37–44 μ diam., in groups under the surface layer.

Rhode Island to Maine; growing on shells, stones, or woodwork, found from the surface, as on floating wharves, to over 25 meters depth; scarce, sporadic, fruiting in summer.

REFERENCES: Algae Exsic. Am. Bor. 17; Phyc. Bor.-Am. 1241 (both as *L. rosea*). Harvey 1849, p. 100 (as *Chrysymenia orcadensis*), 1853, p. 186 (as *Chylocladia rosea*); Farlow 1881, p. 155 (as *L. rosea*); Hoyt 1920, p. 492 (as *L. rosea*); Taylor 1937b, p. 227.

Champia Desvaux, 1808

Plants bushy, tufted from a fibrous base, the branches cylindrical, repeatedly alternately divided, hollow, septate at intervals and contracted at the septa; growth from a group of apical cells; the adult axis composed of an interrupted peripheral layer of small cells, a deeper continuous layer of large thin-walled cells, and a series of longitudinal filaments which run from the apex of the plant down the outer part of the central cavity from the apex, anchored at each septum, and provided with a gland cell in each cavity; sporangia tetrahedral, scattered, numerous in the branches, formed from cortical cells so that they lie just below the cortical layer; spermatangia formed in patches over the surface of the branches of small plants; carpogenic branches each four-celled, much curved and outwardly directed, formed on a primary cortical cell; the auxiliary mother cell cut off from the same support, dividing to form a uninucleate auxiliary beside the carpogonium; after fertilization the auxiliary mother cell enlarges greatly and fuses with others to form a nutritive cell, while the gonimoblast filaments grow out from the auxiliary, only the end cells forming carposporangia; cystocarps surrounded by delicate filaments within ovate sessile ostiolate pericarps.

Champia parvula (C. Agardh) Harvey Pl. 43, figs. 8–10

Plants tufted, the tufts dense, pale dull red, pinkish brown or greenish, crisp-gelatinous in texture, to 3–7 cm. tall; branching alternate, the branches 1–2 (–4) mm. diam., tapering, segmented, the hollow segments cask-shaped, 1.0–1.5 (–5.0) diameters long, the tips of the branches obtuse; tetrasporangia 55–100 μ diam., numerous in the segments; cystocarps relatively few, in prominent superficial pericarps.

From the tropics, Bermuda, Florida to northern Massachusetts; abundant in the sublittoral belt, particularly in quiet water and sheltered coves; fruiting in the summer, but also reported in the winter.

REFERENCES: Phyc. Bor.-Am. 290, 592. Harvey 1853, p. 76; Farlow 1881, p. 156; Bigelow 1887, p. 111; Davis 1892, p. 339, 1896a, p. 109; Hoyt 1920, p. 493; Taylor 1928a, p. 158.

CERAMIALES

Plants generally slenderly filamentous and branched, sometimes coarse, strap-shaped, or membranous; naked or corticated, with the central filament type of structure developing from an apical cell; asexual reproduction by tetraspores formed in sporangia external or more or less covered by the cortex of ordinary branches, or grouped in special sporiferous branchlets or stichidia; sexual reproduction by spermatangia borne on the axial filaments, often in masses, or covering areas of the flat bladelike plants, or covering special colorless branchlets (spermatangial clusters or "antheridia"); and by carpogonia on carpogenic branches borne on the axial or the pericentral cells; auxiliary cells formed after fertilization from the cell supporting the carpogenic branch; gonimoblasts developed from the auxiliary to form a mass of carposporangia, which may be naked, partly enveloped by branchlets, or covered by a pericarp.

KEY TO FAMILIES

1. Foliaceous or, if bushy, with flat branches... DELESSERIACEAE, p. 318
1. Not foliaceous and rarely with flat branches.................... 2

2. Axis clothed with uniseriate branched chromatiferous filaments DASYACEAE, p. 325
2. Axis naked, corticated or covered with branchlets............... 3

3. Axis uniseriate, but the nodes often corticate.... CERAMIACEAE, p. 290
3. Axis polysiphonous .... RHODOMELACEAE, p. 326

CERAMIACEAE

Plants usually bushy; branches in some genera uniseriate, in others corticated, usually terete; growth from the tip producing an axial row of cells; unbranched colorless unsegmented hairs often present, also segmented or unsegmented hyaline extensions from the tips of branches; cortication if present usually developed first about the nodes, consisting of a single ring of cells, or spreading over the internodes in more or less filamentous fashion, later appearing parenchymatous, or else consisting of an investment of rhizoidal filaments; sporangia superficial or stalked on the branches, single or whorled at the nodes, or in corticated species in the cortex, ultimately dividing in tetrapartite or tetrahedral fashion; spermatangia developed on special determinate branchlets, forming small colorless clusters, or covering the cortex of portions of larger species; carpogenic branches of four cells, borne on supporting cells, which may also give rise to sterile cells and after fertilization to one or two auxiliaries near the carpogonia; cystocarps composed of groups of gonimoblast filaments, the outer cells of which produce the carposporangia, naked or enveloped in jelly, or partly enclosed by subtending filaments from below.

KEY TO GENERA

1. Main axes uniseriate, uncorticated.............................. 2
1. Main axes corticated at least at the nodes....................... 8

2. Gland cells usually present; sporangia tetrapartite............... 3
2. Gland cells absent... 4

3. Branching opposite, at least in part........ ANTITHAMNION, p. 292
3. Branching irregular .... TRAILLIELLA, p. 291

4. Cells large and visible to the naked eye; sporangia and cystocarps guarded by involucral cells............ GRIFFITHSIA, p. 303
4. Cells more delicate.. 5

5. Catenate seirospores present.................... SEIROSPORA, p. 296
5. Tetrahedral sporangia, rarely polyspores, produced.............. 6

6. Carpogenic branches subterminal, usually sterile; branching generally opposite .... SPERMOTHAMNION, p. 302
6. Carpogenic branches lateral; branching alternate or dichotomous... 7

7. Cystocarps without involucre; sporangia producing tetraspores CALLITHAMNION, p. 297
7. Cystocarps slightly involucrate; sporangia dividing to polyspores .... PLEONOSPORIUM, p. 301
8. Ultimate branchlets of determinate growth, with naked internodes but small cells about the nodes; branches substantially corticated .... SPYRIDIA, p. 316
8. Ultimate branchlets naked, uniseriate, the branches covered by a spreading cortex............................. PLUMARIA, p. 305
8. Ultimate branchlets with a cortex similar to that of the axes....... 9
9. Branching markedly bilateral and pinnate, axes and branches usually flattened .... PTILOTA, p. 306
9. All axes terete and corticated at the nodes, the internodes if corticated covered by outgrowths from the nodes. CERAMIUM, p. 307

Trailliella Batters, 1896[1]

Plants filamentous, the filaments uniseriate, the primary axes creeping, affixed by disklike holdfasts with filamentous-proliferated margins; erect filaments irregularly alternately branched; irregularly tetrapartite sporangia formed in the erect filaments by segmentation of the axis.

Trailliella intricata (J. Agardh) Batters Pl. 45, figs. 3–5

Plants forming dense entangled tufts, soft in texture, 1–2 (–4) cm. diam., bright reddish rose; primary filaments creeping, 30–40 μ diam., attached by holdfasts; erect filaments 24–45 μ diam., the cells 1.5–2.5 diameters long, little or not contracted at the nodes, irregularly alternately branched above, the forkings distant, the branches rather long, all flexuous, or occasionally somewhat unilateral, little more slender than the primary filaments, though reduced to 20 μ diam. near the tips; cells with many small disklike chromatophores; small refractive colorless gland cells at nearly every node; tetrasporangia in swollen parts of the filaments, 1–6 together, each lateral to and cut off from a flattened vegetative cell, 50–60 μ diam.

Long Island, southern Massachusetts, Nova Scotia, and Newfoundland; in shallow water on various Rhodophyceae; abundant

[1] Researches of Feldmann, Koch, and Harder, summarized by the latter (1948), indicate that Trailliella is the tetrasporophyte of *Asparagopsis hamifera* and so not taxonomically independent.

in the vegetative condition during the summer, but tetrasporangia have seldom been seen, and not along our coast.

REFERENCES: Lewis and Taylor 1928, p. 196; Taylor 1941, p. 73; Harder 1948, p. 24.

Antithamnion Nägeli, 1847

Plants tufted, of uncorticated uniseriate filaments; branches often alternate below, but above repeatedly opposite or whorled, their attachment often considerably below the end of the supporting cell, often one element in a branch pair suppressed, or secund, or in the ultimate branchlets sometimes truly alternate; cells uninucleate, with many small, rounded or bandlike chromatophores; gland cells often present; sporangia tetrapartite, carried on the smaller branchlets, or often replacing branchlets of the last order; spermatangia also in patches on ultimate branchlets; carpogenic branches formed on the lowest cells of such branchlets, these each originating a curved four-celled carpogenic branch and, after fertilization, an auxiliary cell; the cystocarp consisting of a mass of carposporangia upon a stalk of a few sterile cells.

KEY TO SPECIES

1. Branchlets borne on the upper side of the determinate branches.... 2
1. Branchlets usually borne bilaterally.......................... 3

2. Plants obviously flat-spreading, the branchlets conspicuously seriate .... A. plumula, p. 295
2. Plants of tangled habit, irregularly branched; branchlets few, chiefly on the upper side of the determinate branches A. boreale, p. 293

3. Plants forming dense tufts 2–4 cm. diam., rarely more, the branches sparingly divided; lateral branches in 4 rows, the branch tips densely brushlike................ A. cruciatum, p. 294
3. Plants more widely branched, the tips not densely tufted.......... 4

4. Branches of the last order opposite, small and determinate, and very different from the indeterminate branches, or if becoming indeterminate, opposite a determinate branch A. floccosum, p. 293
4. Branches of the last order not distinctively determinate or otherwise, opposite simple branchlets but alternate to other branches 5

5. Secondary branches slender, flexuous; cells in the branchlets 4–8 diameters long........................ A. americanum, p. 294
5. Secondary branches short and thick; cells in the branchlets 1–2 diameters long........................... A. pylaisaei, p. 295

Antithamnion floccosum (Müller) Kleen

Plants bushy, sometimes stoloniferous, branching irregularly below, somewhat alternately above, to 10 cm. tall, soft and delicate throughout; all branches beset with short, simple, strongly tapering opposite branchlets 22–32 μ diam.; segments of the main axes 90–100 μ diam., 5–8 diameters long, and in the branchlets 1–2 diameters long; tetrasporangia on one-celled stalks at the bases of the branchlets.

Northern Massachusetts to Maine, Nova Scotia, Prince Edward Island, and Ile St. Pierre; growing upon various coarse algae and sometimes in tide pools; spring.

REFERENCES: Algae Exsic. Am. Bor. 156 (as *Callithamnion floccosum*); Phyc. Bor.-Am. 495. Harvey 1853, p. 240; Farlow 1881, p. 122 (both as *C. floccosum*); Le Gallo 1947, p. 315.

Antithamnion boreale (Gobi) Kjellman

Plants somewhat entangled in habit; to 20 cm. tall; main branching opposite or alternate, the axis about 150 μ diam., cells 4–6 diameters or more long; indeterminate branches of the outer orders bearing opposite determinate branches, carried laterally on the cells of the axis and about one-fourth below the top; these branches are about 60 μ diam. at the base, each such branch carrying on the upper side of each cell a single branchlet 12–20 μ diam. (rarely two or none, or on the lower side) which tapers from base to apex, with cells 1–3 diameters long; gland cells on the branchlets, attached to a single cell, sometimes absent; tetrasporangia 35–65 (–85) μ long, 22–25 μ diam., replacing the ultimate branchlets; cystocarps large, single or paired.

Northern Massachusetts to Maine, the lower St. Lawrence, James Bay, southeastern Hudson Bay, and Ellesmere Island; sublittoral, growing on shells or coarse algae to 18–36 meters depth; fertile in the summer.

REFERENCES: Phyc. Bor.-Am. 1247, 1346. Kjellman 1883, p. 180; Collins 1896b, p. 461; Croasdale 1941, p. 214.

Antithamnion americanum (Harvey) Farlow in Kjellman

Plants erect, delicately bushy, light rose-pink; occasionally to 20 cm. tall; main branching repeatedly alternate, the branches rather long, to 60–200 μ diam., the cells in the middle of the branches 8–10 (–19) diameters long; branchlets numerous on the outer branches, opposite and in two rows, occasionally a few branchlets forked, spreading, about 10–20 μ diam., the cells 4–8 diameters long; tetrasporangia sessile on the upper side of the penultimate branches, replacing branchlets of the final order; cystocarps found at the bases of the penultimate branchlets, sessile, naked.

New Jersey to Maine, Nova Scotia, Prince Edward Island, and Labrador; growing on wharves, stones, and coarse algae below the low-water mark; from midwinter onward, fruiting in early spring.

REFERENCES: Algae Exsic. Am. Bor. 89 (as *Callithamnion americanum*); Phyc. Bor.-Am. 47 (in some sets), 1100. Harvey 1853, p. 238; Farlow 1881, p. 123 (both as *C. americanum*); Kjellman 1883, p. 184.

Antithamnion cruciatum (C. Agardh) Nägeli

Pl. 44, fig. 3; pl. 45, figs. 6–8

Plants tufted, somewhat intricate below, alternately, above rather sparingly, divided, the branches long, at the tips the branchlets densely tufted or appearing ocellate-congested; height to 5 cm., color dull rose-red; the main axis below 50–90 μ diam., cells 90–300 μ long; lateral branches spreading, short, of rather even length, opposite in alternately placed pairs, or in fours, producing a 4-rowed aspect; branchlets alternate, in 2 rows, or somewhat unilateral; 9–14 μ diam., with cells 4–8 diameters long; gland cells numerous, near the bases of the ultimate branchlets, applied to the upper sides of 3 cells; tetrasporangia 75–85 (–100) μ long, 50–55 (–70) μ diam., replacing the branchlets of the last order, mostly on the upper side of the short branches, sessile or short-stalked; cystocarps to 400 μ diam.

Bermuda, New Jersey to New Hampshire, the common species south of Cape Cod; on stones and coarse algae; fruiting in the later summer.

REFERENCES: Harvey 1853, p. 240; Farlow 1881, p. 122 (both as *Callithamnion cruciatum*); Croasdale 1941, p. 214.

V. **radicans** (J. Agardh) Collins and Hervey (1917, p. 141) shows decumbent primary filaments attached by rhizoids, forming a tangled mass, and these primary filaments producing lax, pinnate branches above; this variety has been reported from southern Massachusetts on woodwork.

Antithamnion pylaisaei (Montagne) Kjellman

Plants erect-bushy, rose-pink, to 15 cm. tall; main branching repeatedly alternate; axes about 250 μ diam., cells 3 diameters long, all branches bearing opposite (sometimes whorled) short branches about 0.5–1.0 mm. long, these repeated or immediately bearing two rows of short, stout branchlets with cells 12–20 μ diam., the cells about 1.0–2.0 diameters long; tetrasporangia sessile, about 60–70 μ long, 38–47 μ diam., replacing the branchlets; cystocarps paired.

Long Island to the lower St. Lawrence, Newfoundland, Labrador, Hudson Strait, and Baffin Island, uncommon in its southern range; growing on wharves and algae below low-water mark; fruiting in the spring.

REFERENCES: Algae Exsic. Am. Bor. 155 (as *Callithamnion pylaisaei;* Phyc. Bor.-Am. 97. Harvey 1853, p. 239; Farlow 1881, p. 123 (all as *C. pylaisaei*); Wilce *in herb.*

Antithamnion plumula (Ellis) Thuret

Plants bushy, bright rose, considering their delicacy rather stiff, the flattened habit of the fronds evident at a glance, to 5–12 cm. tall; alternately to irregularly branched, branches 100–200 μ diam., the cells 3–5 diameters long; all main branches beset with opposite (infrequently 4-ranked) short, more or less recurved branches about 0.5–0.7 mm. long, 50 μ diam., the cells 2–3 diameters long, these bearing on the upper side a series of ultimately spinuliform branchlets which taper from 20 μ at the base to their indurated tips; gland cells present on the upper sides of the branchlets, each attached to a single cell, sometimes seriate; tetrasporangia elliptical to globose, to 35–40 μ long, 25–30 μ diam., sessile or on short stalks, on the lower branchlets of the upper part of the plant; spermatangia forming elliptical patches on the upper sides of the branchlets; cystocarps large, usually 2–4 together, on the upper branches and more or less surrounding the supporting branchlet.

New Jersey, Block Island, Rhode Island, and southern Massachusetts; growing on wharves, shells, or stones in moderately deep water, and occasionally dredged; fruiting in the late summer.

REFERENCES: Phyc. Bor.-Am. 496. Harvey 1853, p. 238; Farlow 1881, p. 124 (all as *Callithamnion plumula*).

Seirospora Harvey, 1846

Plants tufted, delicate, alternately branched below, subdichotomously branched above, the ultimate branches capillary, the lower somewhat corticated; cells cylindrical, uninucleate, with many small chromatophores; sporangia on the upper branches, lateral at the nodes, tetrapartite or tetrahedral; on the asexual plants also occur seirospores, produced in radiating clusters at the ends of the branches, the outermost largest, grading to the point of origin; carpogenic branches three- or four-celled, with two auxiliaries; cystocarps with the gonimoblasts loose, entirely converted into carposporangia.

Seirospora griffithsiana Harvey Pl. 43, fig. 11

Plants erect from basal holdfasts, pyramidal, bushy, 2–7 cm. tall, bright rose, very soft; branching alternate, widely spreading, below the main branches 200–275 μ diam., the cells 1.5–2.5 diameters long, somewhat corticated, mostly devoid of small branches; above more closely branched, toward the tips somewhat corymbose, with the ultimate branchlets 8–14 μ diam., the cells 7–10 diameters long; tetrasporangia scattered on the upper branches, generally short-stalked, irregular in type of division, 60–77 μ long, 42–53 μ diam.; seirospores in corymbose terminal clusters, of several somewhat branched filaments, the outer or mature spores 25–40 μ diam., 30–50 μ long.

New Jersey to Massachusetts; common south of Cape Cod on Zostera and Chorda, or other coarse algae to 18 meters depth; producing seirospores abundantly in the latter part of the summer.

REFERENCES: Phyc. Bor.-Am. 391. Harvey 1853, p. 237; Farlow 1881, p. 129 (both as *Callithamnion seirospermum*); Lewis and Taylor 1928, p. 195.

Callithamnion Lyngbye, 1819 [1]

Plants tufted, erect from a disk, fibrous holdfast or decumbent strands of monosiphonous branching filaments, ecorticate or with rhizoidal cortications, the branching seemingly dichotomous, or alternate; cells plurinucleate, with several to many small rounded to band-shaped chromatophores; sporangia borne on the upper side of the branches, tetrahedral, or rarely transversely bipartite; spermatangia forming small colorless tufts, borne near the bases of the branchlets on the upper side; in preparation for the carpogenic branch a fertile axial segment early forms two large opposite auxiliary mother cells, one serving as the support of the carpogenic branch, which is of four cells and lies partly across the fertile segment; after fertilization the auxiliary mother cells each cut off above large auxiliaries which jointly form the paired gonimoblast masses, of which all cells except the lowest become carposporangia.

KEY TO SPECIES

1. Upper branching subdichotomous, but becoming corymbose-crowded toward the apices of the branches, the ultimate branchlets conspicuously more slender...... C. corymbosum, p. 299
1. Upper branching alternately bilateral.......................... 2
2. Cortication very scanty, limited to the base of the plant.......... 3
2. Cortication obvious, on the main and chief lateral branches........ 4
3. Basal branching not obvious, the divisions slender, spreading widely, the whole plant very pale and delicate.. C. byssoides, p. 297
3. Basal branching obvious; the ultimate branchlets not particularly slender, not corymbose-crowded toward the apices C. roseum, p. 298
4. Only the principal branches much corticated; habit soft though not slippery, light red.......................... C. baileyi, p. 299
4. Nearly all branches densely corticated; habit coarse, dark red C. tetragonum, p. 300

Callithamnion byssoides Arnott

Plants forming very soft globose tufts 3–8 (–12) cm. high, usually light pink; filaments extremely delicate, the lower branching ob-

[1] *Callithamnion polyspermum* C. Agardh, Florida, South Carolina, and New York (Harvey 1853, p. 234; Farlow 1881, p. 126), and *C. tocwottoniensis* Harvey *ms.*, Rhode Island (Farlow 1881, p. 130), are doubtful records.

scured by their slenderness and the involved habit, though but slightly corticated; axes repeatedly pinnately branched, the segments to 40 (–140?) μ diam., the outer secondary branches freer and flexuous, pinnate with many completely pinnately compound smaller branches, the ultimate branchlets very slender, at the base about 10–20 μ diam., at the terminal cell 5–7 μ diam., the cells 6–10 (or more?) diameters long; tetrasporangia sessile on the upper side of the branchlets, oblique-obovate or subglobose, 38–52 μ long, 28–40 μ diam.

Florida, South Carolina, New Jersey to northern Massachusetts, and Nova Scotia, not infrequent south of Cape Cod; growing on Zostera and coarse algae in relatively sheltered places and shallow water; fruiting during the summer.

REFERENCES: Algae Exsic. Am. Bor. 153. Harvey 1853, p. 234; Farlow 1881, p. 127.

This plant is quite variable and often difficult to distinguish from forms of *C. corymbosum,* even though the typical plants are easily differentiated. Harvey (1853, p. 235; Farlow 1881, p. 128) attempted to discriminate between three varieties which, however, are hardly important. V. **fastigiatum** Harvey has its branches fastigiate, with the lesser ones densely beset with branchlets at the tips; it is a common form in southern New England, particularly resembling *C. corymbosum.* V. **unilaterale** Harvey is a small and specially delicate form with unilaterally placed branchlets; it is reported from Long Island and Massachusetts. V. **waltersii** Harvey has the upper divisions repeatedly bilaterally branched.

Callithamnion roseum (Roth) Harvey

Plants erect, tufted, to 4–7 cm. tall, rather deep pink or red and relatively firm; lower branches little entangled, or almost free, the segments 3–4 diameters long, 125–350 μ diam., below eventually a little corticated, repeatedly alternately radially branched, above bilaterally pinnately branched and subcorymbose near the tips; pinnately divided ultimate determinate branches borne on indeterminate branches of all orders, the branchlets 38–42 μ diam., little tapered, cells 2–4 diameters long; tetrasporangia few, generally paired, nearly spherical, mostly unilateral on the inner side of the branchlets, 60–85 μ long, 45–70 μ diam.

New Jersey to southern Massachusetts; growing on coarse algae or, less frequently, on Zostera; common in the late summer, but also found in midwinter.

REFERENCES: Phyc. Bor.-Am. 442. Farlow 1881, p. 125.

Callithamnion corymbosum (J. E. Smith) C. Agardh

Plants forming erect, rounded, soft tufts 2–6 cm. tall, bright pink; below irregularly or pinnately branched, the branches not much entangled, not particularly delicate, 250–450 μ diam., often a little corticate at the base, the cells 4–10 diameters long; above more densely branched, the lesser branchlets dividing dichotomously, forming corymbiform tufts, the tips usually bearing delicate colorless hairs; tetrasporangia 2 or 3 together on the inner side of the branchlets, sparsely produced.

Bermuda, New Jersey to northern Massachusetts, and Nova Scotia, relatively uncommon northward; growing on coarse algae and Zostera to 18 meters depth; fruiting in the late summer and autumn.

REFERENCES: Phyc. Bor.-Am. 444, 1647 (perhaps only in part). Harvey 1853, p. 236; Farlow 1881, p. 128.

V. **secundatum** Harvey (1853, p. 237; Farlow 1881, p. 128) shows the lesser branches frequently secund, the ultimate branchlets irregularly disposed, hardly corymbose; it is reported from Long Island Sound and Massachusetts Bay; the species is quite variable, and this is but one of several ill-defined forms.

Callithamnion baileyi Harvey Pl. 44, fig. 5; pl. 46, figs. 6–9

Plants bushy, from fibrous holdfasts, to 4–7 (–12) cm. tall, dull rose or somewhat bleached, texture softly spongy; branching pyramidal, with long primary axes obviously corticated, to 400 μ diam., the flexuous secondary axes longest near the base of the plant; lesser branches densely enveloping the axes except near the base of old plants, spreading, pinnately alternately branched, somewhat clustered, the ultimate branchlets curved, gradually tapering to acute tips, at the base 20–40 μ diam., the cylindrical cells 3–5 diameters long; tetrasporangia scattered, usually not abundant, sessile on the upper sides of the branchlets, at the outer ends of the lowest 1–3 cells, 70–90 μ long, 55–70 μ diam.; plants dioecious or monoecious, cystocarps bilobed, near the forking of a branch, to about 200 μ across; antheridia in hemispherical clumps near the upper end of the lower cells of smaller branches and branchlets.

New Jersey to Maine, Nova Scotia, and Prince Edward Island; on coarse algae, such as Phyllophora and Chondrus, on Zostera, and on other plants; south of Cape Cod a common plant of the latter part of the summer, fruiting freely, persisting into the winter.

REFERENCES: Phyc. Bor.-Am. 296. Harvey 1853, p. 231; Farlow 1881, p. 127.

A rather variable species; though not sharply distinguishable from *C. tetragonum* it is otherwise easily recognized, but its locally recorded varieties are of quite vague delimitation. V. **boreale** Harvey (1853, p. 231) is more slender in its cells, with those of the branches in the middle of the plant about 4–5 diameters long and those in the branchlets also longer than in the typical form; it is recorded as the commoner form from northern Massachusetts. V. **laxum** Farlow (Phyc. Bor.-Am. 494; Farlow 1881, p. 127, 1893, p. 108; ?*C. dietziae* Hooper in Harvey 1853, p. 236) is more delicate, shows less marked cortications than does the type, and the branchlets are longer and more slender and more fastigiate at the tips; it is recorded from Rhode Island and southern Massachusetts. V. **rochei** Harvey (1853, p. 231) is also a soft and slender form, with a particularly plumose habit, the branchlets long and slender, with divisions spreading, crowded at the ends of the branches; it is reported from northern Massachusetts. V. **squarrosum** Harvey (1853, p. 231) shows the ultimate branches short and little divided, the branchlets short and squarrose; it is recorded from southern Massachusetts.

Callithamnion tetragonum (Withering) C. Agardh

Plants erect, bushy, from disklike holdfasts, 6–20 cm. tall, coarsely spongy and subrigid, dark red; the main branches strong, to 500–750 μ diam., the lateral branches giving a triangular aspect to the main divisions, the main and secondary axes devoid of branchlets; corticated heavily except toward the tips; lateral branches of the lesser orders closely radially beset with branches of the final order, which bear numerous alternate branchlets more or less in two rows, or somewhat clustered, abruptly acute, incurved, at the base 65–90 μ diam., above 75–100 (–140) μ, the cells 1–5 diameters long; tetrasporangia rather scarce, more or less unilateral on the inner side of the branchlets.

New Jersey, Connecticut to southern Massachusetts, New Hampshire, and Maine, not rare south of Cape Cod; on coarse algae in water of moderate depth; apparently to be found throughout the year.

REFERENCES: Phyc. Bor.-Am. 844. Harvey 1853, p. 230; Collins 1901a, p. 134; Croasdale 1941, p. 214.

Pleonosporium Nägeli, 1861

Plants in erect bushy tufts, filamentous, the filaments uniseriate, pinnately branched, corticated at the extreme base only; sporangia on the branchlets dividing to form polyspores; spermatangia in ovoid tufts on the branchlets; cystocarps terminal, with a slight involucre.

Pleonosporium borreri (J. E. Smith) Nägeli

Plants of erect habit, light to brownish red, 3–6 (–12) cm. tall; the axis repeatedly alternately radially branched, the lower segments 100–360 μ diam., uncorticated except at the extreme base, where there is some rhizoidal investment; upper branching becoming bilateral, the final branching producing somewhat triangular stalked pinnate blades; branchlets 25–40 μ diam., cells 2–3 diameters long, somewhat tapered to rounded tips; sporangia sessile, formed on the upper side of the branchlets at the nodes, continuing division to produce polyspores; cystocarps bilobed, terminal on a branchlet, with a slight involucre; spermatangia in oblong masses, on the branchlets.

Florida, South Carolina, New Jersey to southern Massachusetts; fruiting during the summer.

REFERENCES: Algae Exsic. Am. Bor. 193 (as *Callithamnion borreri*); Phyc. Bor.-Am. 342. Harvey 1853, p. 233; Farlow 1881, p. 124 (both as *C. borreri*).

When sterile this is with difficulty distinguished from some callithamnia. The form of the final branching, producing flat or nearly flat triangular plumes with a naked stalk as long as or longer than the blade, is the best distinction. American material referred to this species seems much more slender than that from Europe.

Spermothamnion Areschoug, 1847

Plants tufted, of uniseriate, branched, uncorticated filaments, the basal part stoloniferous, attaching by lobed disciform holdfasts, the erect filaments oppositely or unilaterally branched, the cells multinucleate and with many small chromatophores; sporangia stalked on the branchlets near the base, tetrahedral; spermatangia forming oval or cylindrical clusters on the upper side of the branchlets, or terminal; carpogonia developed near the tips of lateral branchlets, these tips having three small cells, on the middle one of which, with pericentral and sterile cells, is borne the four-celled carpogenic branch ending in the carpogonium with a short trichogyne; after fertilization the two auxiliary cells divide to produce gonimoblasts of which only the outermost cells produce carpospores.

Spermothamnion turneri (Mertens) Areschoug
Pl. 44, fig. 1; pl. 45, figs. 9–11 (*v.*)

Plants with creeping primary filaments which closely invest the host and support a thick bushy tuft of erect filaments about 2–5 cm. tall, the clumps often crowded together, purplish rose, spongy in texture; creeping filaments 30–45 μ diam., the cells 3–6 diameters long, attached by haptera; erect filaments oppositely pinnate, the long branches spreading, slender, with the main filaments 20–80 μ diam., the cells 3–8 diameters long, and the ultimate branchlets 20–45 μ diam.; tetrasporangia spherical, 45–65 μ diam., borne in series on the upper side of the branchlets near the base on stalks one cell long, or the stalk sometimes forked, sometimes absent; procarps subterminal on the main filaments or the main branches, often of tetrasporic plants, the cystocarps becoming surrounded by several small branchlets.

Florida, New Jersey to Nova Scotia; common south of Cape Cod on Chondrus and Phyllophora in shallow water below the low-tide level or deeper; through the year, producing tetrasporangia from early summer, procarps from late summer, but spermatangia and cystocarps infrequently in our area.

REFERENCES: Phyc. Bor.-Am. 197. Harvey 1853, p. 241 (as *Callithamnion turneri*); Farlow 1881, p. 119; Collins 1901b, p. 292; Taylor 1928a, p. 196; Drew 1934, p. 549, 1943, p. 23.

V. **variabile** (C. Agardh) Ardissone has the branches and branchlets alternate or unilateral, instead of opposite; it is found with the typical form. Both forms of branching frequently appear on the same shoot, so that the variety may not be of much consequence.

Griffithsia C. Agardh, 1817

Plants erect, bushy, of notably large-celled falsely dichotomous or laterally dividing branches; cells multinucleate, with many small chromatophores and a large central vacuole; generally bearing delicate colorless repeatedly tri- to polychotomously branched hairs near their upper margins; tetrahedral sporangia whorled at the fertile nodes, partly covered by involucral cells; plants dioecious, the spermatangia very small, in large caplike sori which are usually on the distal ends of the outer cells of the fertile branches; each procarp formed upon an apical cell, which is displaced laterally by growth of the cell below; mature procarp consisting of a broad central cell, certain sterile cells, and the four-celled carpogenic branch, and after fertilization an auxiliary in contact with the carpogonium; mature cystocarps apparently lateral on the branches, partly covered by involucral cells.

KEY TO SPECIES

1. Branching seemingly dichotomous; nodes generally constricted, the cells pyriform in average shape .... G. globulifera
1. Branching alternate, the secondary branches formed laterally near the middle of the supporting cells; cells cylindrical, little constricted at the nodes .... G. tenuis

Griffithsia globulifera Harvey Pl. 43, figs. 1–4

Plants tufted, the holdfasts fibrous, height 4.0–6.5 cm., color bright rose, texture delicate, easily crushed; branching erect and fastigiate, occasionally proliferous, the branches generally moniliform of coenocytic segments which vary from somewhat clavate near the base of the plant to pyriform and nearly spherical at the apex, in length 0.6–3.2 mm., diam. 0.2–0.7 (–1.5) mm., not corticated, except the few lowest segments by descending attaching rhizoids; in sporangial and carposporic plants the filaments somewhat tapering at the apex, but in the male plants with the cells increasingly large and spherical toward the tip; spherical tetrahedral sporangia 8–20, more or less, at a fertile node, their diameter 40–50 μ, attached by short single-celled supports, which may bear secondary lateral sporangia, and enveloped by incurved sausage-shaped involucral cells; dioecious, the carpospores in conspicuous masses single at the fertile nodes, partly covered by similar involucral cells; spermatangia in sori, generally forming a cap over the outer end of the terminal cell of the fertile branch, and sometimes over the outer ends of 1–2 cells below the tip.

From the tropics, Florida, New York to northern Massachusetts; a common annual south of Cape Cod, growing on stones and shells in relatively protected areas below low tide; fruiting abundantly in late summer.

REFERENCES: Algae Exsic. Am. Bor. 88; Phyc. Bor.-Am. 295, 1447 (all as *G. bornetiana*). Harvey 1853, p. 228 (as *G. corallina* v. *globifera*); Harvey in Kützing 1862, p. 10, pl. 30 a–d (as *G. globulifera*); Farlow 1881, p. 131; Lewis 1909, p. 639, 1912, p. 240, 1914, p. 31 (all as *G. bornetiana*); Taylor 1928a, p. 194 (as *G. globifera*).

What may be a very lush form of this species reaching 20 cm. in height is sometimes found in quiet water during the summer. The cells are relatively long and narrow and the branches sometimes arise a bit below the tip of the supporting cell, but the dichotomous habit of branching is not obscured.

Griffithsia tenuis C. Agardh

Plant tufted, 2–7 cm. tall, the branches cylindrical, color light rose, texture soft and gelatinous; habit erect and fastigiate, branching chiefly alternate, the cells forming lateral branches from a point near the middle of the supporting cell; segments coenocytic, nearly cylindrical, diam. 120–300 μ, from 3–6 diameters to 5–10 mm. long, little or not constricted at the nodes; tetrasporangia 50 μ diam., on evident stalk cells, in whorls at the nodes, with slender involucral cells.

Bermuda, Virginia, New Jersey, Long Island, and southern Massachusetts; forming loose tufts in very quiet, warm bays; a summer annual.

REFERENCES: Harvey 1853, p. 228 (as *G. corallina* v. *tenuis*), 1858a, p. 130 (as *Callithamnion tenue*).

In an earlier account the author (1937c, p. 328) confused an attenuate form of *G. globulifera* with this species, but he has seen several apparently correctly identified specimens from our area and the data have been revised accordingly.

Plumaria Stackhouse, 1809

Plants bushy, the branching bilateral, ultimately opposite; the axis and branches corticated, the cortex arising from the basal branchlet cells and from cells cut off from the axis, but the youngest branches and the branchlets of a single series of naked cells; tetrahedral sporangia, or parasporangia of several cells, terminal on the branchlets; spermatangia clustered near the ends of the branchlets; each four-celled carpogenic branch developed on a lateral branchlet; immediately after fertilization the supporting cell cuts off an auxiliary which originates the gonimoblasts, the lowermost cell remaining sterile and fusing with the supporting and auxiliary cells; cystocarp surrounded by incurved involucral branchlets.

Plumaria elegans (Bonnemaison) Schmitz

Pl. 30, figs. 9–10; pl. 52, fig. 2

Plants tufted, the branching in flat sprays, dark red to brownish purple, the main branches fairly stiff, to 550 μ diam., height 5–20 cm.; lower branching apparently alternate, usually free from branchlets, or with small proliferous pinnate branches, but that in the ultimate divisions strictly opposite; main axis, principal branches, and the lower portion of the better-developed lesser, compressed branches corticated, but the young part of the lesser branches and the branchlets uniseriate, naked; younger branches at first strongly incurved, later straighter, those of indefinite growth alternating on opposite sides of the stem, each opposite a neighbor of limited growth, always oppositely bilateral and closely beset with incurved branchlets which are longer on the lower side of the branch than on the incurved upper side, not greatly tapered, the tips rounded, the lower cells 18–22 (–30) μ diam., 0.75–1.50 diameters long; tetrasporangia or parasporangia terminal on branchlets; cystocarps terminal, surrounded by involucral branchlets; spermatangia clustered near the ends of the branchlets.

New Jersey to the lower St. Lawrence; in the northern part of its range becoming intertidal, growing under overhanging Fucus on vertical rock faces, but south of Cape Cod on the basal parts of Laminaria, and on Phyllophora, etc., from moderately deep water; infrequently found in fruit during the summer, probably fruiting earlier in the year.

REFERENCES: Algae Exsic. Am. Bor. 84; Phyc. Bor.-Am. 445. Harvey 1853, p. 224; Farlow 1881, p. 133 (both as *Ptilota elegans*); Collins and Hervey 1917, p. 138 (as *Gymnothamnion sericeum*); Taylor 1937c, p. 330 (as *Plumaria sericea*).

Ptilota C. Agardh, 1817

Plants bushy, the branching abundant, bilateral, ultimately opposite, though members of a pair may be dissimilar; branches and axis flattened, in large part or entirely corticated over a stout uniseriate axis by cells originating from the bases of the branchlets; tetrasporangia terminal on short stalks crowded on somewhat modified branchlets; spermatangia in clusters near the ends of the branchlets; carpogenic branches four-celled, developed terminally on lateral branchlets, the auxiliary cut off beside the carpogonium after fertilization, producing a lobed cystocarp which may become surrounded from below by later-developed involucral branchlets.

KEY TO SPECIES

1. Ultimate branches alternately unequal in size and unlike in form; sporangia crowded together with little sterile filaments on subcylindrical branchlets.......................... P. serrata
1. Ultimate branches alternately unequal in size, but similar in form; sporangia terminal on pinnately placed stalks.... P. plumosa

Ptilota serrata Kützing — Pl. 52, fig. 1

Plants bushy, stiff, dark brownish or purplish red, height to 15 cm.; branched freely, the fronds flat in aspect, the main branching irregularly opposite or one member of an opposed pair suppressed, above tending to become more regular, with all main axes distinctly flattened, the last grade of branching of two generally regularly alternate forms, with undivided serrate branches similar to the branchlets across the stem from each pinnate branch; ultimate branchlets opposite, incurved, often subsimple and large, frequently the lower or outer margin deeply triangularly serrate or the projections partly produced as secondary branchlets, the upper margin entire to serrate; tetrasporangia on reduced subcylindrical branchlets, these usually opposite flattened sterile ones of normal form, the tetrasporangia crowded, to 35–40 μ diam.; associated with short incurved uniseriate filaments; cystocarps terminal on much-reduced branchlets, also opposite sterile ones, surrounded by incurved involucral secondary corticated branchlets with serrate margins.

New Jersey, Connecticut to the lower St. Lawrence, Hudson Bay and Hudson Strait, Labrador, Newfoundland, Baffin Island, and Ellesmere Island; rare and in especially exposed stations south of Cape Cod, but common north of it; growing on stones or coarse algae, especially about the bases of Laminaria plants; perennial, fruiting in the summer.

REFERENCES: Algae Exsic. Am. Bor. 85 (as *Ptilota plumosa* v. *serrata*); Phyc. Bor.-Am. 392 (as *P. pectinata*). Harvey 1853, p. 222; Farlow 1881, p. 133; Kjellman 1883, p. 176; Rosenvinge 1926, p. 33 (both as *P. pectinata*); Taylor 1937c, p. 329 (as *Plumaria pectinata*).

Ptilota plumosa (Hudson) C. Agardh

Plants rather similar to *P. serrata,* to 30 cm. tall; branches of the last grade opposite on the axis, one of each pair large, the other small, these alternating along each side of the stem, both forms pinnated with determinate branchlets; tetrasporangia single at the ends of short stalklike pinnated branchlets, which are sometimes irregularly forked.

Prince Edward Island, Ile Miquelon, Hudson Strait, and the more northern arctic coast.

REFERENCES: Harvey 1853, p. 224; Hariot 1889, p. 195; Wilce *in herb.*

Ceramium Roth, 1797 [1]

Plants erect and bushy, or sometimes partly matted; branching dichotomous or more seldom alternate; branches segmented, the

[1] *Ceramium capri-cornu* (Reinsch) Farlow (*C. youngii* Farlow), New York and Anticosti Island, also *C. corymbosum* C. Agardh (Farlow 1881, p. 138), are records of doubtful value.

axis uniseriate, of relatively large cells, corticated at the nodes by a ring of smaller cells, in some species spreading so as to cover completely the axial internodes; sporangia spherical, tetrahedral, sessile, borne at the nodes or, in the completely corticated species, often between them, immersed, projecting or largely naked; spermatangia formed of minute colorless cells in a layer on the corticated portions; the carpogenic branches borne on a lateral segment of an axial cell acting as a supporting cell, forming a hair rudiment and one or two four-celled carpogenic branches; after fertilization the auxiliary forms a close mass of carpospores which may become involucrate by a few incurving branchlets.

KEY TO SPECIES

1. Cortication limited to the nodes or but slightly extended......... 2
1. Cortication partly or entirely continuous....................... 7

2. Length of naked internodes not more than thrice the length of the corticated node; apices mostly erect.................. 3
2. Naked internodes much longer.................................. 5

3. Plant essentially erect....................................... 4
3. Plant partly decumbent, spreading, with attaching rhizoids and erect branches.......... C. deslongchampii v. hooperi, p. 312

4. Sporangia immersed or exposed slightly on the outer side; the plant purplish red.......................... C. elegans, p. 311
4. Sporangia clearly emergent; the plant dull purplish C. deslongchampii, p. 312

5. Sporangia immersed or a little emergent; the nodal cells becoming irregular, usually several counted lengthwise of the node 6
5. Sporangia emergent or naked; nodal cells in 2–6 fairly distinct transverse series .... C. fastigiatum, p. 309

6. Proliferous branches few or none; cells of the lower margin often 17–20 μ diam........................ C. strictum, p. 310
6. Proliferous branches numerous; cells of the lower margin of the lower nodes usually not averaging over 13 μ diam. C. diaphanum, p. 311

7. Internodes not completely covered in all portions of the plant.... 8
7. Internodes completely covered throughout..................... 10

8. Naked internodes generally less than half as long as broad C. rubriforme, p. 314
8. Naked internodes longer in upper parts of the plant............ 9
9. Nodal growth principally downward; branch tips at first strongly incurved .... C. circinatum, p. 313
9. Nodal growth principally upward; branch tips but little incurved .... C. areschougii, p. 313
10. Older branches showing large cells about the nodes partly covered by small superficial cells........... C. rubriforme, p. 314
10. Older branches showing little difference on the surface between cells of nodal and of internodal cortication...... C. rubrum, p. 315

Ceramium fastigiatum (Roth) Harvey Pl. 47, figs. 3–5, 7; pl. 48, figs. 2–4; pl. 49, figs. 3–4; pl. 50, fig. 4; pl. 51, figs. 6–7

Plants forming dense tufts about 4–8 cm. tall, bright rose-red, very delicately capillary,. the branching regularly dichotomous, fastigiate above, the nodes distinct but not darkly colored, and the internodes somewhat pigmented above; apices divergent, erect or slightly incurved; young nodes with two rows of cells, the upper smaller and more irregular, becoming deeper with age, the intermediate and commonest condition of 3–4 rows of cells, of which the lowest are broader than long, the next higher with larger cells a little sunken, and the upper of about two rows of progressively smaller cells; in oldest parts of 4–6 rows, the lower band of about two irregular rows, the cells often transversely broader, the next band of larger somewhat sunken cells, and the upper band of progressively smaller cells, the marginal ones a little longitudinally elongate; diameter of the older nodes about 60–155 μ, much broader than long (usually 4–6 cells and 55–65 μ long), the margins sharp, the cells moderately differentiated; older internodes 75–150 μ diam., 600–1360 μ long, but the rhizoid-bearing lowest segments less than 200 μ long; tetrasporangia single, or to 4–6 at a node, greatly projecting, somewhat covered below by upgrowth of a few cells, or only the base of the sporangium a little immersed, the sporangia to about 35–65 μ diam., 50–68 μ long, often somewhat oval; cystocarpic plants sometimes with a few lateral branches; cystocarps lateral, sometimes 2–3 together, the involucral branchlets 2–4, short, hardly longer than the cystocarp.

From the tropics, New Jersey to northern Massachusetts, and the Gaspé; on coarse algae and Zostera in shallow, relatively quiet coves; fruiting sparingly in the late summer.

REFERENCES: Phyc. Bor.-Am. 446b, 497 (as *C. tenuissimum non* (Lyngbye) C. Agardh), 895 (as *C. tenuissimum* v. *patentissimum*). Harvey 1853, p. 217 (probably also including *C. arachnoideum non* C. Ag.); Farlow 1881, p. 137 (including also *C. tenuissimum* and vars.); Collins 1905, p. 234.

Two fairly distinct types seem to be referable to this species. One is stouter and more deeply colored and has divergent abrúptly tapered branch tips; the other forms larger paler tufts and has erect or slightly incurved more gradually tapered tips, but there seems to be no difference in nodal structure.

Ceramium strictum (Kützing) Harvey

Pl. 47, fig. 6; pl. 48, figs. 5–6; pl. 51, fig. 5

Plants tufted, the tufts soft, 2–8 cm. tall, dull red, the nodes distinct, the branches rather slender, regularly dichotomous with few lateral proliferations, somewhat spreading or fastigiate above; tips of the branches forcipate; the nodal diameter in older parts about 210–336 μ, length 85–210 μ, sharply delimited at both margins, the cells of somewhat unequal size in the oldest, some of the largest cells of the inner series partly exposed across the center of the node, the cells of the lower margin much smaller than those of the upper margin, 10–22 μ diam.; internodes nearly colorless, the older to about 185–270 μ diam., the segments 0.6–1.5 mm. long; tetrasporangia immersed in the central part of the nodal band, or somewhat exposed above, the sporangia 50–60 μ diam., 60–70 μ long; cystocarps lateral on the upper branches, the 4–6 involucral branchlets considerably overtopping them.

Florida, the Carolinas, New Jersey, Connecticut to Nova Scotia, and Prince Edward Island; common over muddy bottom, on shells, stones, Zostera plants, and various objects, and generally found in relatively quiet bays; fruiting in the summer.

REFERENCE: Farlow 1881, p. 136.

This plant is not always easy to distinguish from *C. diaphanum;* it is more slender, often smaller, with fewer proliferations, the older nodes are less cut up into very small cells, and the older sporangiate nodes are much less irregularly swollen. The numerous plants of intermediate character defy allocation; characteristically differentiated individuals have been used for illustration and measurement.

Ceramium diaphanum (Lightfoot) Roth

Pl. 47, fig. 2; pl. 48, figs. 7–9; pl. 51, figs. 1–4

Plants bushy, 5–10 (–20) cm. tall, dull brownish red, the nodes very sharply demarcate from the colorless internodes; branching slenderly dichotomous, erect-spreading, with numerous lateral adventitious branches, these being in turn often dichotomously forked; branch tips forcipate, later somewhat erect, the outer margins moderately to very little roughened; the nodal diameter in older parts to about 335–450 μ, the nodes broader than long, sharply delimited at both margins and hardly spreading, the cells of unequal size, with very large ones within, which are partly exposed across the center of the band, partly covered and bordered by many small cells which at the lower margin of the older nodes are about 8–13 μ diam.; internodes nearly colorless, to about 200–460 μ diam., the segments 0.50–1.25 mm. long, often somewhat swollen below; tetrasporangia immersed in the upper part of the node, which in the oldest parts may become much lobulate-swollen, the sporangia to about 50–75 μ diam.; cystocarps on short lateral or sublateral branches, near the tip, little overtopped by the 3–4 involucral branchlets; spermatangia in patches on the upper nodes.

Virginia to Maine and Prince Edward Island; common in southern New England on stones, shells, coarse algae and Zostera at one to few meters depth below low tide, in relatively quiet bays; spring to autumn, fruiting during the latter part of the summer.

REFERENCES: Phyc. Bor.-Am. 347, 846 (as *C. strictum non* (Kützing) Harvey), 847 (as *C. strictum* f. *proliferum*). Harvey 1853, p. 215.

Ceramium elegans (Ducluzeau) C. Agardh

Pl. 48, fig. 12; pl. 49, fig. 7; pl. 50, fig. 3

Plants somewhat coarse but not stiff, bushy, to 7–8 cm. tall, dark purplish red, repeatedly dichotomous, sometimes with lateral proliferations; branch apices a little forcipate, soon becoming erect;

internodal segments short, subequal to 2–6 diameters long, toward the base of the plant the internodal cells somewhat swollen just above the nodal band, thence contracted toward the upper end; the nodes to about 280–340 μ diam., and about one half broader than long, structure of the node showing a central zone of large embedded cells, partly exposed and partly covered by the small angular surface cells which form the remainder of the band, the cells along the lower margin about 14 μ diam.; tetrasporangia verticillate, in a single series above, but two series below, nearly immersed across the middle of the swollen cortical zone; cystocarps on the subterminal branches, with an involucre of 4–5 branchlets, which may fork and which greatly overtop the cystocarp.

New Brunswick; summer.

Ceramium deslongchampii Chauvin Pl. 48, fig. 1; pl. 49, figs. 1–2

Plants forming tufts 6–11 cm. tall, somewhat slender, freely and repeatedly dichotomously branched, with numerous lateral proliferous branches, the main forkings distant, with as many as 40 nodal bands between them, the terminal divisions erect-subulate; cortical bands about as long as broad, a little more below the nodal junction than above, cells near the center of the band much bigger than at the margin, largely immersed, the upper margin a little straighter with the cells transversely wider, the lower margin a little less regular and the cells somewhat longer; internodal segments in the lowest parts of the plant 240–320 μ diam., about 3–4 times as long as broad; tetrasporangia few, generally crowded to one side of the swollen node, in a single series and partly emergent.

Maine, with some uncertainty; probably fruiting in late summer or autumn.

Only one or two American specimens seen by the writer seemed to conform with the strict limitations of this species. The commoner American plants seem to be the v. **hooperi** (Harvey) Taylor (Pl. 49, figs. 5–6; pl. 50, fig. 2; Phyc. Bor.-Am. 98; Harvey 1853, p. 214; Farlow 1881, p. 136, all as *C. hooperi;* Taylor 1937c, p. 338), plant forming a dull purplish matted layer attached and consolidated by abundant rhizoid production, sending up erect branches to 5–9 cm. tall which are irregularly dichotomous with lateral proliferations; tips of the branches straight or slightly incurved; segments of the axis as long as broad, or somewhat longer; in the lowest part of the plant the nodes rather short, rhizoidiferous, with subequal naked internodal areas between, the diameter about 150 μ; in the erect branches the diameter lower down about 335 μ, higher to 380 μ, the internodal cells there to 500–590 μ long, the nodes a little wider than long, the upper and lower margins of the band similar, sharply defined, of cells of moderate size, with a zone of largely immersed larger cells between; tetrasporangia in smaller branches nearly external, but in the older immersed at the nodes in a ring causing much swelling, so that the nodes reach 340–380 μ broad, 210 μ long, the sporangia to 80 μ diam., in 1–2 circles; sexual reproduction undescribed; not adhering closely to paper. Connecticut (?), northern Massachusetts to Maine, and the lower St. Lawrence; particularly on vertical rock faces between tide marks, forming a mat overhung by Fucus and other coarse algae.

Ceramium circinatum (Kützing) J. Agardh

Fronds setaceous, dichotomous, fastigiate, divisions erect, spreading, apices forcipate, internodes partly corticated by the cells which are decurrent from the nodes; tetraspores large, projecting in a ring around the upper nodes.

Long Island, Massachusetts, Anticosti Island, and Prince Edward Island; usually growing on coarser algae in shallow water; fruiting in the summer. So far as the writer has seen, specimens from these places attributed to this species all seem to have been incorrectly identified and so all these records are questionable.

REFERENCE: Farlow 1881, p. 135.

Ceramium areschougii Kylin, *prox.*

Pl. 48, fig. 11; pl. 49, figs. 8, 10; pl. 50, figs. 1, 5–6

Plants of small to moderate size, to 1.0 (–2.5) dm. tall, color light red, moderately clearly banded; moderately slender throughout, about 0.5 mm. diam. below, the lesser divisions about 0.15 mm. diam.; branching below irregularly dichotomous with strong lateral branches producing the effect of distinct main axes, the upper divisions rather long and erect; branch tips rather blunt, slightly incurved when young, soon erect or divergent; segments above rather longer than broad, to 2–3 diameters long below and obscured in the oldest parts; nodes very slightly contracted in the uppermost divisions, the internodes in the segments of the middle parts of the plant contracted toward their upper ends; nodal bands nearly touching in the youngest parts, separated by a clear band one-half to as long as broad in the upper forkings, but below disappearing by merging of the spreading margins; nodal outgrowth somewhat delayed, the nodes continuing as little-differentiated well-separated bands, but eventually progressively extended by the outgrowth of the nodes, particularly upward, until the internodes are covered rather thinly, the cells formed above the septum more slender and less close than those below, which may be much wider than long, the upper margin consequently more irregular than the lower; at full development showing larger cells partly emergent about the septum, with small cells over most of the node proper, the internodal extension of elongate cells; tetrasporangia in the nodes of the upper divisions, usually in a single row, immersed or in slender branchlets strongly projecting and emergent.

Southern Massachusetts; growing in a much-protected cove; bearing tetraspores in early summer.

Ceramium rubriforme Kylin, *prox.*
Pl. 48, fig. 10; pl. 49, fig. 9; pl. 50, fig. 7; pl. 52, fig. 4

Plants bushy, of but moderate size and to about 1 (–2) dm. tall, color light red, obscurely banded; rather coarse below, quite slender in the lesser divisions where about 0.1 mm. diam.; branching a little irregular and spreading below, more regularly and closely dichotomous in the upper divisions with occasional secondary lateral branches; branchlet tips very strongly incurved (especially in the carposporic plants), the outer margins markedly dentate in contour and bearing colorless deciduous hairs; segments short above, about two diameters long in the middle parts, obscured below; nodes very slightly constricted in the younger parts, old internodes slightly contracted above the center; nodal bands nearly touching in the youngest parts, separated by a narrow band generally less than one-half as long as broad in the upper forkings, and below merging by spreading of the margins; nodal outgrowth tardily active, the cells above the septum somewhat more slender than those below it, throughout the development showing 1–2 rows of much larger deep-placed cells about the septum and other somewhat smaller cells over them or near by, the whole but partly covered by the small superficial cells which stretch (most strongly upwardly) over the internode in a rather thin layer; sporangia in 1–2 rows, immersed in and nearly covered by the nodal bands of the upper or the middle divisions; cystocarps reported as on the uppermost divisions, with usually two short involucral branchlets.

Southern Massachusetts and New Hampshire; growing in quiet, shallow water; with sporangia in the summer.

REFERENCES: Taylor 1937c, p. 340; Croasdale 1941, p. 214.

Ceramium rubrum (Hudson) C. Agardh

Pl. 47, fig. 1; pl. 52, figs. 5–7

Plants forming large bushy masses, from disklike bases rising to 1–4 dm. tall, color red, of various shades depending on location, obscurely banded; coarse, to 1 mm. diam., and somewhat firm, at least below; branching dichotomous, with, in some forms, irregular habit and more or less abundant lateral branches, tapering to the upper divisions; branch tips forcipate to erect; segments short above, to about two diameters long below, but often obscured in the lowest parts; nodes little swollen, and below indeed more often contracted; nodal rings almost contiguous from the apex, shortly behind which the intervals between the nodal bands may be shown by narrow lines, later obscured; mature axial internodal cells entirely covered by spreading corticating growth from the node, the cortex of small angular to rounded cells, those at the surface not very different about the septa; tetrasporangia 55–80 μ diam., in rows about the nodes, later arising from the spreading cortication, immersed, or causing slight swelling; spermatangia in crowded tufts on younger portions of the plant; cystocarps on lateral branchlets, invested by involucres of 3–6 branchlets which curve up and over them.

From the tropics, Florida, North Carolina to the lower St. Lawrence, Newfoundland, and Baffin Island; from the intertidal zone to water of moderate depth, on stones, shells, woodwork, Zostera, and coarse algae; through the year, fruiting in summer.

REFERENCES: Phyc. Bor.-Am. 345, 646, 894b (*C. botryocarpum p. p.*). Harvey 1853, p. 213; Farlow 1881, p. 135; Hariot 1889, p. 195.

No other alga of our flora when first collected in quantity gives more clearly the impression of belonging to several species, but intergrades between the forms are really commonplace. The following varieties have been recognized in our territory, but intergrades rather than distinctive plants are ordinarily found. V. **corymbiferum** J. Agardh (?Phyc. Bor.-Am. 2249) is closely and regularly dichotomous throughout, the ultimate divisions very slender and corymbose, with the branch tips little incurved; the plant is reported from Connecticut and northern Massachusetts. V. **pedicellatum** Duby (Phyc. Bor.-Am. 645, 894a, as *C. botryocarpum*) is densely corticated at the nodes, but more thinly between them, so that the branches are distinctly banded; branching occurs at somewhat narrow angles, in the cystocarpic plants becoming abundantly forked-proliferous, but more sparsely proliferous in the tetrasporic individuals, with the proliferations pointing irregularly in all directions, rather long, tapering somewhat to base as well as apex, the cystocarps and tetrasporangia primarily on the proliferous branches, but also on the upper main branches; this plant has been reported from Rhode Island and southern Massachusetts and is often considered a distinct species. V. **proliferum** Harvey (1853, p. 214; Farlow 1881, p. 135) shows many short, sometimes forked branchlets usually abundantly covering all orders of branches except the youngest; it occurs in New Jersey, Connecticut to southern Massachusetts and Nova Scotia. V. **secundatum** Harvey (1853, p. 214; Farlow 1881, p. 135) is freely laterally branched, the branchlets generally short and in a row along one side of the supporting branch; it is reported from New Jersey, Connecticut, northern Massachusetts, and Maine. V. **squarrosum** Harvey (1853, p. 214; Phyc. Bor.-Am. 198; Farlow 1881, p. 135) is a small form, fastigiate in habit, regularly but distantly dichotomous below, with but few proliferations; the upper divisions closely succeed each other and spread very widely; this plant is reported from New Jersey, northern Massachusetts, and Nova Scotia.

Spyridia Harvey, 1833

Plants forming erect bushy masses; much alternately branched, the branches corticate by transverse series of longitudinally elongate chromatophore-bearing cells, these later subdivided and covered by rhizoidal downgrowths on the larger axes; ultimate branchlets of limited growth, deciduous, consisting of a uniseriate axis of large cells bearing rings of small chromatophore-bearing cells at the nodes, the elongated internodes naked and translucent, the nodes often, and the tips generally, armed with short, indurated spine cells; sporangia seriate on the upper sides of the branchlets, tetrahedral; spermatangia originating on the nodes of the branchlets; cystocarps near the tips of small branches, often surrounded by slender involucral branchlets.

Spyridia filamentosa (Wulfen) Harvey

Pl. 44, fig. 2; pl. 46, figs. 2–5

Plants bushy, from a rhizoidal disk holdfast, to 1.5–2.0 dm. tall, coarse below, alternately branching, the branches all more or less beset with short, slender deciduous branchlets, or denuded below, the general color dull rose to often bleached, straw-colored or dull brownish, the texture softly spongy; main axes corticated by a layer of small cells surrounding the large ones of the axial row, these formed as alternating transverse nodal and internodal series of longitudinally elongate cells which eventually become irregular; diameter of main axes to 1–2 mm.; branchlets radially inserted, about 0.5–1.5 mm. long, 20–45 μ diam., the segments about 2–4 diameters long, with several cells at each node, at first in a single ring, later some cells dividing transversely and obliquely, the branchlet apex with terminal spine cell; tetrasporangia spherical, sessile, 40–70 μ diam., single or whorled upon the nodes of the branchlets; spermatangia about the nodes of the branchlets and extending over the internodes; cystocarps terminal on short branches, with an involucre of incurved branchlets.

From the tropics, Florida to southern Massachusetts; in protected waters, especially warm coves, on stones and shells below low water, often adrift on the bottom and pale in coloration; fruiting in the summer, and perhaps persisting through the year, as it is known to do toward the south.

REFERENCES: Algae Exsic. Am. Bor. 151; Phyc. Bor.-Am. 393. Harvey 1853, p. 204; Farlow 1881, p. 140; Hoyt 1920, p. 512; Taylor 1928a, p. 197.

This species is particularly variable. In nearly landlocked bays it develops with special luxuriance, the plants large, if exposed to much light pale in color, and very brittle. Where there is more circulation of the water, the plants are more slender and bright rose. V. **refracta** Harvey (1853, p. 205; Farlow 1881, p. 140) is a form with subdichotomous branching, the branches spreading, the terminal ones flexuous, curved to hooked and often entangled; it is common in southern Massachusetts.

DELESSERIACEAE[1]

Plants usually foliaceous, simple, sometimes bushy, alternately or infrequently dichotomously branched, the branches membranous or infrequently subcylindrical; growth from an apical cell producing an axial row which originates connected lateral cell rows of several degrees to produce a membrane and sometimes also a cortex, or the initial cells with age obscured; sporangia tetrahedral, usually in superficial sori; spermatangia in sori; procarps borne on supporting cells which form one or two four-celled carpogenic branches, and groups of sterile cells; auxiliaries developed from the supporting cell after fertilization; cystocarp with a basal fusion cell and branched gonimoblasts, the outer cells of which are carposporangial, the whole invested by an inflated, ostiolate pericarp.

KEY TO GENERA

1. Plants forming large foliar structures 1 dm. long or more.......... 2
1. Plants smaller, or blades narrowly dissected..................... 3

2. Blades ordinarily entire, though occasionally branching from the stalk or proliferating from the surface..... GRINNELLIA, p. 324
2. Blades pinnately lobed when fully developed, or at least crenate PHYCODRYS, p. 322

3. Plants small, grayish purple, segments linear-lanceolate, contracted at the generally dichotomous forkings, and often attaching there by rhizoids.................. CALOGLOSSA, p. 319
3. Plants rose or brighter purple, mostly abundantly alternately branched, the segments not contracted at the forking........... 4

4. Branches with a distinct midrib and thin lateral membrane MEMBRANOPTERA, p. 320
4. Branches more narrow, compressed but without lateral membranes .... PANTONEURA, p. 319

[1] *Nitophyllum laceratum* Greville, Newfoundland (Harvey 1853, p. 104), is a doubtful record.

Caloglossa (Harvey) J. Agardh, 1876

Plants in the form of flat, dichotomously forking, locally constricted blades; each developing from a prominent apical cell a midrib consisting of a broad axial row of large cells covered by a cortex of elongated cells, this midrib bordered by a monostromatic lateral membrane consisting of subhexagonal cells running in oblique series from costa to margin; secondary branching proliferous from the midrib; tetrasporangia spherical, developed in oblique series from cells of the lateral part of the blade, mostly near the upper end; cystocarps sessile on the midribs, each enveloped in a thin pericarp.

Caloglossa leprieurii (Montagne) J. Agardh Pl. 53, figs. 2–3

Plants spreading or tufted, to 4–5 cm. across, color violet; the blades to 2 mm. broad, constricted at the forkings and elsewhere, the individual segments lanceolate, 4–6 mm. long, sometimes linear-attenuate, more rarely ovate, often forming rhizoids at the constrictions; secondary segments or blades formed here or proliferously from the midribs of the blades.

From the tropics, Florida to South Carolina, Maryland, New Jersey to Connecticut; on woodwork, old roots and rocks, in estuaries, sheltered coves, and the edges of salt marshes.

REFERENCES: Algae Exsic. Am. Bor. 66 (as *Delesseria leprieurii*). Harvey 1853, p. 98; Farlow 1881, p. 163 (both as *D. leprieurii*); Collins 1905, p. 232; Post 1936, p. 49.

This plant and *Bostrychia rivularis* are among the most frequent Rhodophyceae in weakly brackish water and even spread to entirely fresh-water streams. Collections from within our area are attributed by Post to f. **pygmaea** (Mart.) Post, which has internodes 0.6 mm. broad or less, but those from the Carolinas southward, and the plants illustrated, belong to the broader typical form.

Pantoneura Kylin, 1919

Plants with a branching habit, branches narrowly linear to filiform, alternately marginally pinnate or occasionally dichotomous; growth apical, the segments isolating a primary axial series the cells of which do not divide further, but become corticated without developing any lateral membranes; sporangia in the upper segments or in subaxillary lateral branchlets; procarp probably with a single carpogenic branch; cystocarps in the middle line in the uppermost segments, or in subaxillary short branchlets.

KEY TO SPECIES

1. Branchlet tips obtuse............................ P. angustissima
1. Branchlets tapering, the tips acute........................ P. baerii

Pantoneura angustissima (Turner) Kylin

Plants bushy, 10–15 cm. tall, branching more or less in a plane; branches linear, compressed, about 0.5–1.0 mm. diam. below, above little broader; branching repeatedly alternate, or in part dichotomous, the apices obtuse; tetrasporangium-bearing segments a little expanded and thickened, terminal or lateral, often forked; cystocarp-bearing segments much swollen, subulate, short.

Northern Massachusetts (with some doubt), and Baffin Island; usually growing attached to the bases of Laminaria plants in moderately deep water.

REFERENCES: Farlow 1881, p. 163 (as *D. alata* v. *angustissima*); Collins 1900b, p. 50 (as *D. angustissima*).

Pantoneura baerii (Postels et Ruprecht) Kylin Pl. 44, fig. 6

Frond from a subconical holdfast, erect, to 15–18 cm. tall, very slenderly pinnate below, the lesser branches somewhat dichotomous, compressed below, to 2 mm. broad and somewhat firm, the upper segments rather erect, linear-acuminate; tetrasporangial sori expanded, in the upper part of the subterminal segments; cystocarps immersed in the terminal segments, the pericarps with prominent short thick beaks.

Labrador, Hudson Strait, and Ellesmere Island.

REFERENCES: Algae Exsic. Am. Bor. 188 (as *Delesseria baerii*). Rosenvinge 1893, p. 806 (as *D. baerii*); Wilce *in herb.*

Membranoptera Stackhouse, 1809

Plants with branching habit, branches linear, growth apical, alternately pinnate to dichotomous, with a conspicuous thin monostromatic wing bordering the midrib; growth from a prominent apical cell; sporangia developed submarginally in spreading sori in the tips of small branches or proliferations, which become corticated; spermatangia developed in marginal sori which occupy the tips of small branchlets or proliferations, no midrib being present on these, but the fertile area becoming corticated; procarps, sometimes in contiguous segments, developed along the midrib of the fertile blade, each from an axial cell which first cuts off laterally a supporting pericentral cell; this developing the four-celled carpogenic branch; the auxiliary cell, formed after fertilization, develops branched gonimoblast filaments of which the outer two to four cells become carposporangia.

KEY TO SPECIES

1. Blades with entire margins and no distinct lateral veins...... M. alata
1. Blades with serrate margins, or lamina reduced; minute lateral veins distinct .... M. denticulata

Membranoptera alata (Hudson) Stackhouse Pl. 53, figs. 4–6

Plants forming tufts 7–20 cm. tall, irregularly branched below, alternately branched above, the apices erect to incurved, the segments linear, 1.5–3.0 (–7.5) mm. wide, with a strong midrib completely corticated by elongated cells, and margined by a distinct membrane of cells averaging about 12.5 μ diam., which is traversed by obscure and immersed opposite veinlets arising from the midrib and is bounded by an entire margin; reproductive organs generally borne in the ultimate segments of the blade or on small lateral branchlets, causing some widening and thickening; tetrasporangial sori covering most of the central part of the fruiting tip with scattered sporangia; cystocarp within a swollen pericarp, borne on the midrib; spermatangia covering the distal lateral part of the fruiting leaflets.

Northern Massachusetts to Maine, the lower St. Lawrence, northern Labrador, Hudson Strait and Baffin Island; a perennial species from the zone below tidal action, growing on coarse algae or on stones; fruiting in the winter.

REFERENCES: Phyc. Bor.-Am. 291 (as *Delesseria alata*). Harvey 1853, p. 95 (as *D. alata*); Farlow 1881, p. 163; Collins 1903b, p. 204; Wilce *in herb*.

Membranoptera denticulata (Montagne) Kylin Pl. 53, figs. 7–8

Plants small, freely branched, the branches flat, narrow, in color rose to purple; segments below subdichotomous, 2–5 mm. broad, above alternately pinnate, 0.5–5.0 mm. wide, entire to minutely serrate, the apices straight; with a distinct midrib eventually corticated by small elongated cells, and with broad, thin lateral membranes, the cells of which are about 23 μ in transverse diameter and are occasionally covered by scattered superficial cells which are about 10–15 μ diam.; membrane traversed by many divergent microscopic veinlets which extend from the midribs to the margin, particularly to the minute marginal teeth, composed of cells 90–110 μ long, 25–35 μ diam., and which in the younger portions of the plant are not immersed; tetrasporangia borne on special lateral branchlets (possibly on the ordinary branches as well); cystocarps borne on the midribs.

Long Island, northern Massachusetts to Maine, the lower St. Lawrence, Labrador, Hudson Strait, and Ellesmere Island; on stones about the bases of Laminaria plants and on *Ptilota serrata;* summer.

REFERENCES: Algae Exsic. Am. Bor. 140; Phyc. Bor.-Am. 291 (both as *Delesseria alata non* (Hudson) Stackhouse), 995 (as *D. denticulata*). Harvey 1853, p. 94; Rosenvinge 1893, p. 802 (as *D. montagnei*), 1926, p. 33; Collins 1903b, p. 204 (as *D. denticulata*).

This plant varies considerably in width. V. **angustifolia** (Lyngbye) Taylor (Collins 1903b, p. 207; Taylor 1937b, p. 231) has the lateral membrane ridgelike or very narrow. It is found through much the same range as the type.

Phycodrys Kützing, 1843

Plants short-stalked, forming pinnately lobed blades monostromatic except at the veins; the lobes growing from a small primary apical cell and later-formed marginal initials; a strong midrib present, corticated lateral veins present and opposite, microscopic veins doubtful or absent; immersed tetrahedral sporangia formed in sori, either submarginal on the ends of lateral veins or in minute marginal ligulate proliferations, the membrane thickened to about five layers of cells; spermatangia formed in submarginal bands; carpogonia scattered over the blade, formed in pairs on the opposite sides of the thallus, the primary membrane cell cutting off segments toward each face as supporting cells each of which forms sterile cells and a four-celled carpogenic branch; previous to fertilization the thallus membrane about the procarp becomes three layers in thickness; after fertilization the support cuts off an auxiliary cell which forms the gonimoblasts, outer cells maturing carpospores about a large fusion cell at the base, enclosed by a distinct pericarp with a pore, but cystocarps do not necessarily mature on both sides of the blade.

Phycodrys rubens (Hudson) Batters [1]

Pl. 30, fig. 8; pl. 42, fig. 1; pl. 46, fig. 1

Plants of stalked, lanceolate-ovate or later deeply lobed blades, bright purple-red, to 10–15 (–30) cm. tall, 2–5 (–12) cm. wide; blade with a midrib and distinct opposite lateral vein systems for each major lobe, the veins disappearing in the margin of the blade, which is somewhat sinuate-serrate; lateral lobes may develop similar to the primary blade in size and form; tetrasporangial sori at the ends of the veinlets near the margins of the primary blades or ultimately occupying little lateral leaflets; spermatangia forming a narrow band just within the margin of the blade; procarps scattered irregularly over the surface of the blade; cystocarps covered with a pericarp, on the primary blades or in old specimens on special lateral leaflets.

New Jersey to the lower St. Lawrence, Newfoundland, northern Labrador, James Bay, southeastern Hudson Bay and Strait, Baffin Island, Devon Island, and Ellesmere Island; common north of Cape Cod, growing in rather deep water on stones and shells; a biennial or a perennial, fruiting at various seasons, possibly throughout the year, but particularly in autumn and winter.

REFERENCES: Phyc. Bor.-Am. 435 (as *Delesseria sinuosa*). Harvey 1853, p. 93; Farlow 1881, p. 162 (both as *D. sinuosa*); Rosenvinge 1926, p. 33; Wilce *in herb.*

Without doubt the most beautiful bladelike red alga of our flora, this plant varies considerably in the general contour of its frond. V. **lingulata** (C. Agardh) Batters (Phyc. Bor.-Am. 1242, in some

[1] It is probable that *Delesseria fimbriata* De la Pylaie from Newfoundland (Harvey 1853, p. 84) belongs to this species.

issues only) is linear lanceolate to almost filiform, but is ill defined and, with other variants, is found with the typical plant in its American range.

Grinnellia Harvey, 1853

Plants of usually simple pink blades; a prominent midrib composed of elongate inner cells covered by rounded outer ones; lateral part of the blade one cell in thicknes and lateral veins absent; growth from an apical cell; sporangia in scattered, elongate, unilaterally projecting nodular sori, more or less elevated, several cells thick, the tetrahedral sporangia formed from cells below the surface; spermatangia in scattered, often confluent sori; procarps individually developed on special islets of cells usually in the plane of the main blade and surrounded by normal blade cells, the supporting cell developing from the axial row of this islet, forming a four-celled carpogenic branch and sterile cells; cystocarps scattered, in each the gonimoblasts composed of large carposporangial cells without and smaller cells within, attaching to a large fusion cell, the whole covered by a thin-walled hemispherical pericarp, which discharges through an apical pore.

Grinnellia americana (C. Agardh) Harvey
Pl. 30, figs. 4–7; pl. 42, fig. 2; pl. 59, fig. 8

Plants more or less gregarious, of large, erect simple translucent pink blades which are occasionally proliferous from the base, and consist of a very short stalk bearing a lanceolate or ovate-oblong blade 1–5 (–10) dm. long, 4–10 cm. or more in width, often decaying at the apex, the strong midrib extending nearly the length of the blade; tetrasporangia in elongate sori about 0.3–1.0 mm. wide and 0.5–2.0 mm. long, scattered over the surface of the blade, the sporangia about 50–65 μ diam. and occasionally germinating in place; cystocarps scattered over similar blades, the pericarps about 0.3–0.6 mm. diam.; spermatangia in separate or confluent sori on much smaller blades which are only about 1–3 cm. long.

The Carolinas, New Jersey to northern Massachusetts; on wharves, stones, or shells in warm quiet water at moderate depths below low tide; fruiting during early summer and midsummer and then disappearing.

REFERENCES: Algae Exsic. Am. Bor. 64; Phyc. Bor.-Am. 593, XXII. Farlow 1881, p. 161; Brannon 1897, p. 1; Lewis 1912, p. 239; Hoyt 1920, p. 495.

DASYACEAE

Plants bushy or with long, terete primary axes; branches radially or dorsiventrally bearing monosiphonous branched filaments of limited growth which may be free or united into a network; growth not continuous from a persisting apical cell, the successive segments before cortication producing laterally a new growing point displacing the preceding apex, which develops into a lateral tuft of filaments; axial cell row in some genera becoming surrounded by a circle of pericentral cells, and in many cases also corticated by the development of rhizoidal downgrowths from the bases of the lateral filaments; these forking filaments in some cases becoming polysiphonous in the lower segments, but monosiphonous above and often ending in colorless filiform extensions; tetrasporangia produced in special swollen polysiphonous branchlets or stichidia; colorless spermatangia borne on lateral branchlets; procarps developed near the bases of the lateral tufts of filaments, a fertile pericentral cell producing sterile cells and the four-celled carpogenic branch, the auxiliary being cut off from the support near the carpogonium; cystocarp enveloped by an ample ostiolate pericarp.

Dasya C. Agardh, 1824

Plants erect, more or less bushy, with stout main branches covered with filiform branchlets; structurally the main branches with five pericentral cells surrounding the axial row, the whole corticated heavily in the older portions of most species by rhizoidal filaments; branchlets crowded on the axes, whorled, spiraled, or scattered, only polysiphonous about the base if at all, and above dividing into delicate, monosiphonous, pseudodichotomously branched chromatophore-bearing filaments; tetrasporangia developed in distinctive siliquose, stalked stichidia; spermatangial clusters lanceolate to subcylindrical, on lateral filaments and often hair-tipped; procarps developed on a fertile segment near the base of a lateral filament tuft; the four-celled carpogenic branch associated with sterile cells; after fertilization the auxiliary cell is cut off from the support beside the carpogonium and produces the gonimoblasts, the outermost cells of which form the carposporangia in moniliform series of about four cells around a large fusion cell; pericarp with a large apical pore, developed immediately after fertilization.

Dasya pedicellata (C. Agardh) C. Agardh Pl. 54, figs. 1–4

Plants from small disklike holdfasts, erect to 2–6 dm. tall, light to deep red-purple, sparingly to freely alternately branched, the lateral branches infrequently redivided, 2–6 mm. diam., sometimes denuded below, but above densely covered with slender monosiphonous chromatophore-bearing filaments, to 4–7 (–14) mm. long, the cells near the base 15–40 μ diam., at the tips 7–12 μ diam.; the stichidia lanceolate to linear-lanceolate, acute, 80–120 μ diam., 0.2–1.25 mm. long, the sporangia tetrahedral, 40–50 μ diam.; sexual plants dioecious; the spermatangial clusters lanceolate to linear-lanceolate, acute, and usually filament-tipped, 60–75 μ diam., 250–550 μ long; the stalked pericarps urceolate, below transversely oval, to 0.75–1.00 mm. diam., above with a narrow ostiolate neck about 200 μ diam.

From the tropics, Florida to northern Massachusetts; growing on shells and stones from 1–4 meters below low tide in protected waters where there is free tidal flow, and occasionally, though smaller, on coarse algae or Zostera; fruiting from midsummer to early autumn.

REFERENCES: Algae Exsic. Am. Bor. 51; Phyc. Bor.-Am. 545, XXIII (all as *Dasya elegans*). Harvey 1853, p. 60; Farlow 1881, p. 171; Lewis 1912, p. 241, 1914, p. 31; Taylor 1928a, p. 173 (generally as *D. elegans*).

RHODOMELACEAE

Plants usually bushy, sometimes sparingly branched; branches often delicate, usually terete, occasionally flat; growth from persisting apical cells producing an axial cell row; branched colorless hairs (trichoblasts) often present; axial cells generally surrounded at least in the fruiting portions by a series of pericentral cells cut off from them by longitudinal walls, producing a typically polysiphonous structure, and sometimes further corticated either by subsequent divisions of these to several degrees or by appressed rhizoidal downgrowths; tetrahedral sporangia formed from internal segments of the pericentral cells, the branchlets bearing them usually little modified, but in extreme cases like stichidia; spermatangial clusters developed from trichoblast rudiments in the form of colorless tufts, cones, or plates of spermatangia; procarps developed from polysiphonous basal trichoblast segments, the fertile pericentral cell as a supporting cell producing sterile cells and the four-celled carpogenic branch; from the supporting cell beside the carpogonium after fertilization an auxiliary is cut off, from which, by means of a fusion cell organized after fertilization, the gonimoblasts are produced; outer cells of the gonimoblasts alone forming carpospores; cystocarps becoming enclosed by an ostiolate pericarp.

KEY TO GENERA

1. Main branches at least in part flattened........ ODONTHALIA, p. 346
1. Branches terete throughout.................................... 2

2. Habit dorsiventral .... BOSTRYCHIA, p. 342
2. Branching radial or bilateral, but not dorsiventral................ 3

3. Pericentral cells in at least the younger branches symmetrically arranged .... POLYSIPHONIA, p. 330
3. Pericentral cells irregular or obscured.......................... 4

4. Plants pale to purplish, coarse and fleshy; pericentral cells remaining intact but forming outwardly a small-celled cortex which invests the axis almost to the apex....... CHONDRIA, p. 327
4. Plants dark reddish purple, branchlets slender; pericentral cells dividing to form a close small-celled investment about the axis RHODOMELA, p. 343

Chondria C. Agardh, 1817

Plants bushy, much and alternately branched, the branches terete, those of the upper divisions constricted at the base and the ultimate spindle- or club-shaped, the branchlets tipped by clusters of branched colorless hairs (trichoblasts); axis with five irregular pericentral cells surrounded by a loose cortex of branching cell series terminating in small cells closely approximated as an epidermal layer; tetrasporangia usually single at a node, in ultimate or penultimate branchlets which are little altered; spermatangial clusters of various shapes, commonly in the form of flat or twisted colorless plates borne near the apices of the branchlets; procarps developed from the second segment of a trichoblast rudiment; immediately after fertilization the active pericentral cell cuts off an auxiliary beside the carpogonium which then originates the branching gonimoblasts, though later these are borne on a fusion cell which includes the auxiliary, the supporting and the inner gonimoblast cells; pericarp in its lower portion becoming several cells thick.

KEY TO SPECIES

1. Branchlets pyriform to club-shaped, short or moderately slender... 2
1. Branchlets slender, spindle-shaped or the end a little blunt........ 3

2. Branching irregular to alternate, subradial to bilateral; ultimate branchlets short and obtuse; trichoblasts not conspicuous, and not staining the mounting paper on drying.. C. sedifolia, p. 330
2. Branching radially alternate; ultimate branchlets moderately short or to 20 mm. long, the ends a little pointed to truncate; trichoblasts very markedly developed, often staining the mounting paper dark brown................ C. dasyphylla, p. 329

3. Plant coarse, generally growing on stones or shells; branchlets tapering both ways from near the center; tetrasporangia toward the upper end of the older ultimate branchlets C. tenuissima, p. 329
3. Plant more slender, generally growing on Zostera or coarse algae; branchlets little tapered distally, the end rather blunt; tetrasporangia in the thickened median portion even of the oldest fertile branchlets..................... C. baileyana, p. 328

Chondria baileyana (Montagne) Harvey Pl. 55, fig. 4

Plants rather densely bushy, 7–10 (–25) cm. tall, gregarious, pale straw-colored or pale purple; slender in aspect and soft in texture; the main axis subsimple or with a few long similar branches, habit generally narrowly pyramidal, less commonly broader; main axis 0.3–1.0 mm. diam., branchlets not congested, to about 5 mm. long, 80–200 μ diam., elongate club-shaped, the distal ends little narrowed, ultimately obtuse; tetrasporangial branchlets with the sporangia 55–105 μ diam., in a circumscribed band well below the apex, and in the oldest branchlets rather near the center.

New Jersey to northern Massachusetts; epiphytic on fuci, less commonly on other coarse algae and Zostera; this plant may occur in very sheltered coves as rounded tufts drifting over the bottom, the branches and branchlets flexuose to arcuate, the habit entangled; fruiting by midsummer.

REFERENCES: Algae Exsic. Am. Bor. 187; Phyc. Bor.-Am. 43 (both as *C. tenuissima* v. *baileyana*). Harvey 1853, p. 20; Farlow 1881, p. 166 (as *Chondriopsis tenuissima* v. *baileyana*); Collins 1905, p. 232.

Chondria tenuissima (Goodenough et Woodward) C. Agardh [1]
Pl. 40, fig. 5; pl. 55, figs. 1–3

Plants bushy, 10–20 cm. tall, pale straw-colored or dull purple; main branches coarse, firm, the lesser segments rather soft; the main axis simple, or with a few similar main branches, 1.0–2.5 mm. diam.; primary axes bearing numerous long, widely divergent secondary axes 5–8 cm. long; ultimate branchlets usually rather distantly placed, spindle-shaped, 2–7 (–15) mm. long, 0.25–0.5 mm. diam., tapering to both ends, the apices pointed, bearing evident tufts of trichoblasts; tetrasporangia 40–60 μ diam., scattered in the upper portion of the fruiting branchlets; pericarps short-stalked or subsessile, scattered along the branchlets.

From the tropics, Florida, New Jersey to northern Massachusetts; on rocks and shells a little below low-tide line in rather open water, seldom on coarse algae; fruiting by midsummer.

REFERENCES: Phyc. Bor.-Am. 42. Harvey 1853, p. 21; Farlow 1881, p. 166 (as *Chondriopsis tenuissima*); Taylor 1928a, p. 171.

This is the common coarse Chondria of southern New England, whereas *C. baileyana* is the yet commoner delicate one. The distinctions between the New England forms are not always evident, and many specimens cause difficulty.

Chondria dasyphylla (Woodward) C. Agardh Pl. 54, figs. 5–6

Plants bushy, 10–20 cm. tall, pale straw-colored or light brownish purple; the main axis with several long similar branches, the habit of each broadly pyramidal; main axis 1–2 mm. diam., erect; branchlets single or clustered, club-shaped, contracted at the base, obtuse

[1] Early reports of *C. littoralis* Harvey and *C. atropurpurea* Harvey (Farlow 1881, p. 167, as *Chondriopsis spp.*) are doubtful, although a specimen ascribed to the latter as derived from New York does seem correctly identified, whatever its origin.

and finally retuse at the apex, with a central apiculus bearing a particularly conspicuous tuft of trichoblasts, the older branchlets generally of uneven, somewhat torulose contour, length 2–3 (–10) mm., diam. 200–500 μ; tetrasporangia formed in the distal portion of the fertile branchlets; spermatangial clusters flat, transversely oval, developed from a portion of a normal trichoblast; pericarps lateral on the branchlets, with a very short stalk, single or 2–3 on one branchlet; showing a considerable tendency to stain when mounted and dried on paper.

From the tropics, Florida, North Carolina, New Jersey to southern Massachusetts; fruiting in summer.

REFERENCES: Algae Exsic. Am. Bor. 186 (?); Phyc. Bor.-Am. 142 (some sets). Harvey 1853, p. 20; Farlow 1881, p. 166 (as *Chondriopsis dasyphylla*); Hoyt 1920, p. 500; Taylor 1928a, p. 170.

Chondria sedifolia Harvey — Pl. 55, figs. 5–6

Plants densely bushy, 10–15 cm. tall, dull reddish purple or somewhat faded, coarse and rather firm in texture, the main axis 1.0–2.5 mm. diam., with several large erect-spreading branches, in habit broadly pyramidal; the branchlets single or clustered, much contracted to the base, short pyriform to more elongated and club-shaped, the top obtuse and with an apical pit bearing a tuft of trichoblasts, the diameter 0.3–0.6 mm., length 1–2 (–5) mm.; tetrasporangia clustered near the tips of the fertile branchlets; pericarps ovate, subsessile, 1–few on a branchlet or lesser branch.

From the tropics, Florida, North Carolina, Long Island, Rhode Island, and southern Massachusetts; from rocks below low-tide level, in rather open water; fruiting in summer.

REFERENCES: Harvey 1853, p. 19; Farlow 1881, p. 166 (as *Chondriopsis dasyphylla* v. *sedifolia*); Hoyt 1920, p. 501; Taylor 1928a, p. 171.

Polysiphonia Greville, 1824

Plants entirely erect, or with decumbent basal filaments giving rise to the erect portions; usually abundantly dichotomously or laterally branched, the branches filamentous, terete, coarse to finely capillary; all branches polysiphonous, with an axial cell series surrounded by four to twenty-four pericentral cells, the main axes and branches (but not the ultimate branchlets) in some species covered by rhizoidal downgrowths which may build up a pseudoparenchymatous cortex; delicate colorless branched hairs (trichoblasts) present in many species, obsolete or absent in others; sporangia in the upper branches, which are slightly thickened, solitary in the segments, often seriate, the fertile pericentral cell dividing lengthwise and the inner half then dividing transversely, the upper portion being the sporangium mother cell and producing tetrahedral spores; spermatangial clusters colorless, ovoid to subcylindrical, developed from trichoblast rudiments on a short stalk, the fertile portion polysiphonous, bearing on the pericentral cells numerous small fertile spermatangia; procarps developed from trichoblast rudiments, where the initial is the upper pericentral cell of the five in the segment; this acting as a supporting cell cuts off a rudiment from which is developed the four-celled carpogenic branch, two lateral sterile cells and a lower sterile cell; after fertilization the support cuts off an auxiliary beside the carpogonium, the auxiliary receives a diploid nucleus and cuts off a gonimoblast rudiment, after which it fuses with the supporting and the axial cells; gonimoblasts developing in branched series, the cells of the second order being the carposporangial rudiments; the large ostiolate pericarp becoming globose, ovate, or suburceolate.

KEY TO SPECIES

1. Pericentral cells four.. 2
1. Pericentral cells more than four.............................. 8

2. Without cortication, except sometimes a little partial cortication near the base of the plant.............................. 3
2. Corticated, except on the branchlets.......................... 5

3. Main axes usually distinct, branching alternate; the base of the plant not creeping...................... P. harveyi, p. 332
3. Without a distinct main axis, branching in part dichotomous; the base of matted creeping filaments........................ 4

4. Trichoblasts present or absent; a plant of brackish water, or of shallow, muddy, and warm coves....... P. subtilissima, p. 334
4. Trichoblasts generally absent; a plant of exposed coasts and deeper, cold waters......................... P. urceolata, p. 337

5. Main branching irregularly dichotomous, the chief branches very coarse, generally obviously segmented.... P. elongata, p. 336
5. Main branching alternate, much more slender.................. 6

6. Main axes strong, with distinct lateral branching, the shape of the plant usually clearly pyramidal and most of the branches corticated 7
6. Axes throughout slender and flexuous, separating, chiefly near the base, into wandlike divisions which are sparingly redivided; only the largest branches corticated.. P. flexicaulis, p. 335

7. Branch rudiments appearing in the axils of trichoblasts; lower axial segments generally longer than broad; main branches beset with short, often recurved adventitious branchlets; usually light brownish red.................. P. fibrillosa, p. 334
7. Branch rudiments appearing in place of trichoblasts on segments near the tips of shoots; lower axial segments generally shorter than broad; without numerous adventitious branchlets, but plants of a dense habit growing in exposed situations; color generally dark reddish purple. P. novae-angliae, p. 336

8. Pericentral cells usually six.................................. 9
8. Pericentral cells more numerous.............................. 10

9. In the slender lower axes the twisted segments over 2 diameters long .... P. arctica, p. 338
9. In the stout lower axes the segments 2 diameters long or less P. denudata, p. 339

10. Stiff, tufted, dichotomously branched epiphytes.... P. lanosa, p. 341
10. Softer plants which are not restrictedly epiphytic............... 11

11. Branchlets tapering to base and apex; pericentral cells usually 12, ordinarily spirally arranged; throughout ecorticate P. nigra, p. 339
11. Branchlets tapering from base to apex; pericentral cells usually 16, ordinarily straight; bases of large plants slightly corticated .... P. nigrescens, p. 340

Polysiphonia harveyi Bailey Pl. 56, fig. 8

Plants bushy, 3–10 cm. tall, basally attached at first, often continuing growth adrift; texture soft to rather stiff, slender, color pinkish when young, commonly light brown to blackish with age; closely irregularly branched below, the primary branches each formed laterally at the base of a trichoblast, which is often deciduous, the main axes to 420–500 μ diam., segments 330–850 μ long, above alternately branched, the branches short, segments with 4 pericentral cells, uncorticated or a very little about the extreme base of the plant, 0.5–2.0 diameters long; branches beset with numerous spreading simple or forked tapering adventitious branchlets about 1–2 mm. long; trichoblasts numerous at the tips, deciduous below; tetrasporangia 65–85 μ diam., in somewhat nodose branchlets; spermatangial clusters short-cylindrical to ellipsoid, clustered near the tips of the branchlets, 100–150 μ long, 45–55 μ diam.; pericarps broadly ovoid, to 420–510 μ diam., short-stalked.

South Carolina to Nova Scotia, and Prince Edward Island, uncommon and local north of Cape Cod; an annual and perhaps occasionally a biennial, starting growth on Zostera or Chorda, but coming adrift in the late summer in the shallow, warm bays which it usually favors; fruiting during summer.

REFERENCES: Algae Exsic. Am. Bor. 133; Phyc. Bor.-Am. 888, 1400. Harvey 1853, p. 41; Reinsch 1875, p. 50 (as *P. americana*); Farlow 1881, p. 171.

In its most distinctive form this plant is typically dark and rather stiff, with the lateral branchlets abundant and very distinct from those of the main branch system. This form eventually becomes particularly rigid, with the adventitious branchlets more or less recurved or hooked, constituting the seasonal phase known as v. **arietina** (Bailey) Harvey (Phyc. Bor.-Am. 889). V. **olneyi** (Harvey) Collins (Pl. 56, fig. 6; pl. 58, figs. 1–5; Phyc. Bor.-Am. 440, 1445; Harvey 1853, p. 40; Farlow 1881, p. 171, all as *P. olneyi;* Taylor 1937b, p. 227) is larger, 7–25 cm. tall, light in color, branching erect-spreading, more evenly distributed and in all orders longer, more cylindrical and much softer; segments below 280–350 μ diam., 0.75–1.25 diameters long, somewhat irregularly formed, above in the branchlets 30–45 μ diam., the segments 2–6 diameters long; spermatangial clusters more scattered, 25–40 μ diam., 100–145 μ long, rounded-cylindrical; pericarps very short-stalked, broadly ovoid, 280–330 μ diam. Though this plant may grow on Zostera, it quite often occurs on Chorda and on stones or shells, and later comes adrift to continue its development in the drifting state in warm bays and salt-marsh pools; it occurs from Virginia to Nova Scotia, and Prince Edward Island.

Polysiphonia fibrillosa Greville Pl. 56, fig. 9; pl. 57, figs. 1–5

Plants bushy, 8–15 (–30) cm. tall, pyramidal, arising from a disk-like holdfast, reddish brown or more commonly bleached and yellowish, repeatedly alternately branched, the branches with many adventitious branchlets; main axes below to 0.6 mm. diam., with few lateral branchlets, the segments 2–3 diameters long, considerably corticated, the cortications continued to the upper subdivisions; the lesser branchlets ecorticate, with segments about as long as broad, and 4 pericentral cells; except in the first divisions usually abundantly beset with adventitious, spreading, often recurved branchlets 40–96 μ diam., the segments 60–85 μ long; primary branches formed in the axils of the abundant trichoblasts; tetrasporangia 60–80 μ diam., in somewhat spiral series in rather distorted branchlets which are 95–105 μ diam.; spermatangial clusters cylindrical to spindle-shaped, grouped near the ends of the branchlets; pericarps short-stalked, ovoid to globose, 200–420 μ diam.

New Jersey to southern Massachusetts, with somewhat unsatisfactory reports northward to New Brunswick; growing on rocks and coarse algae in sheltered locations, particularly warm bays, to several meters below low-tide level; fruiting during summer.

REFERENCES: Algae Exsic. Am. Bor. 181; Phyc. Bor.-Am. 1244 (in some issues). Harvey 1853, p. 43; Farlow 1881, p. 172; Collins 1896b, p. 462 (as *P. vestita non* Kütz., *non* J. Ag.); Schuh 1933g, p. 391.

Polysiphonia subtilissima Montagne

Plants to 15 cm. tall, color blackish purple, texture soft; arising from a creeping base, the erect filaments subdichotomously branched below, alternately branched above, sometimes with numerous subsimple coarse rhizoids, to 90 μ diam. below, the segments 1.5–2.3 (–8) diameters long, with 4 pericentral cells, ecorticate; ordinary branching erect, wandlike to clustered; adventitious branchlets not numerous; above the branchlets 33–45 μ diam., the segments to 130 μ, usually about 1.5 diameters long; apical cells conspicuous; simple or sparingly forked trichoblasts sometimes present, rather persistent, lateral branches replacing certain of these hairs in development from the apex; tetrasporangia seriate in the branches near the central portions of the plant.

From the tropics, Florida, Maryland to northern Massachusetts; a plant of muddy shores, salt-marsh ditches, and estuaries, often extending up large tidal rivers for a considerable distance beyond any marine admixture.

REFERENCES: Phyc. Bor.-Am. 45. Harvey 1853, p. 34 (including v. *westpointensis*); Farlow 1881, p. 170.

Polysiphonia flexicaulis (Harvey) Collins Pl. 56, fig. 7

Plants arising from basal disks to 40 cm. tall, often gregarious, reddish purple and very soft and slippery in texture; main axis flexuous-angled, single or sparingly alternately divided to the sixth order, the secondary branches ascending, similar to and often exceeding the disappearing primary axis; branchlets delicate, scattered and not clustered; trichoblasts abundant, borne in spiral succession unless replaced by a branchlet; the main axes 500–925 μ diam., the segments 1.7–2.5 mm. long or more (to 8 diameters) and corticate, the branchlets 45–85 μ diam., the segments 100–210 μ long, with 4 pericentral cells; tetrasporangia oval to spherical, 90–110 μ long, in long series in branches of the last two orders; often monoecious, pericarps ovoid to globose, sessile or short-stalked; spermatangial clusters short-cylindrical to ovoid, often terminated by a hair, grouped near the tip of a branch.

Long Island to the lower St. Lawrence; growing in the sublittoral on Chondrus, Laminaria, and other coarse algae in somewhat protected localities; common and fruiting in summer.

REFERENCES: Phyc. Bor.-Am. 1144 (as *P. violacea* v. *flexicaulis*), 1245, 1294 (as *P. schuebelerii non* Fosl.). Harvey 1853, p. 44; Farlow 1881, p. 173 (both as *P. violacea* v. *flexicaulis*); Yamanouchi 1906, p. 401; Lewis 1912, p. 239, 1914, p. 31 (all as *P. violacea non* Greville); Taylor 1937b, p. 227.

The greater part of what has passed as *P. violacea* in American algal literature appears to be this plant, although the following has also been called by that name. In *P. violacea* the branches arise at the bases of trichoblasts, instead of replacing them.

Polysiphonia novae-angliae Taylor

Pl. 58, figs. 6–10; pl. 59, figs. 5–6

Plants each from a basal disk, forming more or less pyramidal tufts 8–15 cm. tall, moderately soft in texture but spongy rather than slippery; color light to dark purplish red; alternately branched and with few branchlets below, above densely and finely branched; main branching somewhat coarse, the principal axes erect, straight, with lateral branches wide-spreading, the segments 600–850 μ diam., 250–600 μ long, with 4 pericentral cells about the axis; corticated, the cortications continued to the lesser branches; upper branching alternate to irregular, more dense, so that the branch tips may be somewhat fasciculate; branchlets 85–100 μ diam., segments 40–70 μ long, ecorticate; primary branches probably replacing the evident trichoblasts on the supporting segments; tetrasporangia 70–85 μ diam., oval or subspherical, in straight or somewhat twisted series, or sometimes alternate in the thicker branchlets, which in the fertile part may fork 1–4 times and reach 120–170 μ diam., the segments to 90 μ long; spermatangial clusters cylindrical and pointed at the tip, to spindle-shaped, diameter 40–50 μ, length 135–175 μ, formed near the tips of the branchlets, not terminating in a hair but often with a distinct projecting sterile axial cell; pericarps broadly urn-shaped, 320–400 μ diam., sessile or very short-stalked on the upper branches.

Southern Massachusetts and apparently also from Rhode Island to Nova Scotia and Quebec; growing on Chondrus, barnacles, stones, and woodwork in rather open water below tide level; fruiting during summer.

REFERENCES: Phyc. Bor.-Am. 595 (as *P. violacea non* Greville). Harvey 1853, p. 44; Farlow 1881, p. 173 (as *P. violacea p. p.*); Taylor 1937b, p. 232.

Polysiphonia elongata (Hudson) Harvey

Pl. 56, fig. 1; pl. 57, figs. 11–13

Plants upright, to 30 cm. tall, bushy, each from a basal disk, the primary filaments particularly stout, light red and harsh below, but the ultimate filaments bright and soft; branching dichotomous to irregular below, the main branches spreading, devoid of branchlets, heavily corticated yet perceptibly banded, to 1.3 mm. diam., the segments averaging 0.6 mm. long; above alternately branched, the erect branches smaller, tapering to base and apex, with 4 pericentral cells and heavy cortex; branchlets 60–90 μ diam., the segments 80–120 μ long; adventitious branchlets present in the upper portions of a plant; either a branch or a trichoblast produced from each segment behind the apex, forming during the early part of the season a close tuft of bright red, soft filaments which are lost before summer; tetrasporangia seriate in the branchlets, to 100–130 μ long, 85–100 μ diam., the fertile branchlets often torulose; spermatangial clusters somewhat conical, with a rounded base, grouped near the tips of the branchlets; pericarps transversely oval, 450–510 μ diam., short-stalked near the bases of the branchlets.

New York to New Hampshire, the lower St. Lawrence, and Prince Edward Island; a perennial growing on stones or shells in rather deep water; producing the delicate apical filaments in winter or early spring, this plant is by midsummer generally reduced to the coarser branches, but it fruits at various seasons.

REFERENCES: Algae Exsic. Am. Bor. 182; Phyc. Bor.-Am. 44, 1597. Harvey 1853, p. 42; Farlow 1881, p. 172; Croasdale 1941, p. 214.

Polysiphonia urceolata (Lightfoot) Greville

Plants forming thick, erect tufts from a creeping base, gregarious, height to 30 cm.; color bright to dark wine-red, texture soft; erect axes not much branched, with more or less abundant short, lateral erect branchlets, the erect filaments below to 210 μ diam., the segments to 1.9–2.1 mm. long, with 4 pericentral cells, ecorticate; branchlets alternate, somewhat distant, to corymbose at the tips of the branches, 40–85 μ diam., the segments 80–210 μ long; branches not associated with trichoblasts, which apparently are absent, or present as mere rudiments, or associated with reproductive structures; tetrasporangia 65–90 μ diam., seriate in somewhat nodose often forked branchlets which become 85–110 μ diam.; spermatangial clusters lateral along the tips of the branchlets, short-stalked, long-conical with sterile filiform tips; pericarps short-stalked, urceolate with somewhat flaring apex when fully matured, below to 480 μ diam.

North Carolina (?), New Jersey to the lower St. Lawrence, Newfoundland, James Bay, Hudson Bay, Hudson Strait, and possibly to the arctic; a perennial plant in its creeping matted base, the erect portion in its best vegetative condition in the spring; fruiting and disappearing during summer in the southern portion of its range.

REFERENCES: Phyc. Bor.-Am. 748. Harvey 1853, p. 32; Farlow 1881, p. 170; Kjellman 1883, p. 118.

The distinctive variation of this species falls into two types, of which v. **roseola** (C. Agardh) J. Agardh (Harvey 1853, p. 33; Farlow 1881, p. 170, as *P. urceolata* v. *formosa*) is a spring plant, with particularly soft texture and light color, the segments to 12 diameters long in the main branches; it ranges from New Jersey to Maine, Nova Scotia, and Prince Edward Island, being particularly abundant toward the south. V. **patens** (Dillwyn) Harvey (Phyc. Bor.-Am. 997; Harvey 1853, p. 32; Farlow 1881, p. 170) is by contrast rather stiff, dark in color, the segments relatively short, the branching divergent and the branchlets often recurved; this plant is found from New Jersey to New Brunswick, later in the season than the typical form, being possibly but a seasonal phase.

Polysiphonia arctica J. Agardh

Plants slender, repeatedly branched, the branches erect; color dull red, the texture soft; branching dichotomous below, the axes altogether ecorticate, 190–210 μ diam., the segments 380–550 μ long, with the pericentral cells sometimes much twisted, increasing with age, even to 1–3 spiral turns; above alternately closely laterally branched, fastigiate, the apices acutely tapering, the branchlets 60–85 μ diam., their segments 100–270 μ long, with the pericentral cells nearly straight; usually 6 pericentral cells, but sometimes 4 in small branchlets to 7 below, the segments in the middle portions of the plant 3–8 diameters long; trichoblasts absent (or deciduous?); tetrasporangia seriate in somewhat torulose, sometimes incurved branchlets; pericarps lateral, short-stalked, ovoid or globose.

The lower St. Lawrence, Labrador, Hudson Bay, James Bay, Hudson Strait, Baffin Island, Devon Island, and Ellesmere Island; probably perennial and growing about the bases of Laminaria plants in 18–36 meters of water; fruiting in late summer.

REFERENCES: Phyc. Bor.-Am. 1293. Kjellman 1883, p. 123; Rosenvinge 1926, p. 33; Wilce *in herb.*

Polysiphonia denudata (Dillwyn) Kützing
Pl. 56, fig. 3; pl. 57, figs. 6–10; pl. 59, fig. 1

Plants to 25 cm. tall, the bases disciform, the lower branches strongly divergent (90°–120°) when young, the upper branches more irregular and erect; color dark reddish purple in the upper branches and branchlets, where very soft in texture, but below the branches slender though stiff and usually nearly colorless, or sometimes somewhat brownish; lower portions devoid of branchlets, 340–750 μ diam., the segments 250–500 μ long, nearly or quite ecorticate; secondary branches 100–170 μ diam., the segments 400–900 μ long; branchlets 33–45 μ diam., the segments 45–80 μ long; throughout with 6 pericentral cells, straight or nearly so; trichoblasts absent, rudimentary or moderately developed, maturing well below the apex and not obscuring the prominent apical cells; branchlets formed laterally at the bases of trichoblasts if these are present; tetrasporangia oval to spherical, 60–100 μ long, seriate in the branchlets, which become 75–90 μ diam.; spermatangial clusters long-conical to lanceolate, mostly near the tips of the branchlets, 240–280 μ long, 50–55 μ diam.; pericarps sessile or short-stalked, subglobose to transversely oval, 330–540 μ diam.

From the tropics, Florida to Maine, Nova Scotia, and Prince Edward Island; infrequent north of Cape Cod, but often common south of it; a plant growing on stones, woodwork, and Zostera in warm, well-protected bays; fruiting in the latter part of the summer.

REFERENCES: Algae Exsic. Am. Bor. 135; Phyc. Bor.-Am. 245, 639. Harvey 1853, p. 45; Farlow 1881, p. 173; Taylor 1937c, p. 370 (all as *P. variegata*); Hoyt 1920, p. 503.

Polysiphonia nigra (Hudson) Batters Pl. 56, fig. 5

Plants reddish purple, tufted, gregarious, in texture moderately firm; from a base which is at first disklike, stoloniferous; habit notably virgate, to 20 cm. tall; main axes infrequently dividing, below sparingly to above somewhat more frequently laterally branched, the axes continuing beyond the lateral branches, the shorter branchlets tapering to each end, about 2–10 mm. long; axes 270–335 μ diam. below, segments 0.8–2.0 diameters long, ecorticate, with 12 (8–14) somewhat spiraled pericentral cells; above the vegetative branchlets about 165–210 μ diam., the segments 0.7–1.0 diameter long; trichoblasts numerous, persistent, the new branches arising from their axils; tetrasporangia borne on plants of markedly pyramidal branching with numerous short, spindle-shaped, nodose forked branchlets which are about 85 μ diam., or where sporangia occur reach 130 μ diam.; tetrasporangia seriate, 60–90 μ diam.; spermatangial clusters subconical to cylindrical with a tapering apex; pericarps short-stalked, on plants with relatively few short branchlets, broadly ovoid to globose, 450–550 μ diam.

New Jersey to northern Massachusetts, and Nova Scotia; growing on stones at moderate depths; fruiting in summer.

REFERENCES: Phyc. Bor.-Am. 544 (as *P. atrorubescens*). Harvey 1853, p. 48; Farlow 1881, p. 174 (as *P. atrorubescens*).

Polysiphonia nigrescens (Hudson) Greville

Pl. 56, fig. 2; pl. 58, figs. 11–12; pl. 59, figs. 2–3

Plants erect, to 10–30 cm. tall, with a distinct main axis and abundant lateral branching; color dark purple to black, texture firm to coarse; main branches often rather naked below, above more abundantly laterally branched, near the tips becoming brushlike, corymbose to bilateral, the ordinary branches produced directly from the subapical segments independent of the many deciduous trichoblasts; adventitious branchlets often present; branches with 16 (8–20) pericentral cells, below to 520–850 μ diam., the segments 300–720 μ long, nearly or quite ecorticate; in the middle portions of the plant the segments 0.5–3.0 diameters long, while the branchlets are 60–85 μ diam., the segments 100–150 μ long; tetrasporangia 60–100 μ diam., seriate in somewhat moniliform, distorted, often forked branchlets, these becoming 100–140 μ diam.; antheridia lanceolate, grouped near the tips of the branchlets; pericarps broadly ovoid, short-stalked, 380–420 μ diam.

South Carolina to the lower St. Lawrence; growing in exposed places as well as in sheltered bays from the winter onward; fruiting in spring and summer.

REFERENCES: Harvey 1853, p. 50; Farlow 1881, p. 174.

This plant is quite variable in aspect, but it is hard to maintain varietal distinctions. The most striking phases involve the increased attenuation of the upper branches, the bilateral arrangement of the coarse ultimate branchlets, and the corymbose aggregations of the branchlets near the ends of the main branches. The first group is best represented by v. **affinis** (Moore) Harvey (Phyc. Bor.-Am. 596), which has few branchlets in the lower parts, but above is abundantly ramified to long and slender ultimate divisions without, however, congestion at the tips of the branches; this form is known from New Jersey to Maine, Nova Scotia, and the Gaspé; with it may probably be associated the more pinnate v. *plumosa* Harvey (1853, p. 51) and the more delicate v. *gracillima* Harvey (1853, p. 51). The bilateral habit is represented by v. **durkeei** Harvey (1853, p. 52), in which the upper branches have relatively short, thick, spreading and uniform branches arranged in a flat blade; this variety is commonly met with in young plants in shallow water, and sometimes on a few branches of larger plants otherwise undistinctive; it is known from Connecticut to northern Massachusetts, and probably v. *disticha* Harvey is an intermediate form with similar tendencies less developed. In the third group the generally recognized form is v. **fucoides** (Hudson) Harvey (1853, p. 50), which has stout, often somewhat corticated stems and primary branches with branchlets few below but more numerous above and definitely crowded near the tips; it is known from New Jersey to Maine, the Bay of Fundy, and Prince Edward Island; with it may perhaps be associated the more slender v. *tenuis* Harvey (1853, p. 51).

Polysiphonia lanosa (Linnaeus) Tandy
Pl. 56, fig. 4; pl. 57, figs. 14–15; pl. 59, fig. 4

Plants epiphytic, each with a small prostrate base attached by penetrating rhizoids, supporting stiff, slender, abundantly branched widely spreading filaments, height 2–5 cm.; color black or dark brownish purple, texture firm; branching irregularly dichotomous, uncorticated, the segments with (12–) 20–24 pericentral cells; lower portions to 0.5 mm. diam., the segments 90–170 μ long, the smaller branchlets somewhat fastigiate, 100–160 μ diam., the segments 42–65 μ long, all tapering from the base, the apices incurved and colorless; branches arising from segments behind the apical cell without reference to trichoblasts, which are apparently absent; tetrasporangia seriate or alternate in the branchlets, 200–250 μ diam.; spermatangial clusters ovoid, crowded near the tips of the branchlets; pericarps replacing branchlets of the last order, broadly ovoid, to 480 μ diam.

New Jersey to the lower St. Lawrence, and Newfoundland; growing only as an epiphyte on Ascophyllum, accompanying it in the intertidal zone and much protected by its overhanging masses from drying; fruiting in the summer, the spermatangia being reported as early as May and the cystocarps commonest in July.

REFERENCES: Phyk. Univ. 452; Phyc. Bor.-Am. 145, 1444 (as *P. fastigiata*). Harvey 1853, p. 54; Farlow 1881, p. 175 (both as *P. fastigiata*).

Bostrychia Montagne, 1838

Plant filiform, black or dull purplish; rhizoidal, stoloniferous, and erect branches often distinguishable, even the rhizoidal branches polysiphonous, ordinarily regularly bilaterally branched, the branches near the apex usually incurved; for the most part polysiphonous, several cells of equal length being disposed about the central axis in a circle, or these pericentral cells regularly transversely divided, but the branchlets often monosiphonous at the tips; sporangia whorled in special stichidial branchlets, several tetrahedral sporangia being formed in each segment; pericarps subglobose, terminal on the branchlets.

Bostrychia rivularis Harvey [1]

Frond dull purplish violet, diffuse, becoming erect from creeping stolons attached by holdfasts, tufted to 3 cm. long; repeatedly pinnately branched, the lower branches spreading, forming branchlets bilaterally, the terminal erect, with branchlets subcorymbosely incurved; branchlets 1–2 mm. long, attenuate, polysiphonous almost to their tips but sometimes 5–15 segments monosiphonous; joints of the principal branches ecorticate, half as long as broad, with 6–8 pericentral cells which are transversely divided; branchlets with 4 or fewer pericentral cells; pericarps ovate, formed on branchlets from the lower part of the main branches; stichidia in the middle

[1] Post (1936, pp. 9, 13) refers *B. rivularis* Harv. to *B. radicans* Mont., but as her treatment of these plants is not altogether clear the writer prefers to use Harvey's name (which is based on plants collected near or in our territory) until he is more satisfied as to the need for this change.

portion of the ultimate branchlets, with about 15 whorls of tetrasporangia.

From the tropics, Bermuda, Florida to New Hampshire; growing on mollusks, roots, stones, and woodwork in the intertidal zone, and often in water of reduced salinity, being a plant of warm protected waters and tidal salt-marsh flats, relatively rare in New England.

REFERENCES: Algae Exsic. Am. Bor. 54; Phyc. Bor.-Am. 140. Harvey 1853, p. 57; Farlow 1881, p. 176; Hoyt 1920, p. 507.

Rhodomela C. Agardh, 1822

Plants filiform, coarse or, in the younger growing branches, more slender, alternately branched, dull rose to blackish purple, usually with several axes from a basal disklike holdfast; structurally of an axial cell row surrounded by six or seven pericentral cells, which become much subdivided, unequal and irregular in arrangement, and later with a thick cortex on the main branches; sporangia in two longitudinal series in simple or closely forked branches, tetrahedral; spermatangial clusters conical to cylindrical; procarps originating near the tip of the fertile branchlets, a fertile pericentral supporting initial giving rise to the carpogenic branch and additional sterile cells; subsequently from the support an upper auxiliary is cut off beside the carpogonium; after fertilization the auxiliary receives a diploid nucleus and fuses with the carpogonium and the axial cell, the fusion mass producing the gonimoblasts as chains of cells, the outer cells of which are carposporangial; pericarp eventually much enlarged.

The species of Rhodomela are very difficult to distinguish. The varietal records are of doubtful consequence, so far as our district is concerned.

KEY TO SPECIES

1. Main axis stout, little divided; reproductive organs on special lateral, much-divided branchlets............... R. virgata, p. 345
1. Branching more extensive; reproductive organs on ordinary branches 2

2. Bases of axes naked or nearly so, the upper parts bearing branches of the next order, which are of uniform length or nearly so; penultimate branches short, ending in dense

clusters of branchlets, which are cylindrical or taper to the apex .... R. confervoides, p. 344

2. Branching of nearly uniform density throughout, not manifestly increasing toward the apices; long and short branches of the same order intermingled; short fusiform branchlets more or less numerous .... R. lycopodioides, p. 344

Rhodomela lycopodioides (Linnaeus) C. Agardh

Plants stout, of sprawling, irregular habit, to 25 (–60) cm. tall, much branched in several orders; the branches stiff, those of the same order of unequal length and size, more or less dense along the somewhat persistent main axis; spindle-shaped and often curved adventitious branchlets few to many on the older axes; tetrasporangia in more or less distorted branchlets; pericarps ovoid, short-stalked.

Southern Massachusetts to the lower St. Lawrence, Newfoundland, Labrador, James Bay, Hudson Bay, Baffin Island, and Ellesmere Island; on intertidal rocks or more often in the lower sublittoral; fruiting in spring and early summer.

REFERENCES: Kjellman 1883, pp. 107, 114; Collins 1906c, p. 159.

Among the variants, subf. **compacta** Kjellman, stout in aspect, blackish, the main axes obvious, secondary branches much shorter with more crowded branchlets, has been found in the littoral zone of the Gaspé. Subf. **tenera** Kjellman (1883, p. 107; Phyc. Bor.-Am. 1295) is 15–30 cm. long, slender and limp, much branched throughout, the main axes 5–6 cm. long, with numerous ordinary secondary branches which are short and of uniform length, the ultimate soft and rather crowded, from the end of the main axis bearing a few long slender racemose branch systems; it has been found from Maine to the arctic.

Rhodomela confervoides (Hudson) Silva Pl. 40, fig. 9

Plants to 40 cm. tall, color from red to almost black, texture soft in the younger parts, wiry in the older; the main axis with more or less numerous divisions of the second order; axes of all orders with few branches at the base, densely branched at the apex; lesser branches of even length, with short branchlets crowded at the tips; branchlets cylindrical or tapering, not contracted at the bases; tetrasporangia formed in a double row in the tufted slender branches of the first year's growth.

New Jersey to the lower St. Lawrence, Newfoundland, northern Labrador, Hudson Bay, Hudson Strait, and Baffin Island; on stones in relatively shallow water, in good condition in winter and spring, usually denuded by summer.

REFERENCES: Algae Exsic. Am. Bor. 184; Phyc. Bor.-Am. 93, 1598. Harvey 1853, p. 26; Farlow 1881, p. 169; Hariot 1889, p. 196; Taylor 1937c, p. 377 (all as *R. subfusca*); Silva 1952, p. 269; Wilce *in herb.*

The variants of this species reported in our range are of doubtful taxonomic value but include f. **gracilior** (J. Agardh) n. comb. (Algae Exsic. Am. Bor. 55; Phyc. Bor.-Am. 890, both as v. *gracilis;* Farlow 1881, p. 169, as v. *gracilior;* Harvey 1853, p. 26, as *R. gracilis*), which is more slender and elongate, with longer branchlets in looser apical clusters, moniliform tetrasporangial branchlets, racemose and stalked pericarps; it is found from Long Island to Maine, Nova Scotia, and the lower St. Lawrence. F. **rochei** (Harvey) n. comb. (Phyc. Bor.-Am. 1296; Harvey 1853, p. 27, both as *R. rochei;* Taylor 1937b, p. 227 (as *R. subfusca* f. *rochei*) is slender, loosely branching with a bilateral tendency, up to 60 cm. tall, the ultimate branchlets capillary, with abundant trichoblasts; color bright rose; pericarps long-stalked, in long racemose series; reported from New York to Massachusetts, it is typically a spring plant.

Rhodomela virgata Kjellman

Plants to 20 cm. tall, main axis coarse, little divided, bearing more or less numerous short, slender racemose branches, tapering somewhat to base and apex, with secondary branches and very slender branchlets, all much contracted at their bases; during summer denuded by loss of the finer divisions, this state persisting until winter when fruiting occurs on short densely compound branch systems about 3 mm. long issuing from the main axis; tetrasporangia few in each division of the distorted branchlets; spermatangial clusters conical to cylindrical; pericarps lateral, short-stalked.

Rhode Island, southern Massachusetts; fruiting in winter.

REFERENCE: Kjellman 1883, p. 110.

Odonthalia Lyngbye, 1819

Plants erect, somewhat bushy, with flattened shoots alternately branching from the margin; axis of a longitudinal series of large cells surrounded by four pericentral cells which become much subdivided, eventually strongly corticated and, in the broader bandlike forms, show as an indistinct midrib, bordered by a membranous margin; reproductive organs borne on branchlets formed in tufts on the margins of the branches; tetrasporangia developed from the upper segments of the pericentral cells, covered externally; antheridial branchlets small, ligulate, short-stalked, the spermatangia formed by the outer cells of the cortex; procarps originating near the tips of the branchlets; after fertilization the auxiliary forms the gonimoblasts, the enveloping pericarp eventually becoming much swollen.

KEY TO SPECIES

1. Somewhat compressed, the subulate branches without evident midrib; habit bushy .... O. floccosa
1. Flattened throughout, habit and branching notably plane; midrib evident .... O. dentata

Odonthalia floccosa (Esper) Falkenberg

Plants bushy, erect, brownish black, alternately pinnately compound, the main axis somewhat bare below, much branched above; nearly cylindrical, the lower sterile branchlets evidently compressed to the base, bilaterally pinnate from the margins, those higher up elongate, somewhat cylindrical, pinnately branched, the lesser branchlets compressed-subulate, 4.0–6.5 mm. long, diverging, or below somewhat recurved, above subfasciculate or corymbosely incurved, becoming tetrasporiferous; cystocarps obovate-globose, pedicellate.

Ile St. Pierre and southeastern Hudson Bay; dredged in 20 meters; late summer.

REFERENCE: Howe 1927, p. 24B.

Odonthalia dentata (Linnaeus) Lyngbye Pl. 60, figs. 4–6

Plants to 30 cm. tall, deep reddish purple, sparingly branched, the branches flat and serrate; the main axes often denuded below, above branched to 1–3 degrees, the branches marginal in origin, spreading, membranous, to 2–5 mm. broad, with the flexuous midrib obscure or apparently lacking, at their bases contracted, coarsely serrate above by reason of alternate, ascending, chiefly determinate branchlets which may remain about 3–10 mm. long with dentate tips, or grow out into branches of the next order; reproductive organs on minute marginal branchlets.

Nova Scotia, the Gaspé, the lower St. Lawrence, Labrador, Hudson Strait, James Bay and southwestern Hudson Bay, Foxe Basin, Baffin Island, and Ellesmere Island; growing just below extreme low tide; fruiting until summer.

REFERENCES: Algae Exsic. Am. Bor. 56. Harvey 1853, p. 14; Farlow 1881, p. 168; Kjellman 1883, p. 105.

There has also been reported f. **angusta** Harvey (Pl. 60, figs. 7–8; Phyc. Bor.-Am. 1297; Harvey 1852, p. 49), the plants smaller, narrower throughout, the branches 1–3 mm. wide, the branchlets relatively long, acute, erect and somewhat incurved near the apices, the midrib and membranous portions less distinguishable; this plant is known from stations on the lower St. Lawrence.

BIBLIOGRAPHY

AGARDH, J. G. 1870. Om de under Korvetten Josephines expedition sistliden sommar, insamlade Algerne. Öfvers. Kungl. Vetensk.-Akad. Förhandl., 1870(4) : 359–366.

ASHMEAD, S. 1854. Marine Algae. Proc. Acad. Nat. Sci. Phila., 6 : 147–148. 1852–53.

—— 1856. Catalogue of Marine Algae Discovered at Beesley's Point during the Past Summer, with Some Remarks Thereon. *Ibid.,* 7 : 410–413.

—— 1857. Catalogue of the Marine Algae Discovered at Beesley's Point during the Summer of 1855. Geol. Surv. of New Jersey. Geology of the County of Cape May, pp. 152–154.

—— 1859. Note on *Griffithsia setacea.* Proc. Acad. Nat. Sci. Phila., 1858 : 8.

—— 1864. Algae.—Enumeration of Arctic Plants Collected by Dr. J. J. Hayes in his Exploration of Smith's Sound between Parallels 78th and 82nd during the Months of July, August and Beginning of September 1861. *Ibid.,* 1863 : 96.

—— 1878. Algae. *In* O. R. Willis, Catalogus plantarum in Nova Caesarea repertarum. xxviii + 88. New York.

BACHELOT DE LA PYLAIE, A. J. M., *see* De la Pylaie, A. J. M. B.

BAILEY, J. W. 1847–48. Notes on the Algae of the United States. Am. Journ. Sci. and Arts, II, 3 : 80–85, 399–403. 1847; 6 : 37–42. 1848.

BARTON, B. W. 1893. On the Origin and Development of the Stichidia and Tetrasporangia in *Dasya elegans.* Stud. Biol. Lab., Johns Hopkins Univ., 5 : 279–282.

BELL, H. P. 1927. Seasonal Disappearance of Certain Marine Algae. Trans. Nova Scotian Inst. Sci., 17 (1) : 1–5.

—— AND MACFARLANE, C. 1933a. The Marine Algae of the Maritime Provinces of Canada, I. List of Species with Their Distribution and Prevalence. Canadian Journ. Res., 9 : 265–279.

—— —— 1933b. The Marine Algae of the Maritime Provinces of Canada, II. A Study of Their Ecology. *Ibid.,* 9 : 280–293.

BELL, H. P. AND McFARLANE, C. 1933c. Marine Algae from Hudson Bay. Contrib. Canadian Biol. and Fish., 8(3) : 65–68.

BIGELOW, R. P. 1887. On the Structure of the Frond of *Champia parvula* Harv. Proc. Am. Acad. Arts and Sci., 23 : 111–120.

BLUM, J. L., AND CONOVER, J. T. 1953. New or Noteworthy Vaucheriae from New England Salt Marshes. Biol. Bull., 105(3) : 395–401.

BØRGESEN, F., AND JÓNSSON, H. 1905. The Distribution of the Marine Algae of the Arctic Sea and of the Northernmost Part of the Atlantic. Botany of the Faeröes, 3 : Appendix, I–XXVIII. Copenhagen.

BORNET, E., AND FLAHAULT, C. 1889. Sur quelques plantes vivant dans le test calcaire des mollusques. Bull. Soc. Bot. France, 36: cxlvii-clxxvi.

BORY DE SAINT VINCENT, J. B. 1822–1831. Dictionnaire classique d'histoire naturelle. Vols. 1–17. Paris. (Laminarias discussed on pp. 188–191 of vol. 9, 1826.)

BRANNON, M. A. 1897. The Structure and Development of *Grinnellia americana* Harv. Ann. Bot., 11 : 1–28.

BROWN, H. J. 1929. The Algal Family Vaucheriaceae. Trans. Am. Microsc. Soc., 48(1) : 86–117.

BURNHAM, S. H., AND LATHAM, R. A. 1914, 1917. The Flora of the Town of Southold, Long Island, and Gardiners Island. Torreya, 14 : 201–225; 17 : 111–122.

CHAPMAN, V. J. 1939. Some Algal Complexities. A. *Rhizoclonium tortuosum* (Dillw.) Kütz. and *Chaetomorpha tortuosa* Kütz. B. Marsh Forms of *Fucus spiralis* and *F. vesiculosus*. Rhodora, 41 : 10–28.

CHESTER, G. D. 1896. Notes concerning the Development of *Nemalion multifidum*. Bot. Gaz., 21 : 340–347.

CHRISTENSEN, T. 1952. Studies on the Genus Vaucheria. I. A List of Finds from Denmark and England with Notes on Some Submarine Species. Bot. Tidsskr., 49(2) : 171-188.

CLELAND, R. E. 1919. The Cytology and Life History of *Nemalion multifidum* Ag. Ann. Bot., 33 : 323–351.

COLLINS, F. S. 1880. A Laminaria New to the United States. Bull. Torr. Bot. Club, 7 : 117–118.

—— 1882. Notes on the New England Marine Algae, I. *Ibid.*, 9 : 69–71.

Collins, F. S. 1883. Notes on the New England Marine Algae, II. *Ibid.*, 10 : 55–56.

—— 1884a. Notes on the New England Marine Algae, III. *Ibid.*, 11 : 29–30.

—— 1884b. Notes on the New England Marine Algae, IV. *Ibid.*, 11: 130–132.

—— 1888a. Algae from Atlantic City, N. J., Collected by S. R. Morse. *Ibid.*, 15 : 309–314.

—— 1888b. Algae. *In* Maria L. Owen, A Catalog of Plants Growing without Cultivation in the County of Nantucket, Massachusetts. 87 pp. Northampton, Mass.

—— 1890. *Brachytrichia Quoyii* (C. Agardh) Born. & Flah. Bull. Torr. Bot. Club, 17: 175–176.

—— 1891. Notes on New England Marine Algae, V. *Ibid.*, 18 : 335–341.

—— 1894. Algae. *In* E. L. Rand and J. H. Redfield, A Preliminary Flora of Mount Desert Island, Maine. 286 pp. Cambridge, Mass.

—— 1896a. Notes on New England Marine Algae, VI. Bull. Torr. Bot. Club, 23 : 1–6.

—— 1896b. Notes on New England Marine Algae, VII. *Ibid.*, 23 : 458–462.

—— 1899a. A Seaweed Colony. Rhodora, 1 : 69–71.

—— 1899b. To Seaweed Collectors. *Ibid.*, 1 : 121–127.

—— 1900a. Notes on Algae, II. *Ibid.*, 2 : 11–14.

—— 1900b. Preliminary Lists of New England Plants, V. Marine Algae. *Ibid.*, 2 : 41–52.

—— 1900c. Seaweeds in Winter. *Ibid.*, 2 : 130–132.

—— 1900d. The New England Species of Dictyosiphon. *Ibid.*, 2 : 162–166.

—— 1900e. The Marine Flora of Great Duck Island, Maine. *Ibid.*, 2: 209–211.

—— 1901a. Notes on Algae, III. *Ibid.*, 3 : 132–136.

—— 1901b. Notes on Algae, IV. *Ibid.*, 3 : 289–293.

—— 1902a. An Algologist's Vacation in Eastern Maine. *Ibid.*, 4 : 174–179.

—— 1902b. The Marine Cladophoras of New England. *Ibid.*, 4 : 111–127.

—— 1903a. The North American Ulvaceae. *Ibid.*, 5 : 1–31.

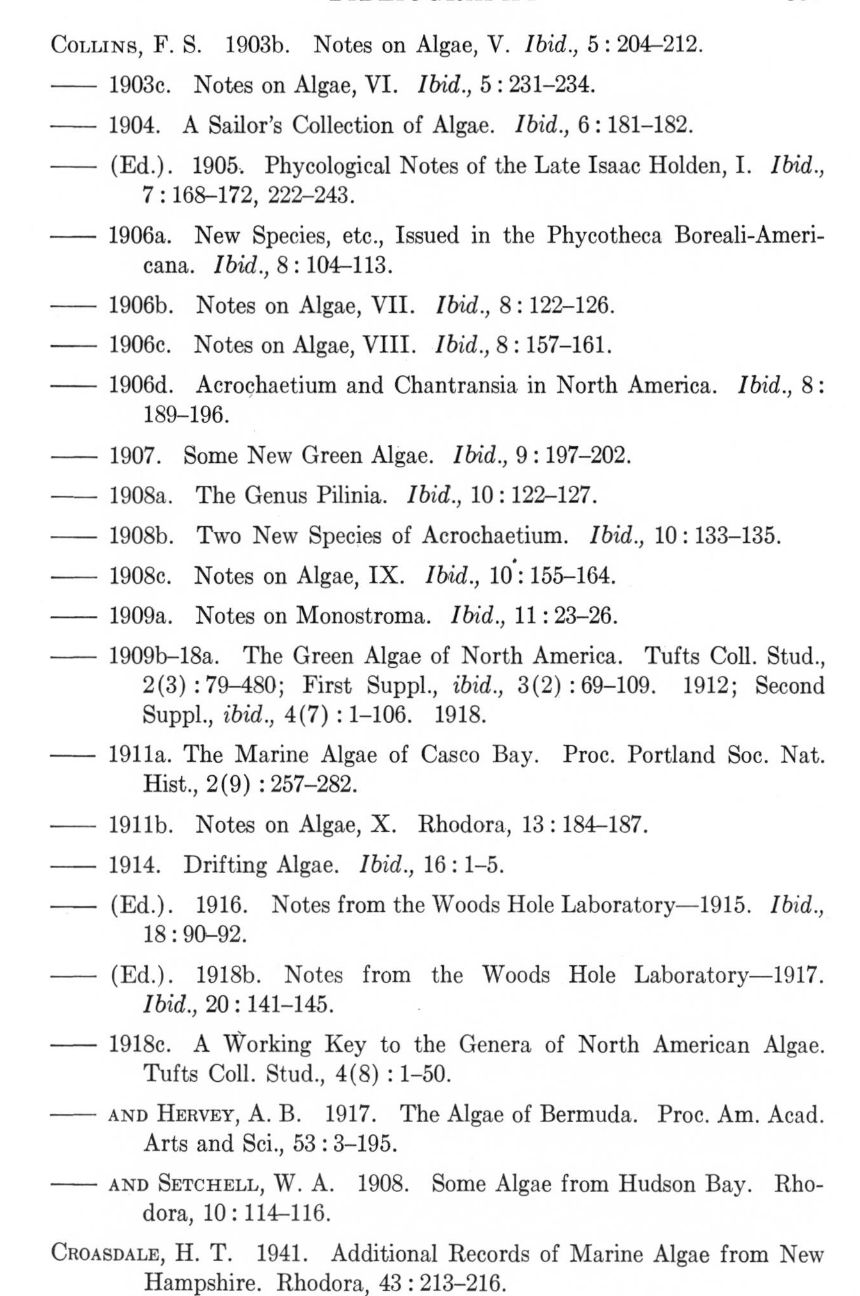

Collins, F. S. 1903b. Notes on Algae, V. *Ibid.*, 5 : 204–212.

—— 1903c. Notes on Algae, VI. *Ibid.*, 5 : 231–234.

—— 1904. A Sailor's Collection of Algae. *Ibid.*, 6 : 181–182.

—— (Ed.). 1905. Phycological Notes of the Late Isaac Holden, I. *Ibid.*, 7 : 168–172, 222–243.

—— 1906a. New Species, etc., Issued in the Phycotheca Boreali-Americana. *Ibid.*, 8 : 104–113.

—— 1906b. Notes on Algae, VII. *Ibid.*, 8 : 122–126.

—— 1906c. Notes on Algae, VIII. *Ibid.*, 8 : 157–161.

—— 1906d. Acrochaetium and Chantransia in North America. *Ibid.*, 8 : 189–196.

—— 1907. Some New Green Algae. *Ibid.*, 9 : 197–202.

—— 1908a. The Genus Pilinia. *Ibid.*, 10 : 122–127.

—— 1908b. Two New Species of Acrochaetium. *Ibid.*, 10 : 133–135.

—— 1908c. Notes on Algae, IX. *Ibid.*, 10 : 155–164.

—— 1909a. Notes on Monostroma. *Ibid.*, 11 : 23–26.

—— 1909b–18a. The Green Algae of North America. Tufts Coll. Stud., 2(3) : 79–480; First Suppl., *ibid.*, 3(2) : 69–109. 1912; Second Suppl., *ibid.*, 4(7) : 1–106. 1918.

—— 1911a. The Marine Algae of Casco Bay. Proc. Portland Soc. Nat. Hist., 2(9) : 257–282.

—— 1911b. Notes on Algae, X. Rhodora, 13 : 184–187.

—— 1914. Drifting Algae. *Ibid.*, 16 : 1–5.

—— (Ed.). 1916. Notes from the Woods Hole Laboratory—1915. *Ibid.*, 18 : 90–92.

—— (Ed.). 1918b. Notes from the Woods Hole Laboratory—1917. *Ibid.*, 20 : 141–145.

—— 1918c. A Working Key to the Genera of North American Algae. Tufts Coll. Stud., 4(8) : 1–50.

—— and Hervey, A. B. 1917. The Algae of Bermuda. Proc. Am. Acad. Arts and Sci., 53 : 3–195.

—— and Setchell, W. A. 1908. Some Algae from Hudson Bay. Rhodora, 10 : 114–116.

Croasdale, H. T. 1941. Additional Records of Marine Algae from New Hampshire. Rhodora, 43 : 213–216.

Curtis, M. A. 1867. Geological and Natural History Survey of North Carolina, III, Botany; Containing a Catalogue of the Indigenous and Naturalized Plants of the State. Algae listed on pp. 155–156. Raleigh, N. C.

Curtiss, A. H. 1899. Mrs. Floretta A. Curtiss, a Biographical Sketch by Her Son. 14 + IV. Jacksonville, Fla.

Davis, B. M. 1892. Development of the Frond of *Champia parvula* Harv. from the Carpospore. Ann. Bot., 6 : 339–354.

—— 1894a. Notes on the Life History of a Blue-Green Motile Cell. Bot. Gaz., 19 : 96–102.

—— 1894b. Euglenopsis, a New Alga-like Organism. Ann. Bot., 8 : 377–390.

—— 1896a. The Development of the Cystocarp of *Champia parvula* Harv. Bot. Gaz., 21 : 109–117.

—— 1896b. Development of the Procarp and Cystocarp in the Genus Ptilota. *Ibid.*, 22 : 353–376.

—— 1898. Kerntheilung in der Tetrasporenmutterzelle bei *Corallina officinalis* L., var. *mediterranea*. Ber. d. d. bot. Gesell., 16 : 266–272.

—— 1899a. Recent Work on the Life History of the Rhodophyceae. Bot. Gaz., 27 : 315–320.

—— 1899b. Translation of C. Sauvageau, The Sexuality of the Tilopteridaceae (Résumé). *Ibid.*, 28 : 213–214.

—— 1905. The Sexual Organs and Sporophyte Generations of the Rhodophyceae. *Ibid.*, 39 : 64–66.

—— 1910. Nuclear Phenomena of Sexual Reproduction in Algae. Am. Nat., 44 : 513–532.

—— 1913a. General Characteristics of the Algal Vegetation of Buzzards Bay and Vineyard Sound in the Vicinity of Woods Hole. Dept. Comm. and Labor, Bull. (U. S.) Bur. Fisheries, 31(1) : 443–544. 1911.

—— 1913b. A Catalogue of the Marine Flora of Woods Hole and Vicinity. *Ibid.*, 31(2) : 795–833. 1911.

—— 1916. Life Histories in the Red Algae. Am. Nat., 50 : 502–512.

De la Pylaie, A. J. M. B. 1824. Quelques observations sur les productions d'île de Terre-Neuve, et sur quelques Algues de la côte de France appartenant au genre Laminaire. Ann. Sci. Nat., 4 : 174–184.

De la Pylaie, A. J. M. B. 1829. Flore de l'Ile de Terre-Neuve et les Iles Saint Pierre et Miclon. 128 pp. Paris.

Derick, C. M. 1899. Notes on the Development of the Holdfasts of Certain Florideae. Bot. Gaz., 23 : 246–263.

De Toni, G. B. 1889–1924. Sylloge algarum omnium hucusque cognitarum. 1 (Sylloge Chlorophycearum) : 1–12, I–CXXXIX, 1–1315. 1889. 3 (Sylloge Fucoidearum) : I–XVI, 1–638. 1895. 4 (Sylloge Floridearum) (1) : I–XX, I–LXI, 1–388. 1897; (2) : 387–776. 1900; (3) : 775–1525. 1903; (4) : 1523–1973. 1905. 6 (Sylloge Floridearum) (5, Additamenta) : I–XI, 1–767. 1924. Patavii.

Dickie, G. 1852. Notes on the Algae. *In* P. C. Sutherland, Journal of a Voyage in Baffins Bay and Barrow Straits in the Years 1850–1851; 2 : cxci–cc. London.

—— 1867. Notes on a Collection of Algae Procured in Cumberland Sound by Mr. James Taylor, and Remarks on Arctic Species in General. Journ. Proc. Linn. Soc. London, 9 : 235–243.

Doty, M. S. 1947. The Marine Algae of Oregon. Part I. Chlorophyta and Phaeophyta. Farlowia, 3(1) : 1–65. Part II. Rhodophyta. *Ibid.,* 3(2) : 159–215.

—— 1948. The Flora of Penikese, Seventy-four Years After. I. Penikese Island Marine Algae. Rhodora, 50 : 253–269.

—— and Newhouse, J. 1954. The Distribution of Marine Algae into Estuarine Waters. Amer. Journ. Bot., 41(6) : 508–515.

Drew, K. M. 1928. A Revision of the Genera Chantransia, Rhodochorton and Acrochaetium, with Descriptions of the Marine Species of Rhodochorton (Näg.) gen. emend. on the Pacific Coast of North America. Univ. Cal. Publ. Bot., 14(5) : 139–224.

—— 1934, 1943. Contributions to the Cytology of *Spermothamnion Turneri* (Mert.) Aresch. I. The Diploid Generation. Ann Bot. 48 : 549–573. II. The Haploid and Triploid Generations. *Ibid.,* N.S., 7 : 23–30.

—— 1949. *Conchocelis*-Phase in the Life-History of *Porphyra umbilicalis* (L.) Kütz. Nature, 164 : 748.

—— 1955. Phycology and the British Phycological Society. Brit. Phycol. Bull., 3 : 1–10.

Dunn, G. A. 1916. A Study of the Development of *Dumontia filiformis.* Plant World, 19 : 271–281.

DUNN, G. A. 1917. The Development of *Dumontia filiformis*. II. Development of Sexual Plants and General Discussion. Bot. Gaz., 63 : 425–467.

DURANT, C. F. 1850. Algae and Corallines of the Bay and Harbor of New York. 43 pp., and specimens. New York.

EATON, A. 1822. A Manual of Botany for the Northern and Middle States of America. Algae listed on pp. 130–138, and described, in alphabetical order, through the text. Ed. 3, Albany.

EATON, D. C. 1873. List of Marine Algae Collected near Eastport, Maine, in August and September, 1873, in Connection with the Work of the U. S. Fish Commission under Prof. S. F. Baird. Trans. Conn. Acad. Arts and Sci., 2 : 343–350.

ENGLER, A., AND PRANTL, K. 1890–97. Die natürlichen Pflanzenfamilien, I, 2 : i–xii, 1–580. Leipzig.

—— 1927. *Ibid.*, 2. Aufl., 3 : 1–463. Leipzig.

FARLOW, W. G. 1873. List of the Seaweeds of Marine Algae of the South Coast of New England. U. S. Comm. Fish and Fisheries, Report on the Condition of the Sea Fisheries of the South Coast of New England, in 1871–1872, 1 : 281–294.

—— 1875. List of the Marine Algae of the United States with Notes of New or Imperfectly Known Species. Proc. Am. Acad. Arts and Sci., 10(11) : 351–380.

—— 1876. List of the Marine Algae of the United States. Rep. U. S. Comm. Fish and Fisheries for 1873–4 and 1874–5 : 691–718.

—— 1877. On Some Algae New to the United States. Proc. Am. Acad. Arts and Sci., 12 (N. S. 4) : 235–245.

—— 1879. List of Algae Collected at Points in Cumberland Sound during the Autumn of 1877. Contribution to the Natural History of Arctic America Made in Connection with the Howgate Polar Expedition, 1877–1878, by Ludwig Kumlien. Bull. U. S. Nat. Mus., No. 15 : 169.

—— 1881. The Marine Algae of New England. Rep. U. S. Comm. Fish and Fisheries for 1879, Appendix A-1 : 1–210. 1882. Separate copies of the independently circulated appendix carry the title "Marine Algae of New England and Adjacent Coast" and the date 1881 on the title page, but the completed volume when issued was dated 1882. The work was reissued with the shorter name on the title page in 1891, with the same pagination.

—— 1882. Notes on New England Algae. Bull. Torr. Bot. Club, 9 : 65–68.

FARLOW, W. G. 1886. Notes on Arctic Algae; Based Principally on Collections Made at Ungava Bay by Mr. L. M. Turner. Proc. Am. Acad. Arts and Sci., 31 (N.S. 13) : 469–477.

—— 1889. On Some New or Imperfectly Known Algae of the United States, I. Bull. Torr. Bot. Club, 16 : 1–12.

—— 1893. Notes on Some Algae in the Herbarium of the Long Island Historical Society. *Ibid.*, 20(3) : 107–109.

—— 1895. Algae, p. 214. *In* H. E. Wetherill, Botany. List of Plants Obtained on the Peary Auxiliary Expedition of 1894. Bull. Geog. Club Phila., 1(5), Appendix C : 208–215.

FOSLIE, M. 1905. Remarks on Northern Lithothamnia. K. Norske Vidensk. Selsk. Skr. 1905(3) : 1–138.

—— 1929. Contributions to a Monograph of the Lithothamnia. Edited by H. Printz. 60 pp., 75 pl. Trondhjem.

FRITSCH, F. E. 1935, 1945. The Structure and Reproduction of the Algae, I. xvii + 791. 1935. II. xiv + 939. 1945. Cambridge.

GARDNER, G. 1937. Liste annotée des espèces de pteridophytes, de phanérogames et d'algues récoltées sur la côte du Labrador, à la baie d'Hudson et dans le Manitoba nord, en 1930 et 1933. Bull. Soc. Bot. France, 84 : 19–51.

GARDNER, N. L. 1917. New Pacific Coast Marine Algae I. Univ. Cal. Publ. Bot., 6(14) : 377–416.

GRIER, N. M. 1925. The Native Flora of the Vicinity of Cold Spring Harbor, L. I., N. Y., Am. Midland Nat., 9(6, 7, 9–11) : 245–256, 283–318, 384–437, 513–527, 550–563.

GRUNOW, A. 1915, 1916. Additamenta ad cognitionem Sargassorum. Verhandl. d. K. K. Zool-Bot. Gesellsch. 65: 329–448, 1915. *Ibid.*, 66 : 1–48, 136–185, 1916.

HALL, F. W. 1876. List of Algae Growing in Long Island Sound within 20 Miles of New Haven. Bull. Torr. Bot. Club, 6 : 109–112.

HAMEL, G. 1930a–32. Chlorophycées des côtes françaises. Rev. Algol., 5 : 1–54, 383–430. 1930b–31. *Ibid.*, 6 : 9–73. 1931–32.

—— 1931–39. Phaeophycées de France: pp. 1–80 (Ectocarpacées). 1931; pp. 81–176 (Myrionematacées—Spermatochnacées). 1935; pp. 177–240 (Spermatochnacées—Sphacelariacées). 1937; pp. 241–336 (Sphacelariacées—Dictyotacées). 1938; pp. 337–432, I–XLVII (Dictyotacées—Sargassacées). 1939.

HARDER, R. 1948. Einordnung von *Trailiella intricata* in den Generationswechsel der Bonnemaisoniaceae. Nachr. Acad. Wiss. Göttingen, Math.-Phys. Kl., Biol.-Physiol.-Chem. Abt. 1948 : 24–27.

HARIOT, P. 1889. Liste des algues recuillies à l'île Miquelon par M. le docteur Delamarre. Journ. de Bot., 3(9–11) : 154–157, 181–183, 194–196.

HARVEY, W. H. 1848. Directions for Collecting and Preserving Algae. Am. Journ. Sci. and Arts, II, 6 : 42–45.

—— 1849. A Manual of British Marine Algae. lii + 252. London.

—— 1850. Observations on the Marine Flora of the Atlantic States. Proc. Am. Assn. Adv. Sci., pp. 79–80.

—— 1852–58a. Nereis Boreali-Americana. Part I, Melanospermae. Smithsonian Contrib. to Knowledge, 3(4) : 1–150, Pl. 1–12. 1852; Part II, Rhodospermae. *Ibid.*, 5(5) : 1–258, Pl. 13–36. 1853. Part III, Chlorospermae. *Ibid.*, 10 : ii + 1–140, Pl. 37–50. 1858 (including Supplements).

—— 1858b. List of Arctic Algae, Chiefly Compiled from Collections Brought Home by Officers of the Recent Searching Expeditions. *Ibid.*, Part III, Suppl. 2 : 132–134.

HAY, G. U. 1887. Marine Algae of the Maritime Provinces. Bull. New Brunswick Nat. Hist. Soc., 1(6) : 62–68.

—— AND MACKAY, A. H. 1886. Marine Algae of Bay of Fundy. *Ibid.*, 1(5) : 32-33.

—— —— 1888. Marine Algae of New Brunswick: with an Appendix containing a List of the Marine Algae of the Maritime Provinces of the Dominion of Canada. Trans. Roy. Soc. Canada, 5(4) : 167–174. 1887.

HAZEN, T. E. 1902. The Ulotrichaceae and Chaetophoraceae of the United States. Mem. Torr. Bot. Club, 11 : 135–250.

HERVEY, A. B. 1881. Sea Mosses. A Collector's Guide and an Introduction to the Study of Marine Algae. xii + 281. Boston.

—— 1882. *Arthrocladia villosa* Duby. Bull. Torr. Bot. Club, 9 : 126–127.

HITCHCOCK, E. 1835. Catalogues of the Animals and Plants of Massachusetts. (From the author's second (1833) edition of his Report on the Geology of Massachusetts.) 142 pp. Amherst.

HOLDEN, I. 1899. Two New Species of Marine Algae from Bridgeport, Connecticut. Rhodora, 1 : 197–198.

HOOPER, J. 1850. Introduction to Algology: with a Catalogue of American Algae, or Seaweeds, according to the Latest Classification of Professor Harvey. 34 pp. Brooklyn.

HOWE, M. A. 1902. An Attempt to Introduce a Seaweed into the Local Flora. Journ. New York Bot. Gard., 3 : 116–118.

—— 1905. Phycological Studies, II. New Chlorophyceae, new Rhodophyceae and Miscellaneous Notes. Bull. Torr. Bot. Club, 32: 563–586.

—— 1914a. The Marine Algae of Peru. Mem. Torr. Bot. Club, 15 : 1–185.

—— 1914b. Some Midwinter Algae of Long Island Sound. Torreya, 14 : 97–101.

—— 1920. Algae, pp. 553–618. *In* N. L. Britton and C. F. Millspaugh, The Bahama Flora. viii + 695. New York.

—— 1927. Hudson Bay Algae. Rep. Canadian Arctic Exped. 1903–1918, 4(Botany) : 18–29.

HOYT, W. D. 1920. Marine Algae of Beaufort, N. C., and Adjacent Regions. Bull. [U. S.] Bur. Fisheries, 36 : 367–556 + V.

HUMM, H. J. 1950. Notes on the Marine Algae of Newfoundland. Journ. Tenn. Acad. Sci., 25 : 229.

HUS, H. T. A. 1902. An Account of the Species of Porphyra found on the Pacific Coast of North America. Proc. Cal. Acad. Sci., iii, 2(6) : 171–240.

HYLANDER, C. J. 1928. The Algae of Connecticut. [Conn.] State Geol. and Nat. Hist. Surv., Bull., 42 : 1–245.

JAO, C.-C. 1936. New Rhodophyceae from Woods Hole. Bull. Torr. Bot. Club, 63 : 237–258.

JELLIFFE, S. E. 1904. Additions to the "Flora of Long Island." Torreya, 4 : 97–100.

JENKINS, E. H., AND STREET, J. P. 1917. Manure From the Sea. Conn. Agr. Exp. Sta. Bull., 194 : 1–13.

JOHNSON, D. S. 1937. A New England Station for *Laminaria faeroensis* Børg., forma. Bull. Torr. Bot. Club, 64 : 103.

—— AND SKUTCH, A. F. 1928. Littoral Vegetation on a Headland of Mount Desert Island, Maine. Ecology, 9 : 118–215, 307–338.

—— AND YORK, H. H. 1912. The Relation of Plants to Tide Ievels. A Study of the Distribution of Marine Plants at Cold Spring Harbor. Johns Hopkins Univ. Circ., 1912(2) : 111–116.

Jónsson, H. 1904. The Marine Algae of East Greenland. Medd. om Grønl., 30 : 1–73.

Jordan, D. S. 1874. A Key to the Higher Algae of the Atlantic Coast between Newfoundland and Florida. Am. Nat., 8 : 398–403.

—— 1875. *Fucus serratus* and *Fucus anceps*. *Ibid.*, 9 : 309–310.

Jorde, I. 1933. Untersuchungen über den Lebenzyklus von Urospora Aresch. und Codiolum A. Braun. Nyt Mag. Naturvidensk., 73 : 1–20.

Kemp, A. F. 1860. A Classified List of Marine Algae from the Lower St. Lawrence, with an Introduction for Amateur Collectors. Canadian Nat. and Geol., 5 : 30–42.

—— 1862. On the Shore Zones and Limits of Marine Plants on the Northeastern Coast of the United States. *Ibid.*, 7 : 20–34.

—— 1870. Notice of *Fucus serratus* Found in Pictou Harbor. Canadian Nat. and Quart. Journ. Sci., N.S., 5 : 349–350.

Kjellman, F. R. 1883. The Algae of the Arctic Sea. Kongl. Svensk. Vetensk. Akad. Handl., 20(5) : 1–349.

Klugh, A. B. 1917. The Marine Algae of the Passamaquoddy Region, New Brunswick. Contrib. Canadian Biol., Suppl. 6th Ann. Rep. Dept. Naval Serv., Fisheries Bur., pp. 79–85.

—— 1922. Ecological Polymorphism in *Enteromorpha crinita*. Rhodora, 24 : 50–55.

Kolderup Rosenvinge, L. *See* Rosenvinge, L. Kolderup.

Kraemer, H. 1899. Some Notes on Chondrus. Amer. Journ. Pharm., 71 : 479–483.

Kuntze, O. 1891–98. Revisio generum plantarum . . . , Pars I: I–CLVI + 1–376. 1891; Pars II: 377–1011. 1891; Pars III (1): CLVII–CCCXXII. 1893; (2): I–VI + **1–202** + 1–576. 1898. Leipzig.

Kützing, F. T. 1845–71. Tabulae Phycologicae. Vols. 1–20. Nordhausen.

Kylin, H. 1932. Die Florideenordnung Gigartinales. Lunds Univ. Årsskr. N. F., Avd. 2, 28(8) : 1–88.

—— 1933. Über die Entwicklungsgeschichte der Phaeophyceen. Lunds Univ. Årsskr. N. F., Avd. 2, 29(7) : 1–102.

—— 1935. Über einige kalkbohrende Chlorophyceen. Kungl. Fysiogr. Sällsk. i Lund, Förhandl., 5(19) : 1–19.

KYLIN, H. 1944. Die Rhodophyceen der Schwedischen Westküste. Lunds Univ. Årsskr. N. F., Avd. 2, 40(2) : 1–104.

—— 1947. Die Phaeophyceen der Schwedischen Westküste. *Ibid.*, 43(4) : 1–99.

—— 1949. Die Chlorophyceen der Schwedischen Westküste. *Ibid.*, 45(4) : 1–79.

LAMBERT, F. T. 1930. On the Structure and Development of Prasinocladus. Zeitschr. Bot., 23 : 227–244.

LAWSON, G. 1864. Botanical Science — Record of Progress. 5. Gulfweed at Cape Sable. Canadian Nat. and Geol., II, 1 : 3.

—— 1870a. On the Laminariaceae of the Dominion of Canada and Adjacent Parts of British America. Trans. Nova Scotian Inst. Nat. Sci., 2(4) : 109–111.

—— 1870b. North American Laminariaceae. Canadian Nat. and Quart. Journ. Sci., N.S., 5 : 99–101.

LE GALLO, C. 1947. Algues Marines des Iles Saint-Pierre et Miquelon. Le Nat. Canad., 74(11, 12) : 293–318. Reissued in: Esquisse Générale de la Flore Vasculaire des Iles Saint-Pierre et Miquelon, suivie d'un supplement sur les algues marines. Contr. Inst. Bot. Univ. Montréal, 65 : 1–84. 1949.

LEVRING, T. 1940. Studien über die Algenvegetation von Blekinge, Südschweden. Akad. Abh. Lund. vii + 179.

LEWIS, I. F. 1909. The Life History of *Griffithsia Bornetiana.* Ann. Bot., 23 : 639–690.

—— 1912. Alternation of Generations in Certain Florideae. Bot. Gaz., 53 : 236–242.

—— 1914. The Seasonal Life-Cycle of Some Red Algae at Woods Hole. Plant World, 17 : 31–35.

—— 1924. The Flora of Penikese, Fifty Years After. Rhodora, 26 : 181–195, 211–219, 222–229.

—— AND TAYLOR, W. R. 1921. Notes from the Woods Hole Laboratory, 1921. *Ibid.*, 23 : 249–256.

—— —— 1928. Notes from the Woods Hole Laboratory, 1928. *Ibid.*, 30 : 193–198.

—— —— 1933. Notes from the Woods Hole Laboratory, 1932. *Ibid.*, 35 : 147–154.

LIFE, A. C. 1905. Vegetative Structure of Mesogloia. Missouri Bot. Gard., Ann. Rep., 16 : 157–160.

LUND, S. 1933. The Godthaab Expedition, 1928. The Marine Algae. Medd. om Grønl., 82(4) : 5–17.

MACY, A. M. 1882. Botany, Conchology and Geology of Nantucket (Algae, pp. 46, 47, by F. S. Collins). *In* E. K. Godfrey, The Island of Nantucket, What It Was and It Is. VI + 365 + map. Boston.

MARTINDALE, I. 1889a. Marine Algae of the New Jersey Coast and Adjacent Waters of Staten Island. Mem. Torr. Bot. Club, 1 : 87–109.

—— 1889b. Marine Algae. *In* N. L. Britton, Catalogue of Plants. Geol. Surv. New Jersey, 2(1) : 384–430, 602–615 (*p. p.*).

MCCULLOUGH, M. S. 1900. Algae. *In* Daily Union History of Atlantic City and County. Not seen; quoted by Richards, H. G.

MOORE, G. T. 1900. New or Little Known Unicellular Algae. I. *Chlorocystis Cohnii.* Bot. Gaz., 30 : 100–112.

NIEUWLAND, J. A. 1917. Critical Notes on New and Old Genera of Plants.—IX. Am. Midland Nat., 5 : 30.

OLNEY, S. T. 1847. Rhode Island Plants, 1846, or Additions to the Published Lists of the Providence Franklin Society. Proc. Providence Franklin Soc., 1(2) : 25–42 (Algae, pp. 38–42).

OSTERHOUT, W. J. V. 1896. On the Life History of *Rhabdonia tenera* J. Agardh. Ann. Bot., 10 : 403–427.

OVERTON, J. B. 1913. Artificial Parthenogenesis in Fucus. Science, N.S., 37 : 841–844.

PAPENFUSS, G. F. 1933. Note on the Life-Cycle of *Ectocarpus siliculosus* Dillw. Science, 77 : 390–391.

—— 1935. Alternation of Generations in *Ectocarpus siliculosus.* Bot. Gaz., 96(3) : 421–446.

—— 1945. Review of the Acrochaetium-Rhodochorton Complex of Red Algae. Univ. Cal. Publ. Bot., 18 : 299–344.

—— 1947. Further Contributions Toward an Understanding of the Acrochaetium-Rhodochorton Complex of the Red Algae. *Ibid.*, 18(19) : 433–447.

—— 1950. Review of the Genera of Algae Described by Stackhouse. Hydrobiol., 2(3) : 181–208.

PARKE, M. W. 1953. A Preliminary Check-List of British Marine Algae. Journ. Marine Biol. Assoc. United Kingd., 32 : 497–520.

PERRY, G. W. 1883. *Arthrocladia villosa* Duby. Bull. Torr. Bot. Club, 10(9) : 106.

PIKE, N. 1886. Check List of Marine Algae. Based on Specimens Collected on the Shores of Long Island, from 1839 to 1885. Bull. Torr. Bot. Club, 13 : 105–114.

POST, E. 1936. Systematische und pflanzengeographische Notizen zur Bostrychia-Caloglossa-Assoziation. Rev. Algol., 9 : 1–84.

PRAT, H. 1933. Les Zones de végétation et les facies des rivages de l'estuaire du St. Laurent, au voisinage de Trois-Pistoles. Le Nat. Canad., 60 : 93–136.

REINSCH, P. F. 1875. Contributiones ad Algologiam et Fungologiam, I. xii + 103 + i. Leipzig.

RICHARDS, H. G. 1931. Notes on the Marine Algae of New Jersey. Bartonia, 13 : 38–46.

RICHARDS, H. M. 1891. On the Structure and Development of *Choreocolax Polysiphoniae*. Proc. Am. Acad. Arts and Sci., 26 : 26–63.

ROBINSON, C. B. 1903. The Distribution of *Fucus serratus* in America. Torreya, 3 : 132–134.

—— 1907. The Seaweeds of Canso. Being a Contribution to the Study of Eastern Nova Scotia Algae. 39th Ann. Rep. Dept. Marine and Fisheries, 22a (Further Contrib. Canadian Biol., 1902–1905 [7]) : 71–74.

ROSCOE, M. V. 1931. The Algae of St. Paul Island. Rhodora, 33 : 127–131.

ROSENVINGE, L. KOLDERUP. 1893. Grønlands Havalger. Medd. om Grønl., 3 : 765–981.

—— 1894. Les Algues marines du Groenland. Ann. Sci. Nat. Bot., VII, 19(2) : 53–164.

—— 1926. Marine Algae Collected by Dr. H. G. Simmons during the 2nd Norwegian Arctic Expedition in 1898–1902. Norske Vidensk.-Akad. i Oslo, Report of the Second Norwegian Arctic Expedition in the "Fram" 1898–1902, 37 : 1–40.

RUSSELL, J. L. 1856. Contributions to the Cryptogamic Flora of Essex County. Proc. Essex Instit., 1 : 191–194.

—— 1870. The Sea-Weeds at Home and Abroad. Amer. Nat., 4(5) : 274–297. 1871.

SCHMITT, J. 1904. Monographie de l'Ile d'Anticosti. Thèses présentées a la Faculté des Sciences de Paris no. 1195 (sér. A, no. 486) : vi + 1–370 (Algae, pp. 140–145).

SCHUH, R. E. 1900a. Rhadinocladia, a New Genus of Brown Algae. Rhodora, 2 : 111–112.

—— 1900b. Notes on Two Rare Algae of Vineyard Sound. *Ibid.*, 2 : 206–207.

—— 1901. Further Notes on Rhadinocladia. *Ibid.*, 3 : 218.

—— 1914a. The Discovery of the Long-Sought Alga, *Stictyosiphon tortilis*. *Ibid.*, 16 : 105.

—— 1914b. *Kjellmannia sorifera* Found on the Rhode Island Coast. *Ibid.*, 16 : 152.

—— 1933a. *Pylaiella fulvescens* (Schousb.) Bornet. *Ibid.*, 35 : 63.

—— 1933b. *Ectocarpus paradoxus* in New England. *Ibid.*, 35 : 107.

—— 1933c. *Myriotrichia densa* in New England. *Ibid.*, 35 : 256–257.

—— 1933d. A Second Station for Isthmoplea in North America. *Ibid.*, 35 : 293.

—— 1933e. Dumontia in Maine. *Ibid.*, 35 : 315–316.

—— 1933f. On the Distribution of Sorocarpus. *Ibid.*, 35 : 347.

—— 1933g. On *Polysiphonia fibrillosa* in New England. *Ibid.*, 35 : 391.

—— 1937a. On *Ectocarpus granulosus*. *Ibid.*, 39 : 50–51.

—— 1937b. On *Ectocarpus ovatus*. *Ibid.*, 39 : 148–149.

SETCHELL, W. A. 1891. Concerning the Life History of *Saccorhiza dermatodea* (De la Pyl.) J. Ag. Proc. Am. Acad. Arts and Sci., 26 : 177–217.

—— 1900. Critical Notes on the New England Species of Laminaria. Rhodora, 2 : 115–119, 142–149.

—— 1905. *Gymnogongrus torreyi* (C. Ag.) J. Ag. *Ibid.*, 7 : 136–138.

—— 1914. The Scinaia Assemblage. Univ. Cal. Publ. Bot., 6(5) : 79–152.

—— 1922. Cape Cod in Its Relation to the Marine Flora of New England. Rhodora, 24 : 1–11.

—— 1923. *Dumontia filiformis* on the New England Coast. *Ibid.*, 25 : 33–37.

—— AND COLLINS, F. S. 1908. Some Algae from Hudson Bay. *Ibid.*, 10 : 114–116.

SILVA, P. C. 1952. A Review of Nomenclatural Conservation in the Algae from the Point of View of the Type Method. Univ. Cal. Publ. Bot., 25(4) : 241–324.

SIMONS, E. B. 1906. A Morphological Study of *Sargassum filipendula.* Bot. Gaz., 41 : 161–182.

SMITH, A. A. 1896. The Development of the Cystocarp of *Griffithsia bornetiana.* Bot. Gaz., 22 : 35–47.

SMITH, G. M. 1933, 1950. Fresh-Water Algae of the United States. xi + 716. 1933. New York. *Ibid.;* Ed. 2, vii + 719. 1950.

—— (Ed.). 1951. Manual of Phycology. viii + 375. Waltham, Mass.

SPALDING, V. M. 1890. Development of the Sporocarp of *Griffithsia bornetiana* (Abstract). Proc. Am. Assn. Adv. Sci., 39 : 327.

STEPHENSON, T. A., AND A. 1954. Life Between Tide-Marks in North America. III. Nova Scotia and Prince Edward Island. Journ. Ecol., 42(1) : 14–70.

SVERDRUP, H. U., JOHNSON, M. W., AND FLEMING, R. H. 1942. The Oceans, Their Physics, Chemistry and General Biology. x + 1087. New York.

TAYLOR, W. R. 1921. Additions to the Flora of Mount Desert, Maine. Rhodora, 23 : 65–68.

—— 1922. Recent Studies on Phaeophyceae and Their Bearing on Classification. Bot. Gaz., 74(4) : 431–441.

—— 1928a. The Marine Algae of Florida, with Special Reference to the Dry Tortugas. Papers Tortugas Lab., Carnegie Inst. Wash., 25 : 1–219.

—— 1928b. A Species of Acrothrix on the Massachusetts Coast. Am. Journ. Bot., 15 : 577–583.

—— 1929. Notes on the Marine Algae of Florida. Bull. Torr. Bot. Club, 56 : 199–210.

—— 1936. Phaeophycean Life Histories in Relation to Classification. Bot. Review, 2 : 554–563.

—— 1937a. General Botanical Microtechnique, pp. 155–245. *In:* C. E. McClung (Ed.), Handbook of Microscopical Technique, Ed. 2. New York.

—— 1937b. Notes on North Atlantic Marine Algae. I. Papers Mich. Acad. Sci., Arts, and Letters, 22 : 225–233.

TAYLOR, W. R. 1937c. Marine Algae of the Northeastern Coast of North America. ix + 427. Ann Arbor, Michigan.

—— 1940. Marine Algae from Long Island. Torreya, 40 : 185–195.

—— 1941. The Reappearance of Rare New England Marine Algae. Rhodora, 43 : 72–74.

—— 1945. The Collecting of Seaweeds and Freshwater Algae. Suppl. to Company D Newsletter. Instructions to Naturalists in the Armed Forces for Botanical Field Work. Ed. 2. 18 pp. Ann Arbor, Michigan.

—— 1950. Field Preservation and Shipping of Biological Specimens. Turtox News, 28(2) : 42–43.

—— 1954. Cryptogamic Flora of the Arctic. II. Algae: Non-Planktonic. Bot. Review, 20(6, 7) : 363–399.

—— AND BERNATOWICZ, A. J. 1952. Bermudian Marine Vaucherias of the Section Piloboloideae. Papers Mich. Acad. Sci., Arts, and Letters, 37 : 75–86.

—— *See* Lewis, I. F., and Taylor, W. R.

TERRY, W. A. 1901. Causes of Variation in Color in Some Red Algae. Rhodora, 9 : 90–91.

THIVY, F. 1942. A New Species of Ectochaete (Huber) Wille, from Woods Hole, Massachusetts. Biol. Bull., 83(1) : 97–110.

—— 1943. New Records of Some Marine Chaetophoraceae and Chaetosphaeridiaceae from North America. Biol. Bull., 85(3) : 244–264.

TONI, G. B. DE. *See* De Toni, G. B.

TRANSEAU, E. N. 1913. The Vegetation of Cold Spring Harbor, Long Island, I. The Littoral Successions. Plant World, 16 : 189–209.

VERRILL, A. E. 1874. Brief Contribution to Zoölogy from the Museum of Yale College. XXVI. Results of Recent Dredging Expeditions on the Coast of New England, 4. Am. Journ. Sci. and Arts, III, 7 : 38–46.

WEBBER, H. J. 1891. On the Antheridia of Lomentaria. Ann. Bot., 5 : 226–227.

WHELDEN, R. M. 1928. Observations on the Red Alga *Dumontia filiformis.* Maine Nat., 8 : 121–130.

—— 1947. Algae. Pp. 13–127. *In:* Polunin, N. (Ed.), Botany of the Canadian Eastern Arctic. Part II. Thallophyta and Bryophyta. Nat. Mus. Canada, Bull. 97, Biol. Ser. 26 : 1–573.

WILLEY, H. 1872. A Seaweed New to Our Coast. Am. Nat., 6 : 767.

WOLFE, J. J. 1904. Cytological Studies on Nemalion. Ann. Bot., 18 : 607–630.

WOOD, R. D. 1954. Macroscopic Algae of the Coastal Ponds of Rhode Island. Amer. Journ. Bot., 41 (2) : 135–142.

YAMANOUCHI, S. 1906. The Life History of Polysiphonia. Bot. Gaz., 42 : 401–449.

YENDO, K. 1919. A Monograph of the Genus Alaria. Journ. Coll. Sci., Imp. Univ. Tokyo, 43 (1) : 1–145.

PLATES AND DESCRIPTIONS

PLATE 1

	PAGE
FIGS. 1–2. *Chaetomorpha linum:* Fig. 1, habit of portion of plant mass, × 0.45. Fig. 2, detail of filament, showing cell form and chromatophores, × 38	78
FIG. 3. *Rhizoclonium riparium:* portions of three filaments, showing rhizoidal branches, × 150	81
FIGS. 4–7. *Prasiola stipitata:* Fig. 4, habit of a small group of plants, × 11. Figs. 5–7, portions from base, middle and upper margin of plant, showing structure and akinetes, × 76	75
FIG. 8. *Percursaria percursa:* small portion of filaments, showing biseriate arrangement of cells, × 120	61
FIG. 9. *Ulothrix flacca:* small portion of filament, showing cell form and chromatophores, × 450	45
FIGS. 10–12. *Chaetomorpha aerea:* Fig. 10, habit of small group of plants, × 0.45. Fig. 11, base of filament, showing modification of basal cell as a holdfast, × 38. Fig. 12, cells at summit of filament, showing, below, vegetative structure; above, emptied sporangia cells, × 38	79
FIGS. 13–14. *Gomontia polyrhiza:* Fig. 13, group of vegetative filaments, showing form and branching, × 320. Fig. 14, sporangium, showing rhizoidal appendages, which project toward the surface of the shell, × 320	59

For an additional figure of Gomontia see Plate 7.

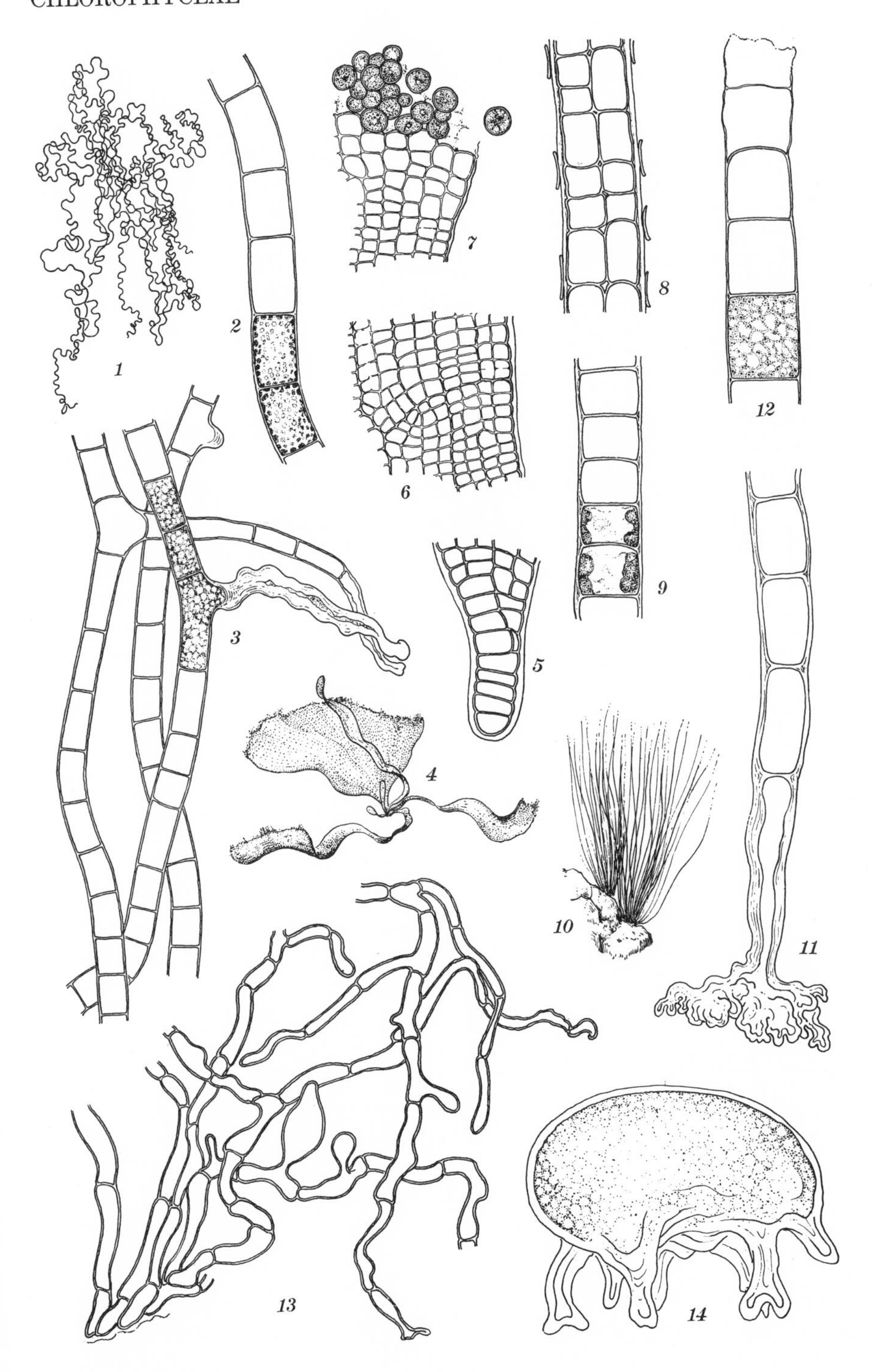
1
2
3
4
5
6
7
8
9
10
11
12
13
14

PLATE 2

PAGE

FIGS. 1–2. *Entocladia viridis:* Fig. 1, habit of a considerable portion of a plant, × 115. Fig. 2, a small part from the older region of a plant, showing chromatophores and sporangial cells, × 340....... 54

FIG. 3. *Ostreobium quekettii:* habit of a small group of filaments from a shell, × 90... 95

FIGS. 4–5. *Phaeophila dendroides:* Fig. 4, habit of a considerable group of filaments, showing branching, cell form, hairs, and sporangial cells, × 90. Fig. 5, details of the structure of three cells, particularly to show chromatophores, × 270.................... 51

FIGS. 6–8. *Blastophysa rhizopus:* Fig. 6, habit of a group of filaments, showing vegetative cells, slender connecting tubes, and hair groups, × 135. Fig. 7, details of a sporangial cell, × 270. Fig. 8, details of a vegetative cell, particularly to show chromatophores, × 680.. 92

1
2
3
4
5
6
7
8

PLATE 3

For additional figures of Enteromorpha see Plate 4.

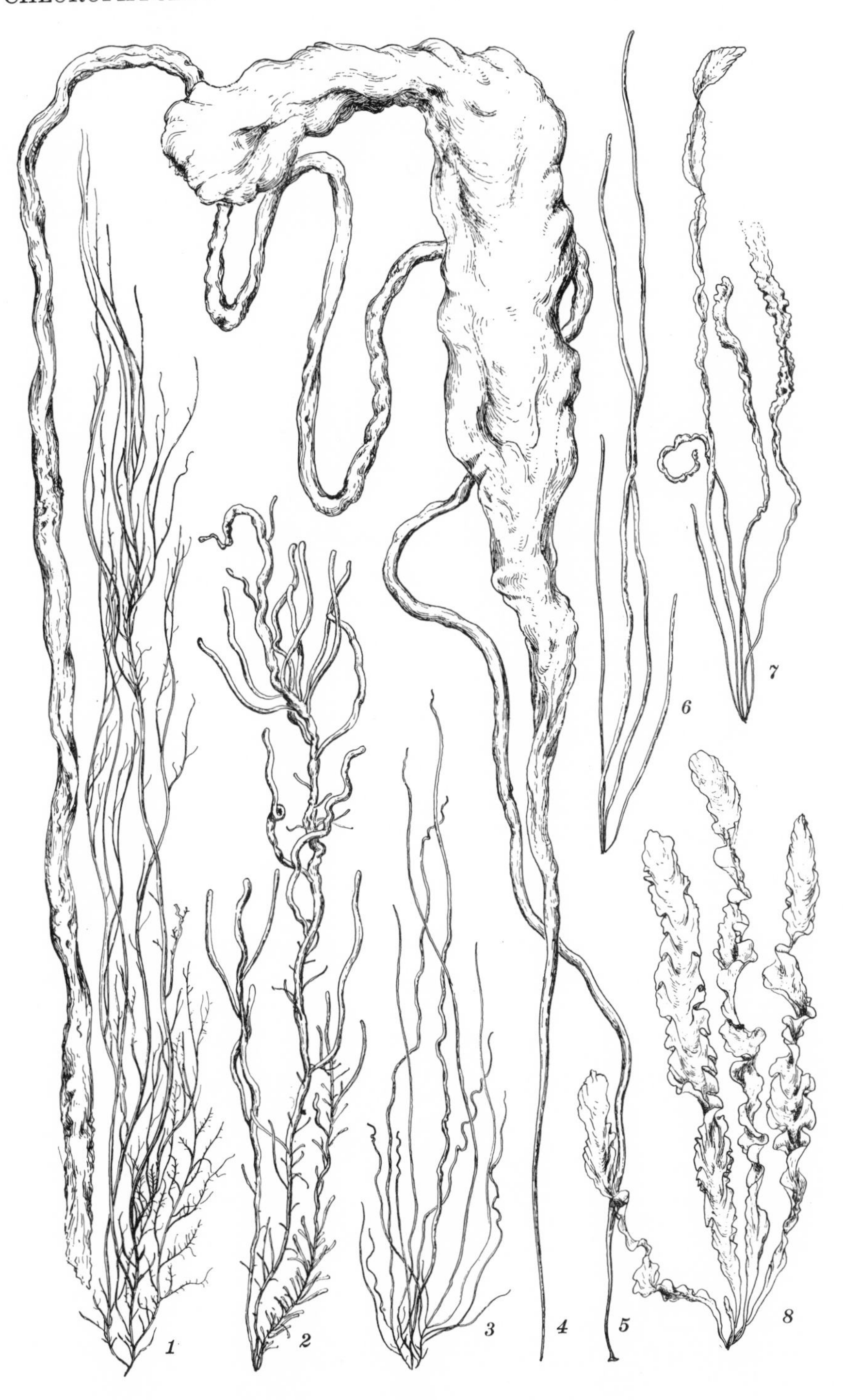
1
2
3
4
5
6
7
8

PLATE 4

PAGE

FIGS. 1–3. *Monostroma oxyspermum:* Fig. 1, habit of plant, × 0.45 Fig. 2, surface view, showing cell arrangement and structure, × 450. Fig. 3, sectional view, showing cell form and structure, × 450 72

FIGS. 4–5. *Enteromorpha intestinalis:* Fig. 4, sectional view of portion of the wall of the upper part of the tubular thallus, showing cell form and structure, × 540. Fig. 5, surface view, showing cell arrangement and structure, × 540.......... 66

FIG. 6. *Ulva lactuca:* habit of rather narrow specimen, × 0.45...... 74

FIGS. 7–8. *Ulva lactuca* v. *rigida:* Fig. 7, surface view, showing cell arrangement and structure, × 320. Fig. 8, sectional view of the thallus in the middle region, showing cell form and structure, × 320 74

FIG. 9. *Ulva lactuca v. latissima:* sectional view, showing cell form and structure, × 320.......... 74

For additional figures of Enteromorpha see Plate 3.

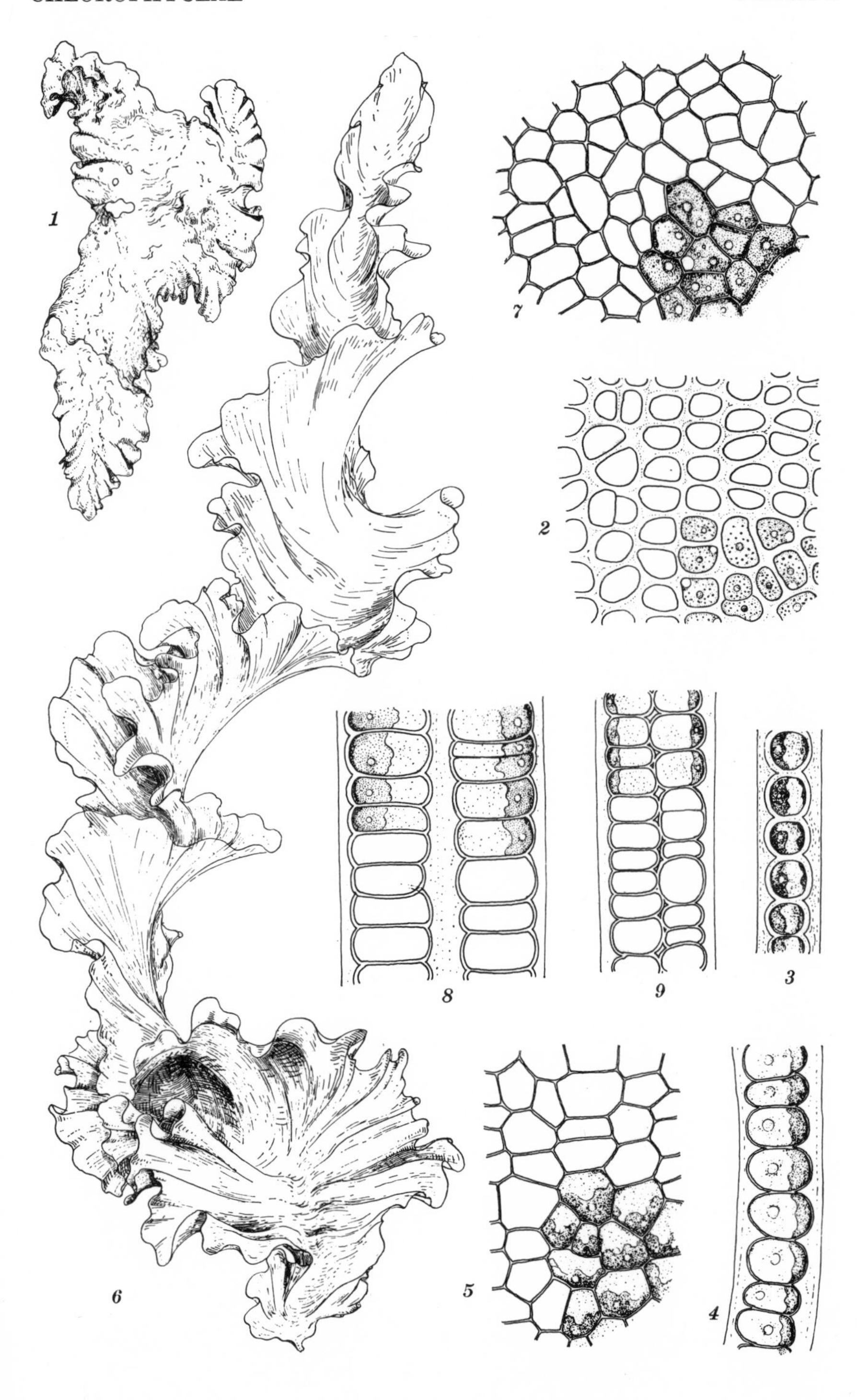
1
7
2
8
9
3
6
5
4

PLATE 5

	PAGE
FIG. 1. *Cladophora rupestris:* habit of branching near tip of a main shoot, × 15	88
FIG. 2. *Cladophora gracilis:* detail of branching, × 34	86
FIG. 3–4. *Cladophora rudolphiana:* Fig. 3, habit of branching near tip of a main shoot, × 2.3. Fig. 4, detail of branching, × 45	84
FIG. 5. *Cladophora expansa:* detail of branching from an old filament in the proliferous state, × 45	85
FIGS. 6–7. *Spongomorpha lanosa:* Fig. 6, habit of a twisted group of branches, × 1.8. Fig. 7, details of branching, showing the interweaving rhizoidal filaments, × 32	89

For additional figures of Cladophora and Spongomorpha see Plate 6. Figure 1 drawn by the author.

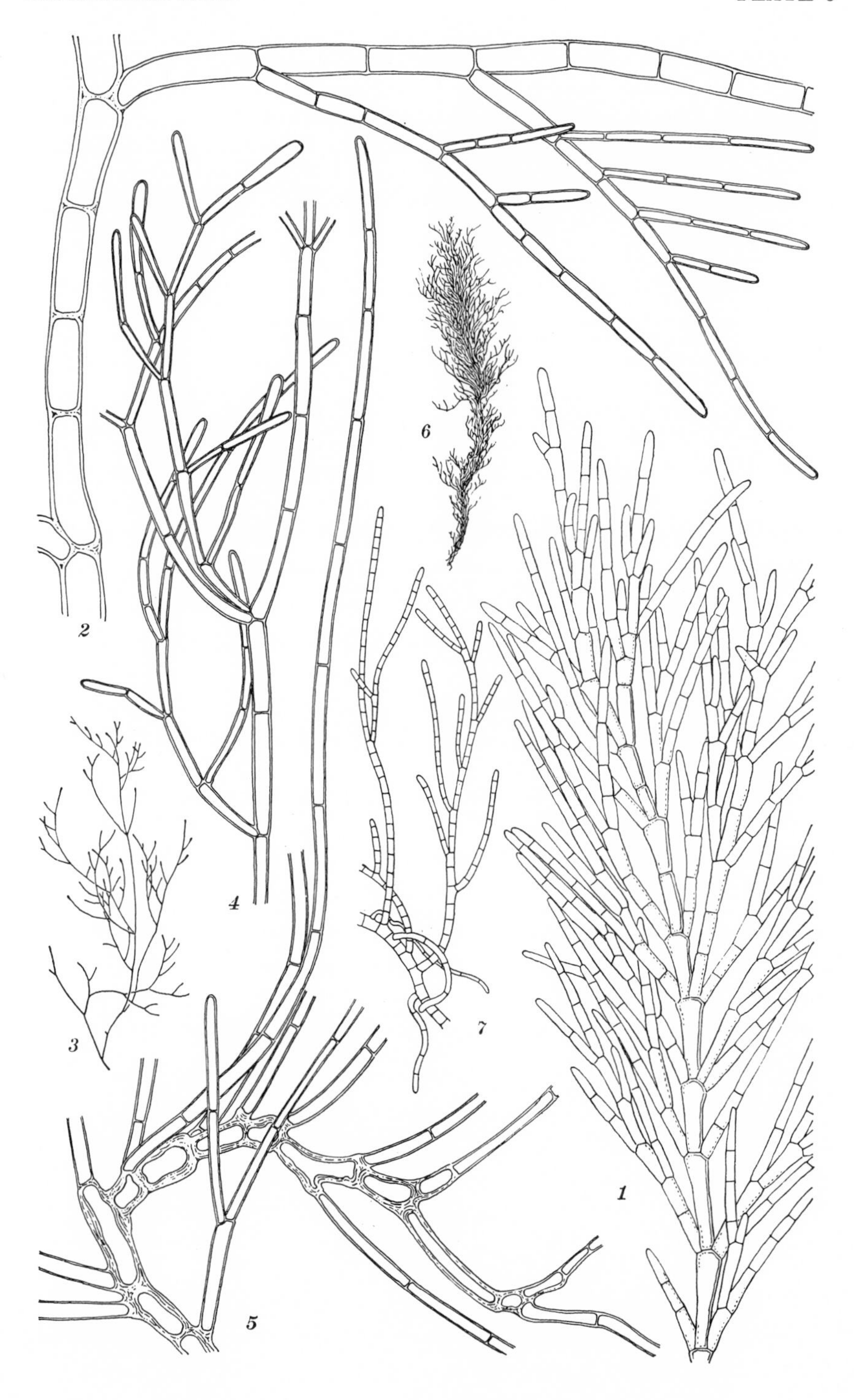
1
2
3
4
5
6
7

PLATE 6

For additional figures of Cladophora and Spongomorpha see Plate 5.

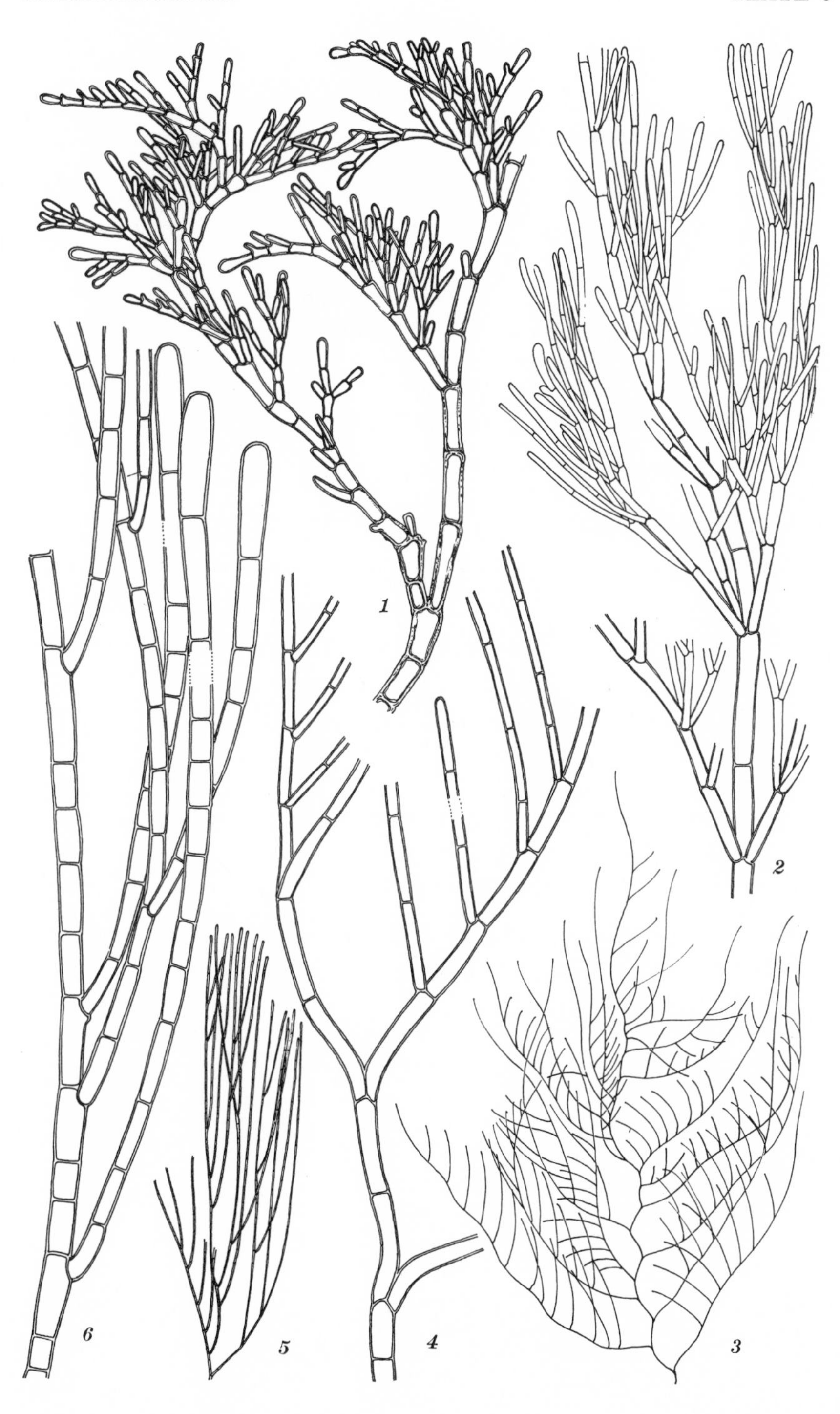
1
2
3
4
5
6

PLATE 7

PAGE

FIGS. 1–3. *Bryopsis plumosa:* Fig. 1, habit of a tuft, × 0.9. Fig. 2, detail of base of a plant, showing rhizoidal downgrowths, × 11. Fig. 3, tip of a branch, × 11 94

FIG. 4. *Gomontia polyrhiza:* sporangium of the narrow type, × 320... 59

FIGS. 5–6. *Ectocarpus subcorymbosus:* Fig. 5, base of a plant, × 150. Fig. 6, tips of several filaments, with gametangia, × 150.......... 109

FIGS. 7–9. *Giffordia granulosa:* Fig. 7, lateral branch from the upper part of a plant, showing branchlets, hairs and several gametangia, × 28. Fig. 8, detail showing two maturing gametangia, × 150. Fig. 9, cell in a main branch, showing the chromatophores, × 150. 112

For additional figures of Ectocarpus see Plate 8; for Gomontia, Plate 1. Figures 4–9 were drawn by the author.

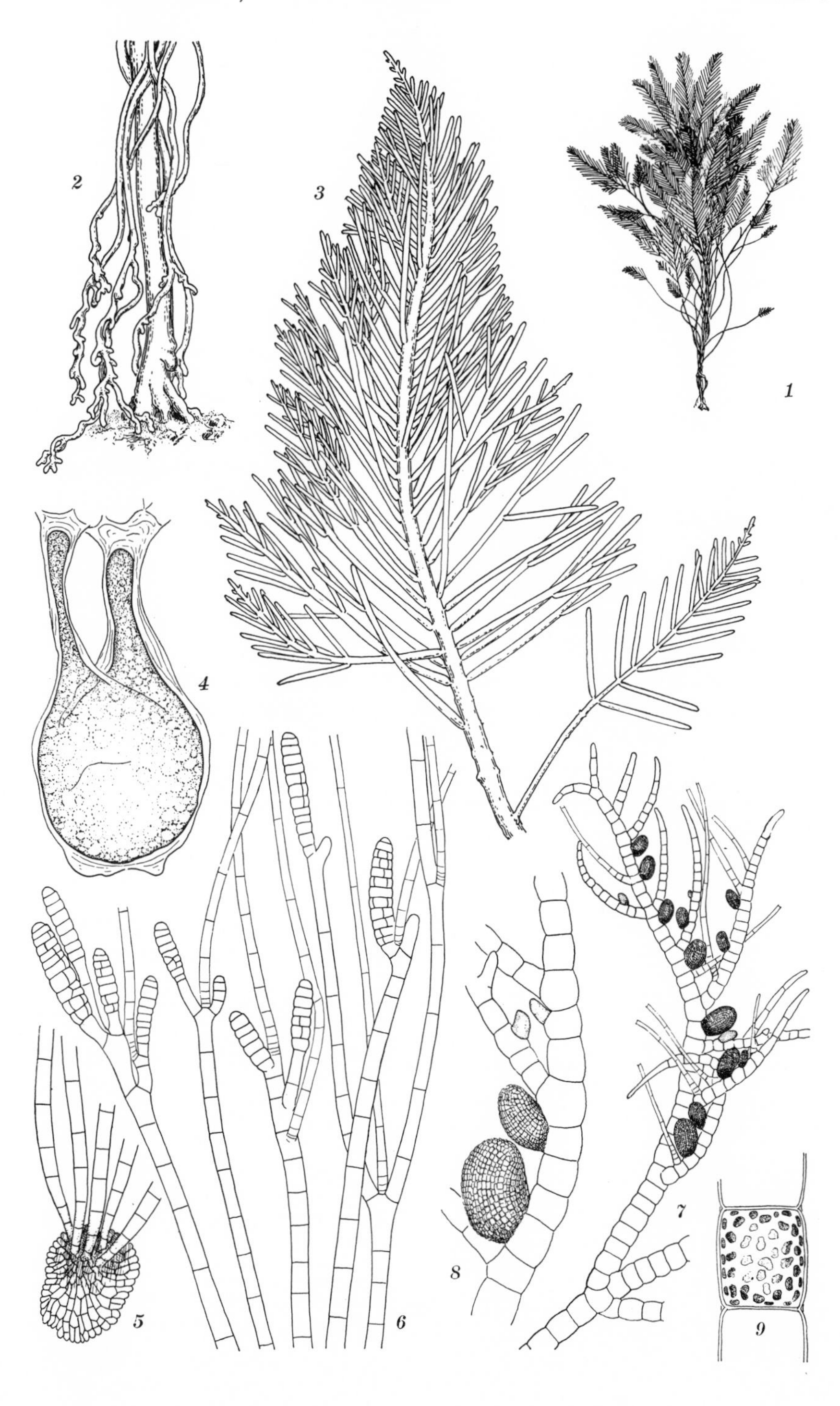
2
3
1
4
5
6
7
8
9

PLATE 8

PAGE

FIGS. 1–3. *Ectocarpus confervoides:* Fig. 1, small portion of upper branch system, showing branching, cell form, and gametangia, × 76. Fig. 2, portion of branch, showing sporangia, × 76. Fig. 3, cell with chromatophore, × 320 106

FIGS. 4–5. *Ectocarpus siliculosus:* Fig. 4, small portion of upper branch system, showing branching, cell form, and gametangia with hair tips, × 76. Fig. 5, with chromatophores, × 320 105

FIGS. 6–8. *Ectocarpus tomentosus:* Fig. 6, portion of tuft of plant, showing twisted, cordlike branch systems, × 0.45. Fig. 7, cells with chromatophores, × 320. Fig. 8, small portion of branch system, showing branching, recurved branchlets, and stalked, recurved, often falcate gametangia, × 150 108

For additional figures of Ectocarpus see Plate 7.

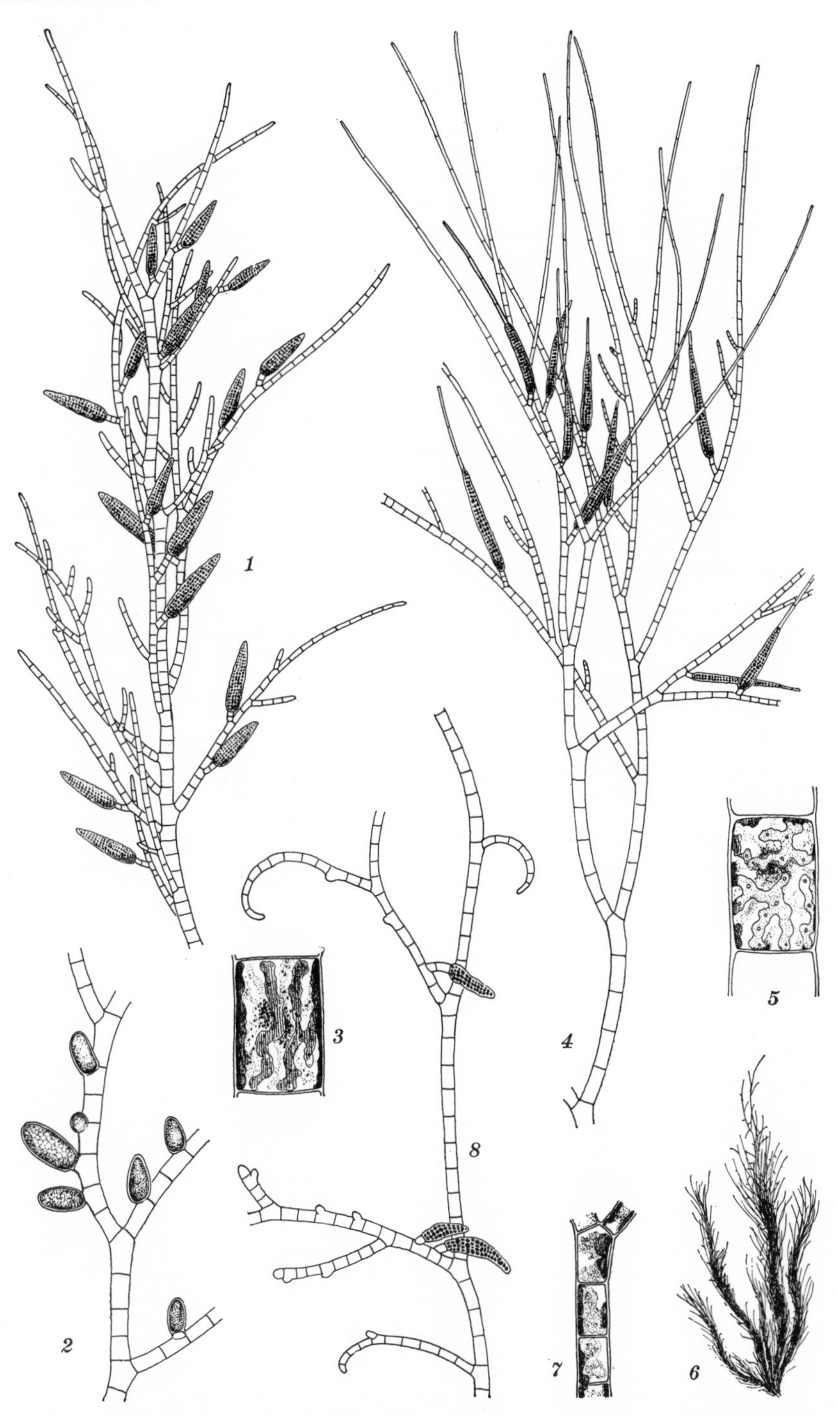
1
2
3
4
5
6
7
8

PLATE 9

PAGE

FIGS. 1–3. *Pylaiella littoralis:* Fig. 1, small portion of the upper branch system, showing branching, cell form, and gametangia, with the vegetative branches extending beyond the gametangia, but in the form figured, rather briefly, × 76. Fig. 2, small portion of the branch system, showing seriate sporangia intercalary in the lateral branchlets, × 76. Fig. 3, a few segments from a lower part of the axis, showing occasional longitudinal walls and chromatophores, × 320 102

FIGS. 4–5. *Isthmoplea sphaerophora:* Fig. 4, small portion of the lower part of a main branch, showing multiseriate arrangement of cells. Fig. 5, small part of upper branch, showing cell arrangement, branching, chromatophores, and spherical sporangia, × 150..... 156

FIG. 6. *Sorocarpus micromorus:* small portion of the upper branch system, showing forking, cell form, groups of gametangia, and colorless hairs, × 200.. 117

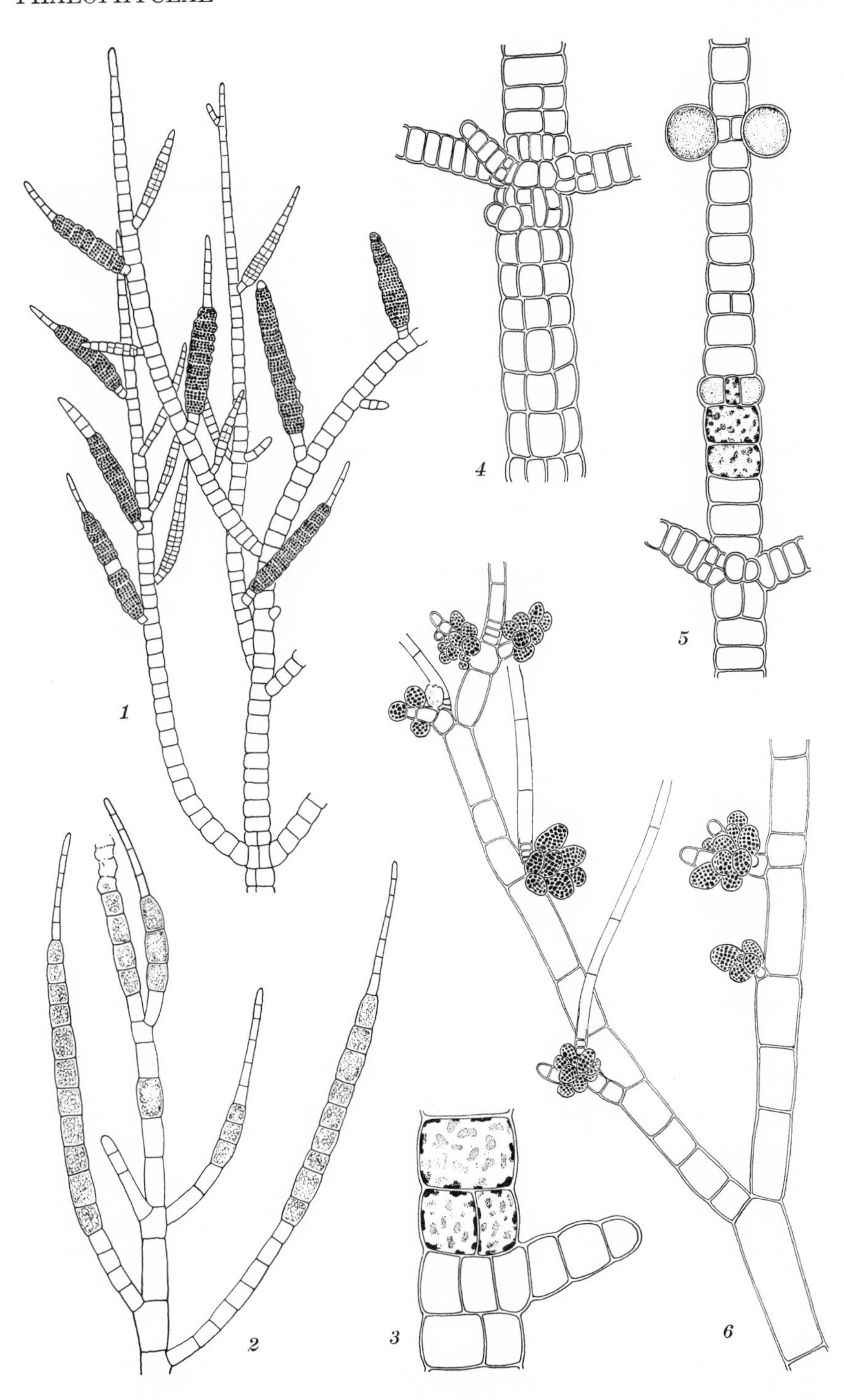
1
2
3
4
5
6

PLATE 10

PAGE

FIGS. 1–3. *Elachistea fucicola:* Fig. 1, fragment of Fucus plant with several tufts of Elachistea, × 0.45. Fig. 2, somewhat diagrammatic section of an Elachistea tuft, showing penetrating base, crowded sporangia, short paraphysal filaments, and long exserted assimilatory filaments. Fig. 3, basal part of an assimilatory filament, showing transition between the vegetative part and the colorless base, with attached branching system of paraphyses and several sporangia in various stages of development, × 150. 140

FIGS. 4–7. *Myriotrichia clavaeformis:* Fig. 4, single individual, × 22. Figs. 5–7, three portions of a plant, successively from the upper portion to the base, showing short lateral branchlets with hairs and spherical sporangia, the increase in diameter of the axial cells, the beginning development of longitudinal walls, and the base with evidences of rhizoidal development, × 68. 161

FIGS. 8–9. *Myriactula minor:* Fig. 8, portion of a tuft from a Sargassum cryptostoma, showing uniseriate gametangia, sporangia, the bases of three exserted assimilators, and a colorless hair, × 79. Fig. 9, the upper portion of an assimilator laterally bearing short gametangia, × 79. 142

FIGS. 10–11. *Eudesme zosterae:* Fig. 10, portion of axis with a tuft of assimilators, showing vegetative filaments, gametangia and hair, × 165. Fig. 11, portion of a tuft of assimilators, showing a sporangium, × 165. 146

For additional figures of Eudesme see Plate 12.

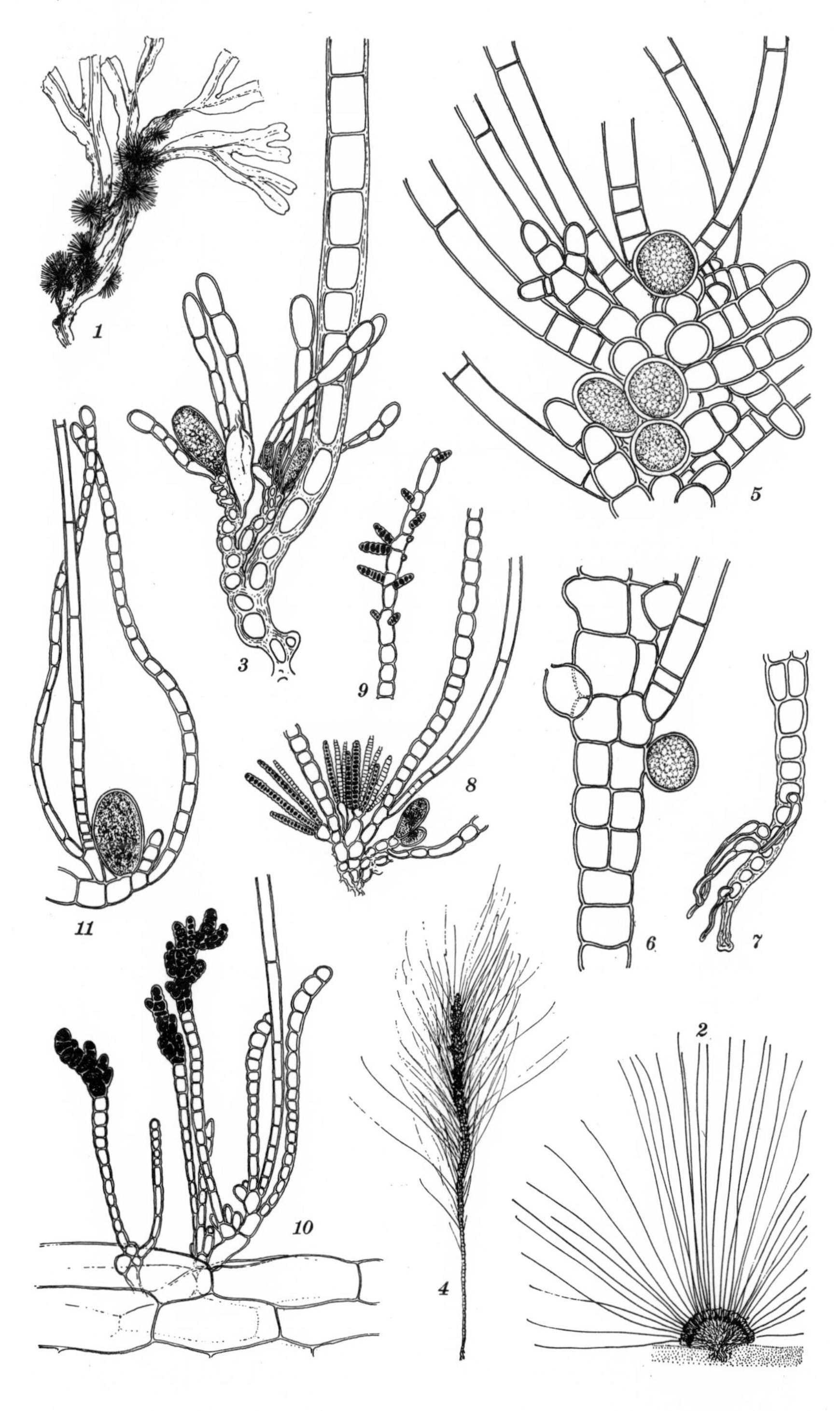
1
3
5
9
6
7
8
11
10
4
2

PLATE 11

PAGE

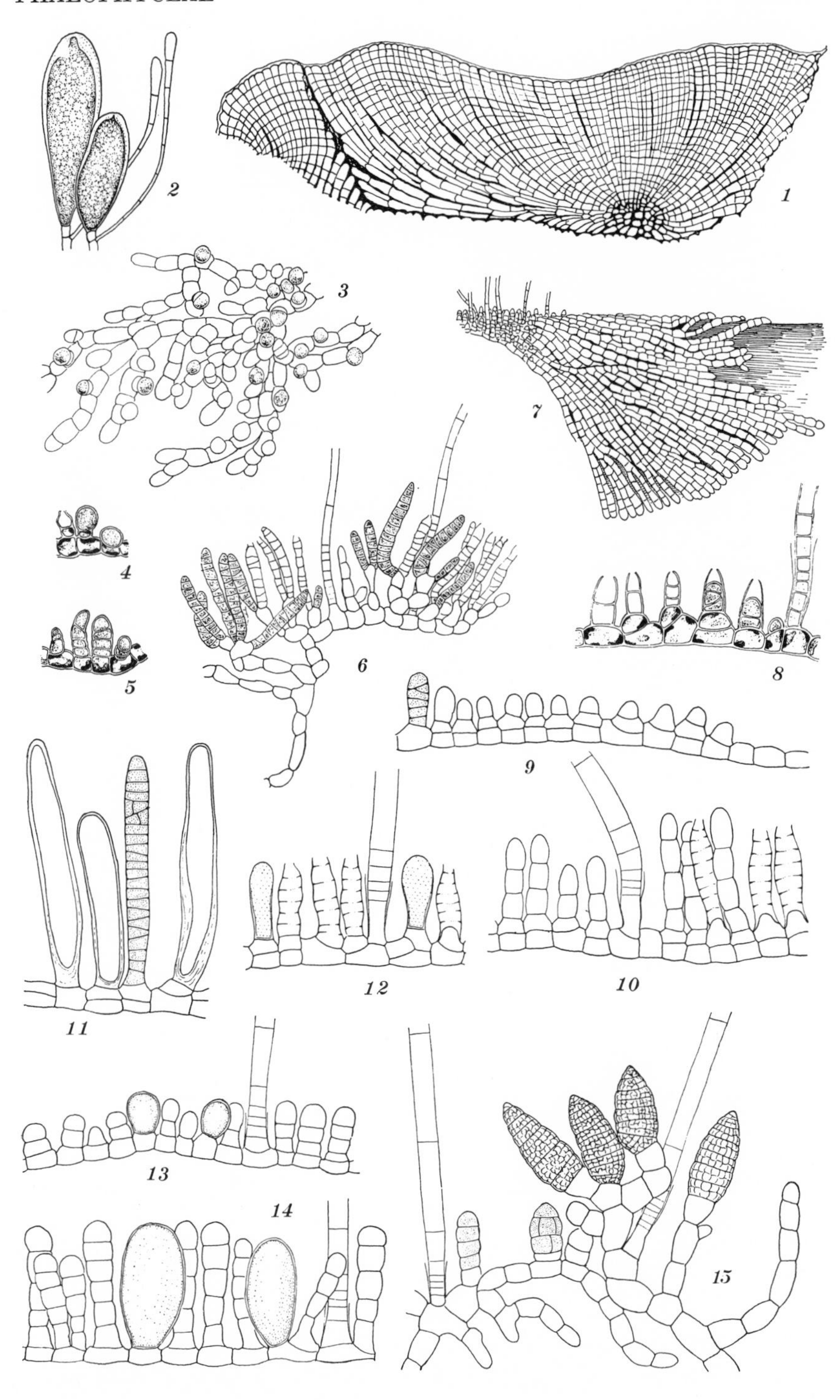
2
1
3
7
4
6
5
8
9
11
12
10
13
14
15

PLATE 12

PAGE

FIG. 1. *Sphaerotrichia divaricata:* final portion of a branch system, × 0.45 147

FIG. 2. *Eudesme zosterae:* leaf of Zostera, with several individuals epiphytic upon it, × 0.45.......... 146

FIG. 3. *Eudesme virescens:* individual plant, × 0.45.......... 145

FIG. 4. *Dictyosiphon foeniculaceus:* individual plant, × 0.45.......... 172

FIG. 5. *Leathesia difformis:* branches of Chondrus, with several Leathesia plants growing upon it, × 0.45.......... 149

FIG. 6. *Chordaria flagelliformis:* individual plant, × 0.45.......... 148

For additional figures of Eudesme see Plate 10; for Chordaria, Dictyosiphon, Leathesia, and Sphaerotrichia, Plate 14.

1
2
3
4
5
6

PLATE 13

PAGE

FIG. 1. *Acrothrix novae-angliae:* individual plant. This specimen is more flexuous in the branches than typical ones, × 0.45 150

FIG. 2. *Arthrocladia villosa:* portion of a main branch, with principal lateral branches and branchlets and the tufted assimilators, × 0.45. 152

FIG. 3. *Desmarestia viridis:* holdfast and basal portion of a small plant, showing opposite branching, × 0.45 153

FIGS. 4–5. *Desmarestia aculeata:* Fig. 4, portion of a plant in the summer condition, showing spinuliform ultimate branchlets, × 0.45. Fig. 5, very small portion of a branch system in early spring condition, showing the tufted assimilators, × 0.45 154

FIG. 6. *Stilophora rhizodes:* portion of a plant, showing habit of branching and the characteristic rough surface, × 0.45 151

For additional figures of Arthrocladia see Plate 17; for Desmarestia and Stilophora, Plate 14.

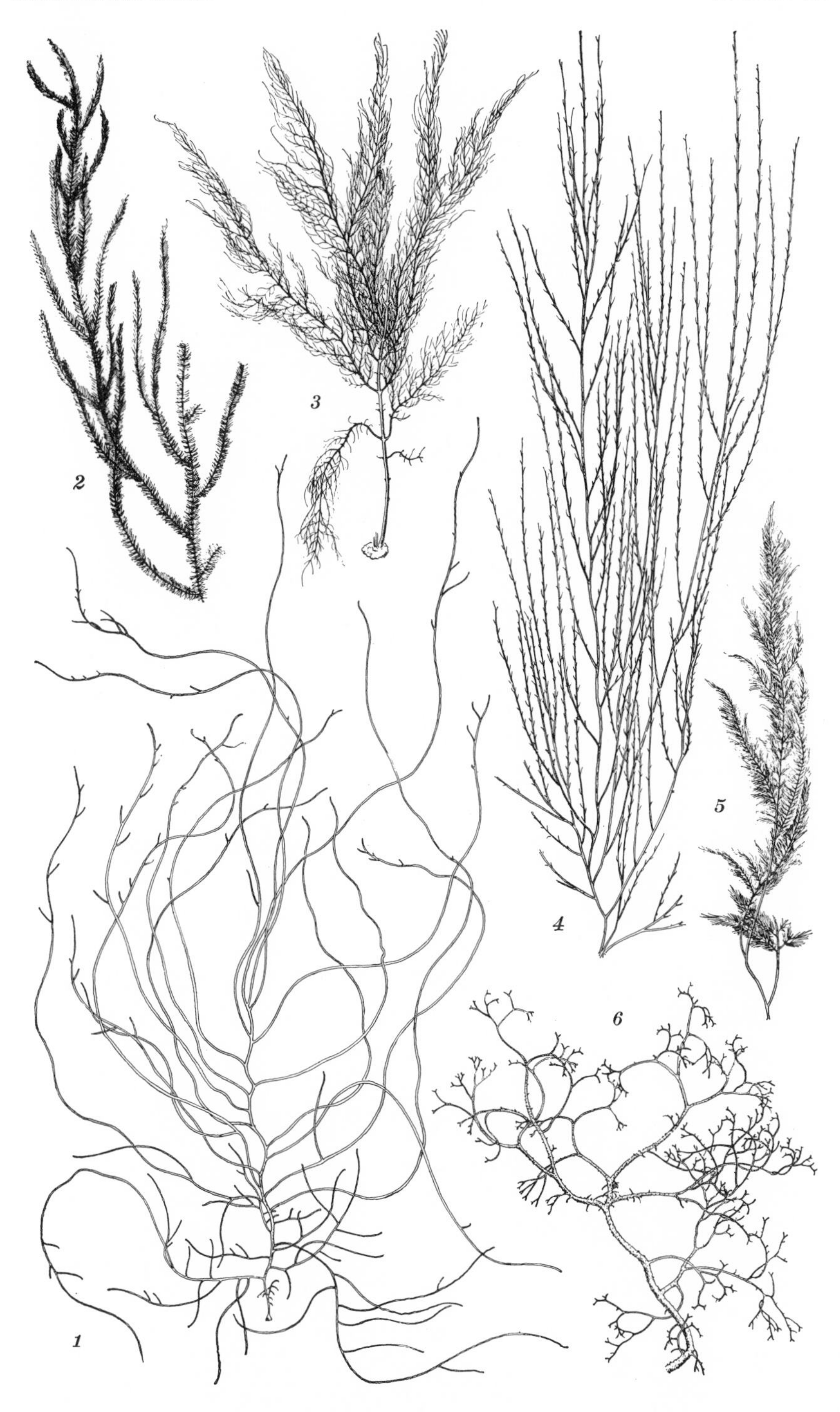
1
2
3
4
5
6

PLATE 14

PAGE

Fig. 1. *Sphaerotrichia divaricata:* portion of a small branch in longitudinal section, showing peripheral assimilators and colorless hairs, × 150 147

Fig. 2. *Dictyosiphon foeniculaceus:* portion of a hollow main axis in transverse section, showing hairs and young sporangia, × 120.... 172

Fig. 3. *Chorda filum:* portion of a hollow plant in transverse section, showing hairs, paraphyses, and young sporangia, × 150.......... 176

Fig. 4. *Chordaria flagelliformis:* portion of a plant in transverse section, showing hairs and assimilators, × 150.................... 148

Fig. 5. *Petalonia fascia:* portion of a blade in transverse section, showing hairs and extensive development of gametangia, × 320... 167

Fig. 6. *Stilophora rhizodes:* transverse section through one sorus on a branch, showing hair, superficial assimilators, sporangia, and young gametangia, × 240.. 151

Fig. 7. *Desmarestia aculeata:* transverse section of a main axis showing axial filament and general structure, × 120.................. 154

Fig. 8. *Leathesia difformis:* portion of a transverse section of a plant, showing hairs, assimilators, and sporangia, × 320............... 149

For additional figures of Chorda and Petalonia see Plate 15; for Chordaria, Dictyosiphon, Leathesia and Sphaerotrichia, Plate 12; for Desmarestia and Stilophora, Plate 13.

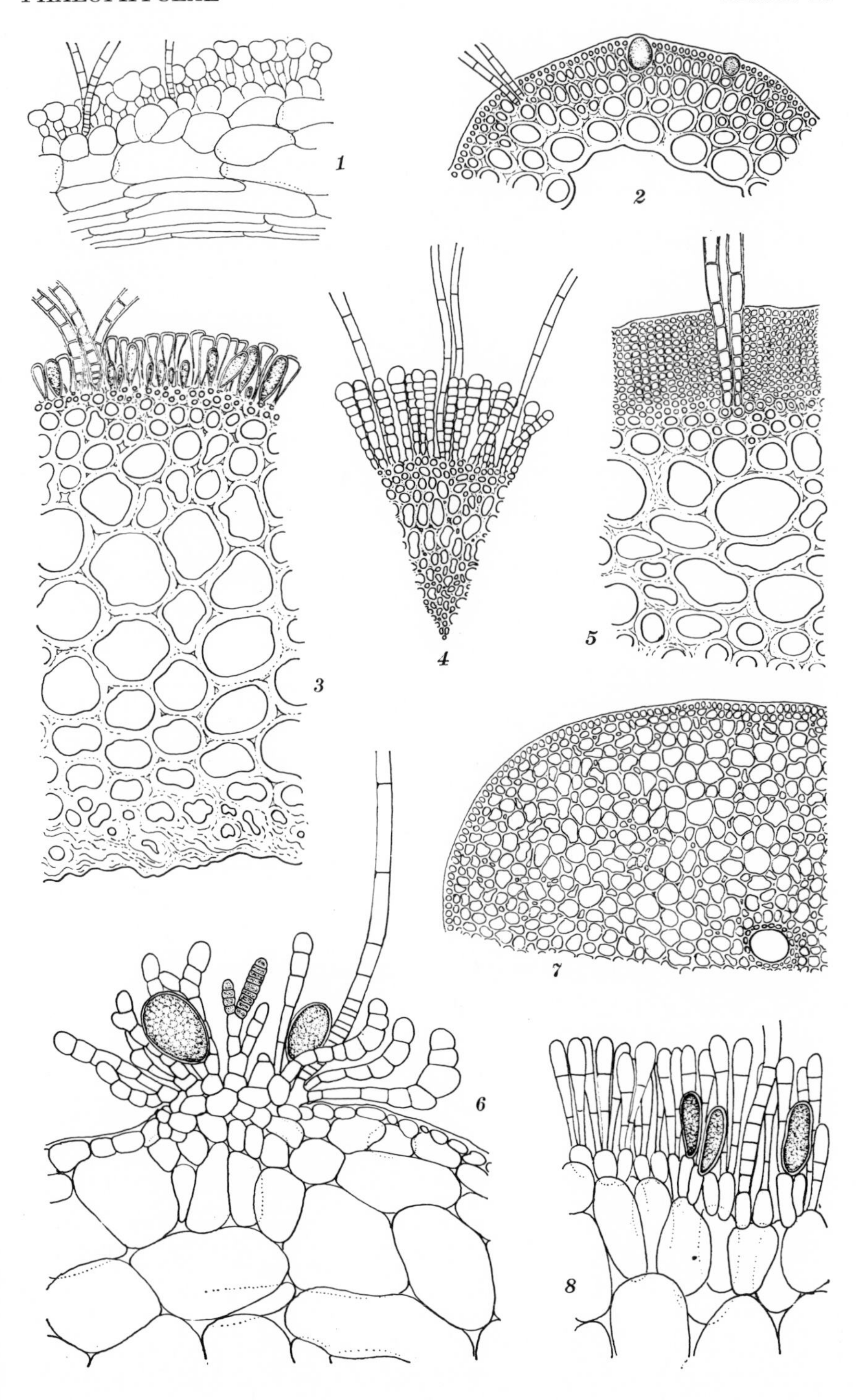
1
2
3
4
5
6
7
8

PLATE 15

PAGE

FIG. 1. *Chorda filum:* group of three plants from crowded holdfasts, × 0.45 176

FIG. 2. *Scytosiphon lomentaria:* group of four plants from crowded holdfasts, × 0.45.. 168

FIG. 3. *Petalonia fascia:* group of several plants from crowded holdfasts, × 0.45... 167

FIG. 4. *Punctaria plantaginea:* individual with eroded tip, × 0.45... 166

FIG. 5. *Punctaria latifolia:* two individuals, × 0.45................. 166

FIG. 6. *Desmotrichum undulatum:* tip of Zostera leaf, with tufts of plants epiphytic upon it, × 0.45............................. 165

FIG. 7. *Asperococcus echinatus:* fragment of Fucus, with a crowded group of plants on it, × 0.45................................ 169

For additional figures of Asperococcus, Desmotrichum, Punctaria and Scytosiphon see Plate 16; for Chorda and Petalonia, Plate 14.

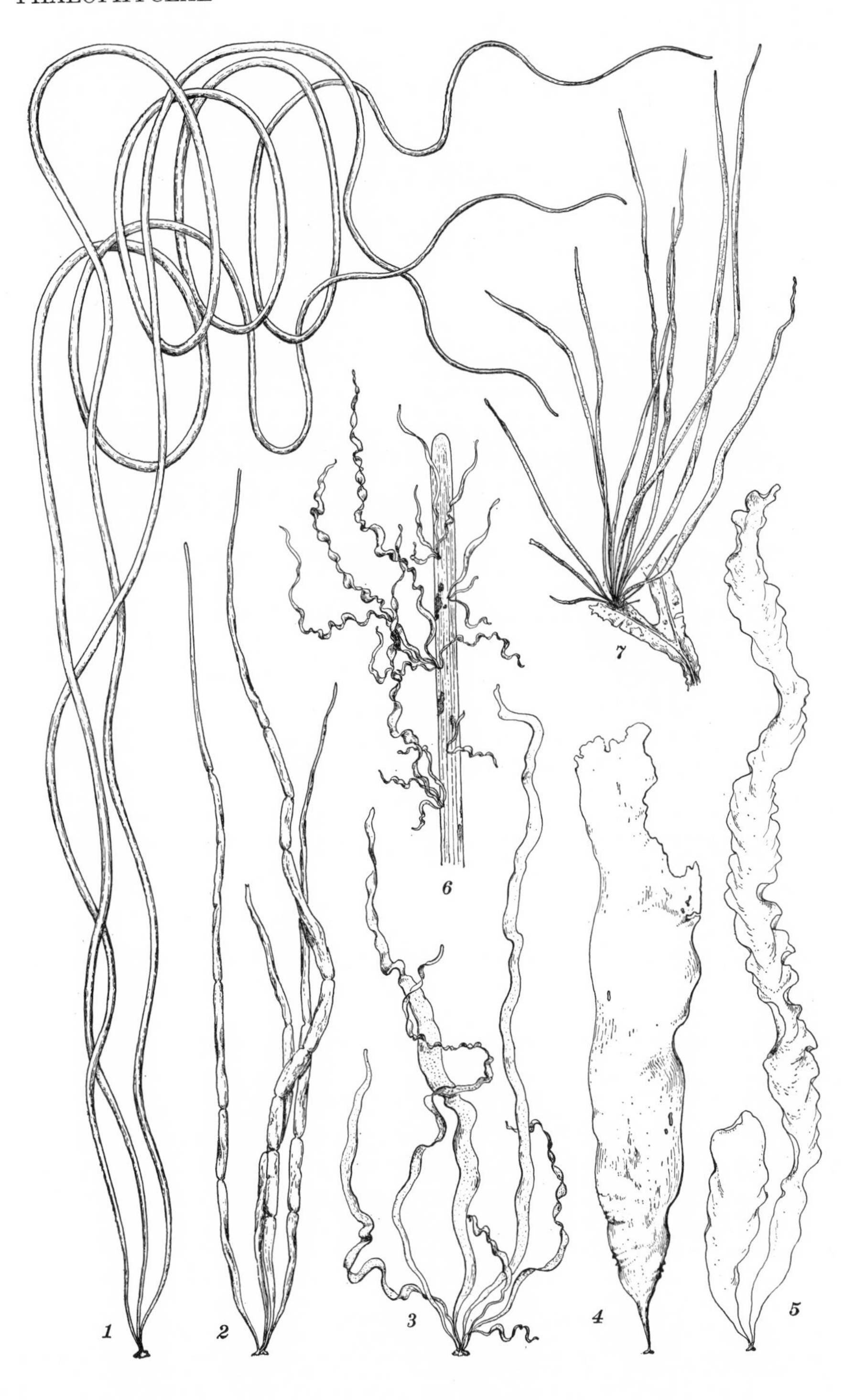
1
2
3
4
5
6
7

PLATE 16

PAGE

FIGS. 1–2. *Desmotrichum undulatum:* Fig. 1, transverse section of blade, showing gametangia, × 320. Fig. 2, marginal portion of a blade, showing hairs and arrangement of cells, × 320........... 165

FIG. 3. *Scytosiphon lomentaria:* portion of a transverse section of the hollow plant, showing hairs, paraphyses, and gametangia, × 540.. 168

FIG. 4. *Punctaria plantaginea:* portion of a transverse section, showing hairs, gametangia and (?) a sporangium, × 150............. 166

FIGS. 5–6. *Asperococcus echinatus:* Fig. 5, portion of a transverse section of a small plant, showing a sorus with chromatophore-bearing paraphyses, young hairs, and sporangia, × 150. Fig. 6, portion of a small plant in surface view, showing sori of paraphyses and sporangia, × 150.. 169

For additional figures of Asperococcus, Desmotrichum, Punctaria, and Scytosiphon see Plate 15.

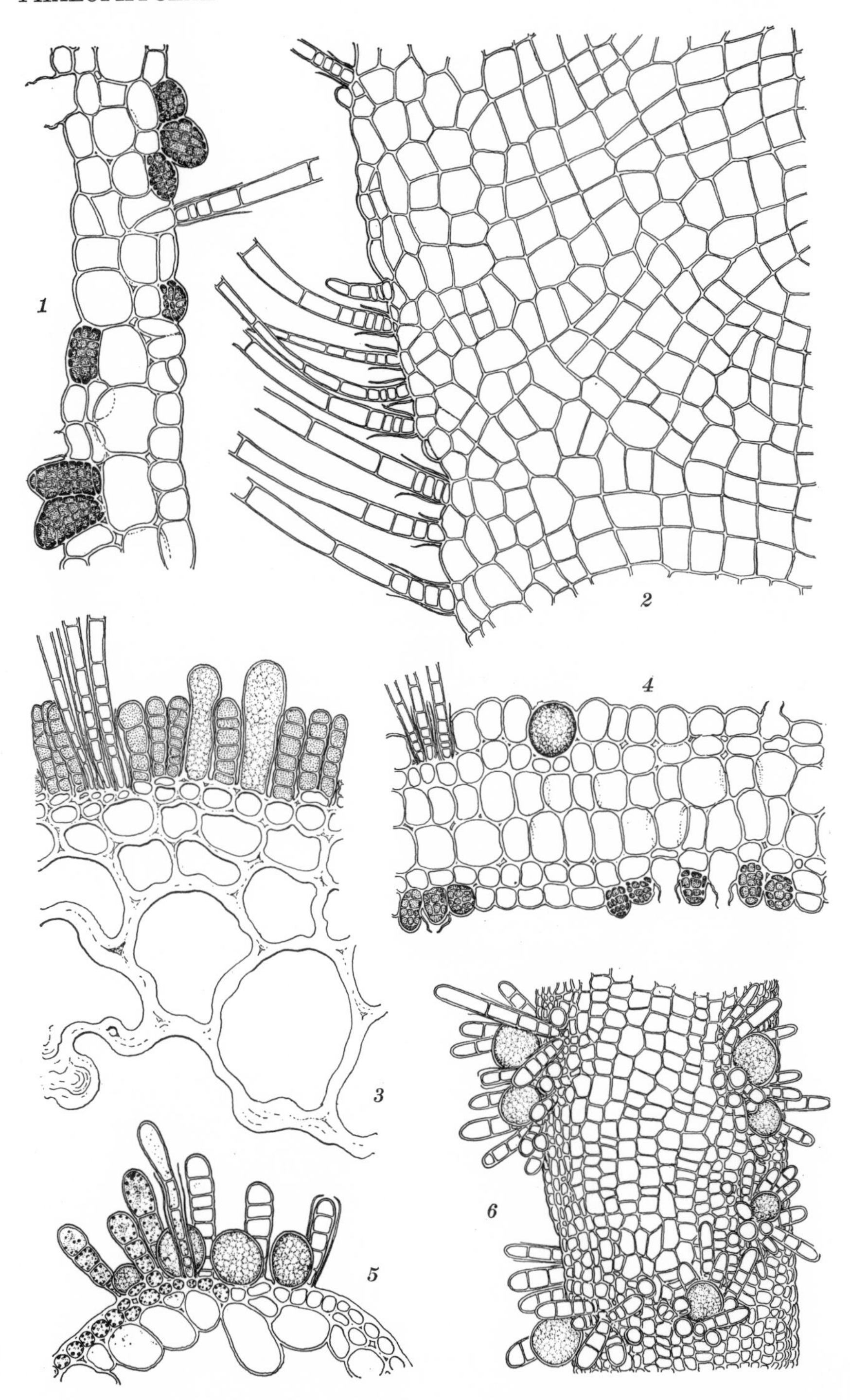
1
2
3
4
5
6

PLATE 17

PAGE

FIGS. 1–6. *Sphacelaria cirrosa:* Fig. 1, branches of Fucus, with tufts of epiphytic Sphacelaria, × 0.45. Fig. 2, individual plant, showing branching, × 8.1. Fig. 3, portion of branch axis, showing segmentation and attachment of branchlets, × 160. Fig. 4, apex of branch, showing apical cell and hair, × 160. Fig. 5, base of plant, showing basal disk, erect axes, rhizoids and a branchlet forming a terminal disk, × 70. Fig. 6, propagulum on a branch, × 70 121

FIGS. 7–8. *Arthrocladia villosa:* Fig. 7, small portion of a small branch in surface view, showing tufts of branched uniseriate assimilators, × 38. Fig. 8, portion of an assimilator, showing seriate sporangia laterally attached, × 170.................................... 152

FIGS. 9–11. *Cladostephus verticillatus:* Fig. 9, upper portion of primary branchlet, showing subsidiary branchlets and hairs, as well as cellular structure, × 150. Fig. 10, lower portion of a secondary branchlet from base of a fertile plant, showing three pedicellate gametangia, × 150. Fig. 11, transverse section of a well-developed stem, showing the embedded bases of several branchlets, × 150 124

For additional figures of Arthrocladia see Plate 13.

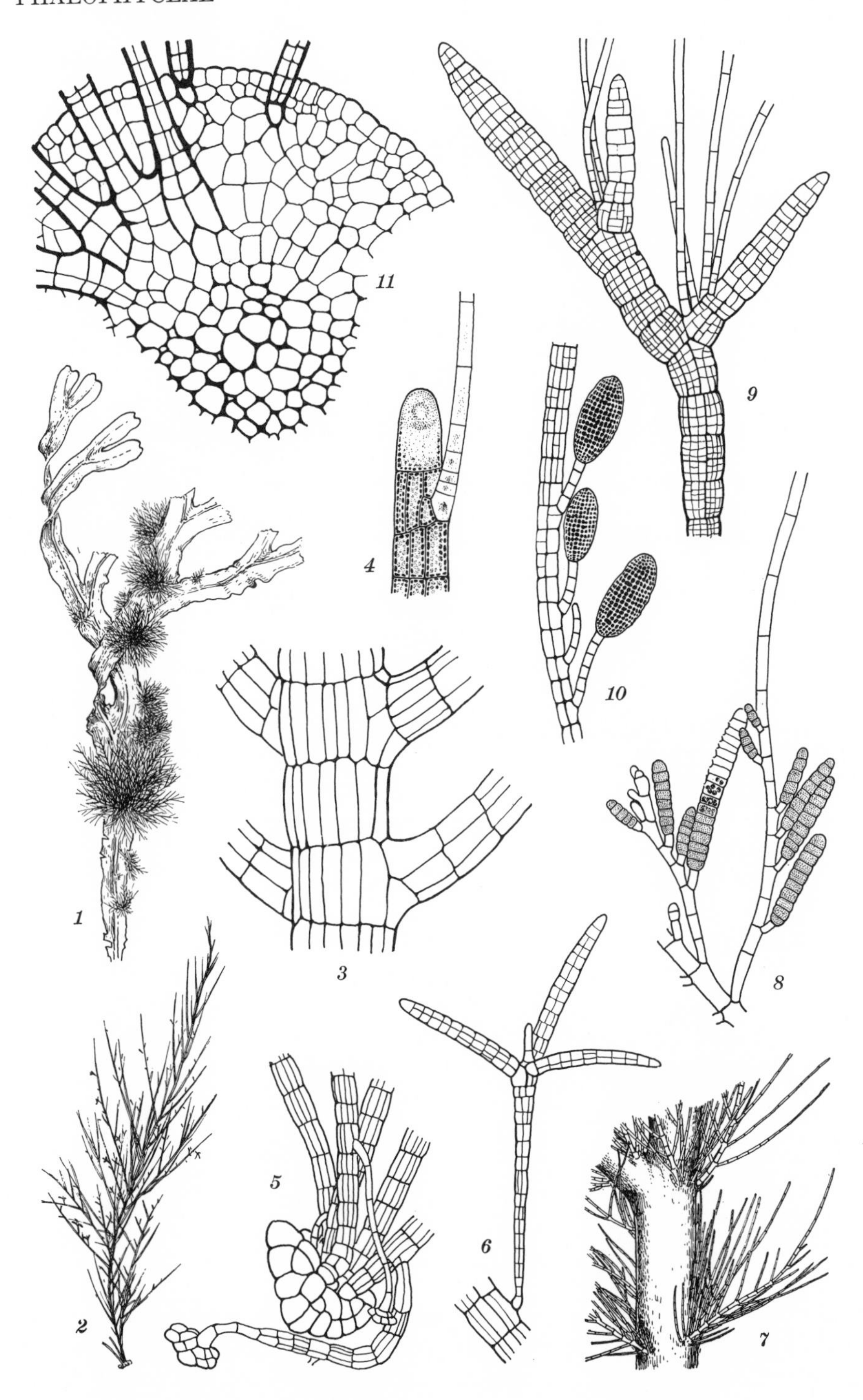
11
9
4
10
1
3
8
5
6
2
7

PLATE 18

PAGE

FIG. 1. *Laminaria digitata:* individual plant, well-developed specimen, showing habit, × 0.23 184

FIG. 2. *Laminaria agardhii:* young and old plants, in summer condition, showing habit, × 0.14 179

For additional figures of Laminaria see Plates 19–21.

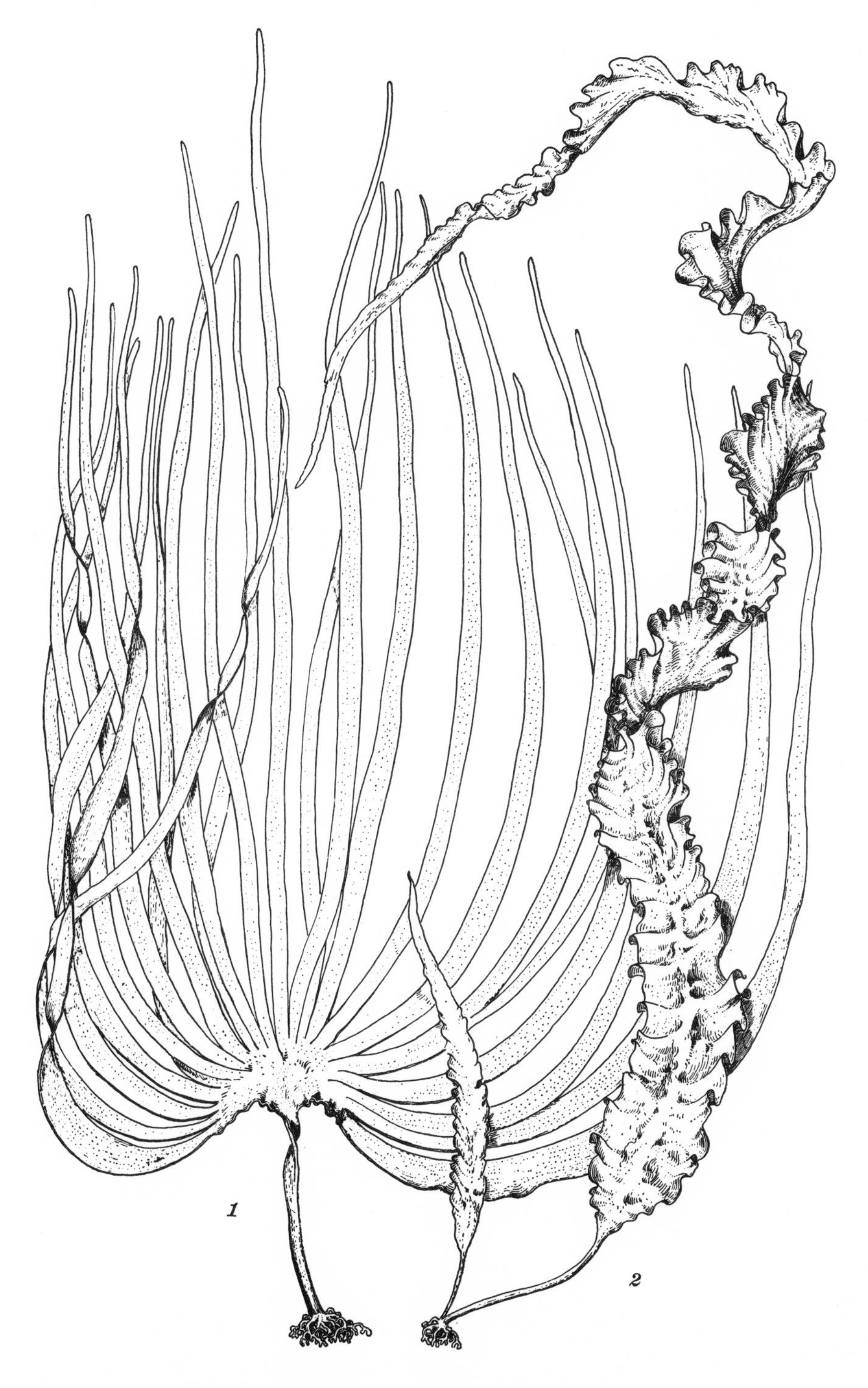
1
2

PLATE 19

For additional figures of Laminaria see Plates 18, 20–21.

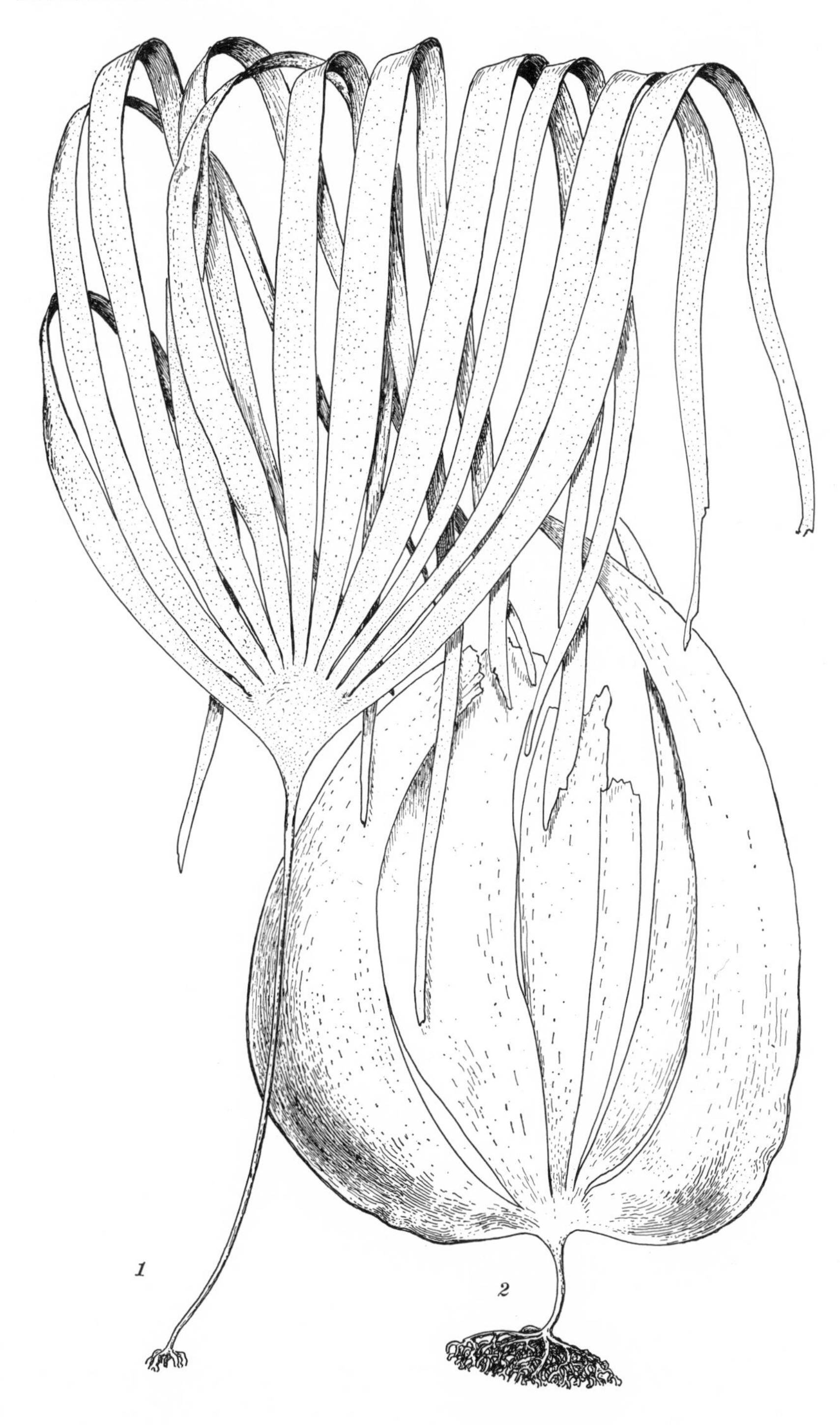
1
2

PLATE 20

PAGE

FIGS. 1–2. *Laminaria longicruris:* individual plants. The specimen shown in fig. 1 probably came from shallow water where there was a strong tidal current, that in fig. 2 from somewhat deeper, quieter water, × 0.14. 181

For additional figures of Laminaria see Plates 18–19, 21.

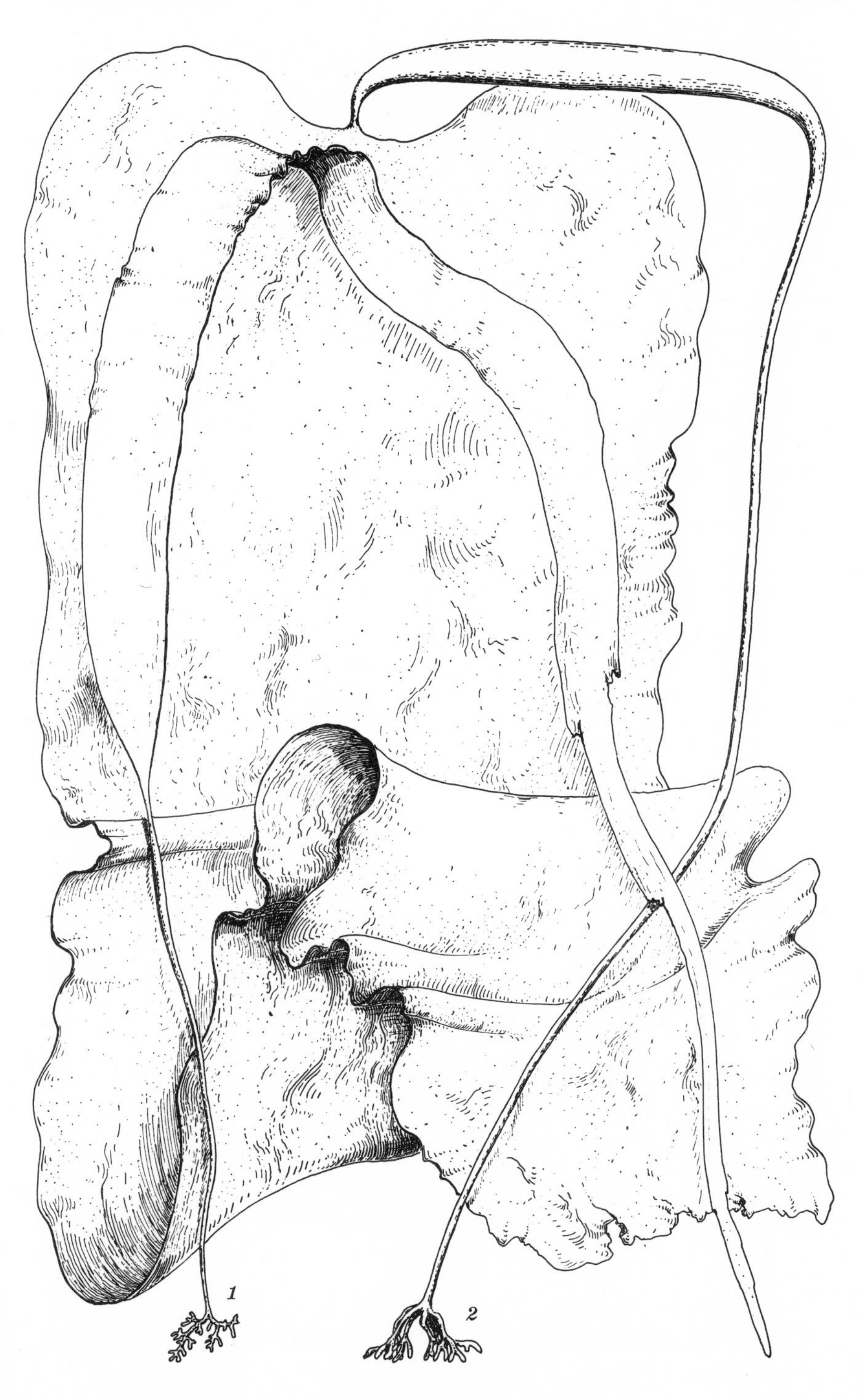
1
2

PLATE 21

PAGE

FIG. 1. *Laminaria platymeris:* individual plant, × 0.23............. 183

FIG. 2. *Alaria esculenta:* individual plant, with the old distal part of the vegetative blade partly eroded away and with sporophylls fully developed, × 0.23...................................... 186

For additional figures of Laminaria see Plates 18–20.

1
2

PLATE 22

PAGE

Fig. 1. *Agarum cribrosum:* individual plant, very well developed, showing characteristic form and perforations, × 0.45 185

Fig. 2. *Saccorhiza dermatodea:* individual plant, × 0.45 177

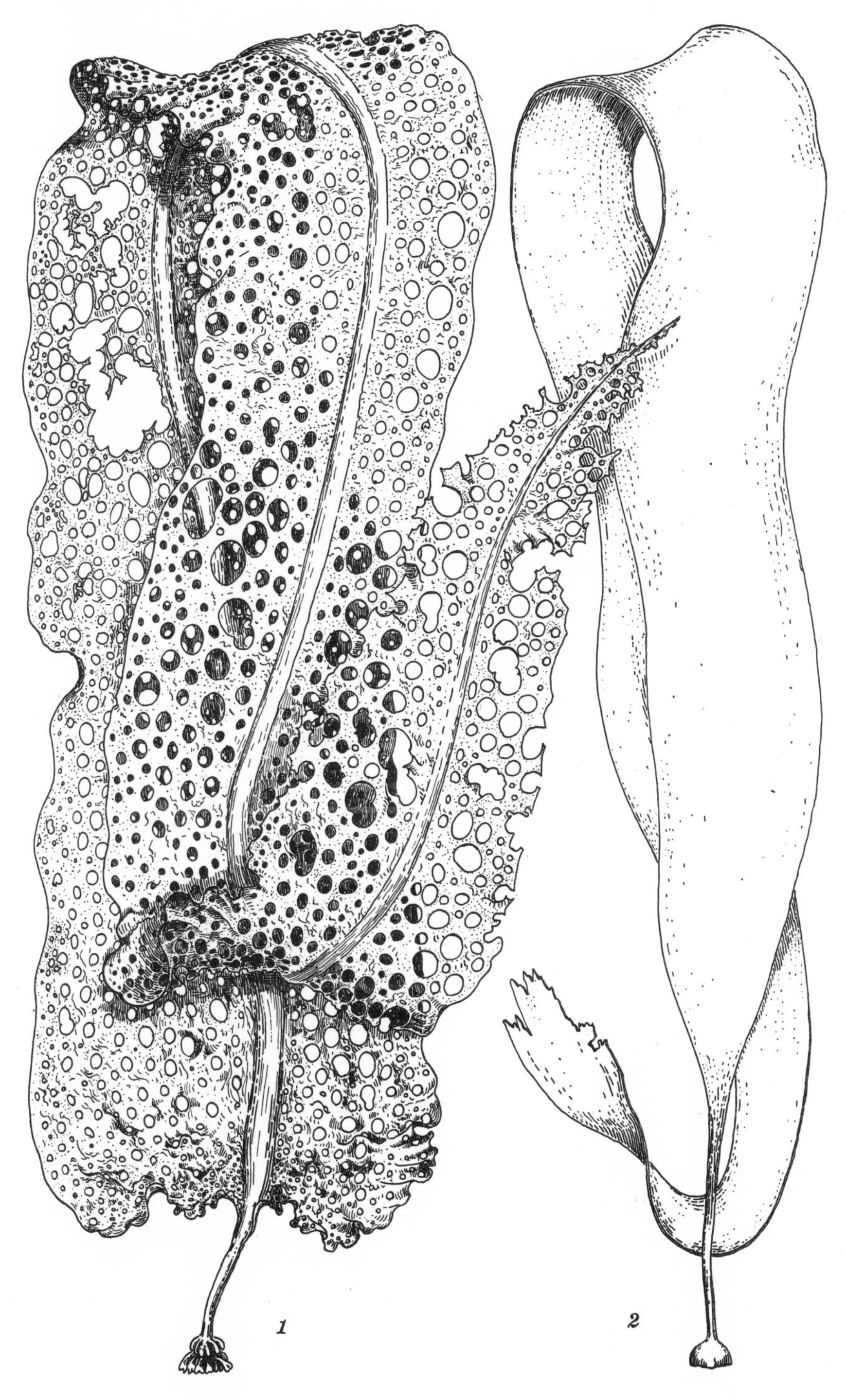
1
2

PLATE 23

PAGE

FIG. 1. *Fucus edentatus* f. *angustior:* portion of plant, showing the branching and receptacles, × 0.45 191

FIG. 2. *Fucus filiformis:* portion of plant, showing the branching and receptacles, × 0.45 190

FIG. 3. *Fucus edentatus:* basal holdfast, main branching system, and part of ultimate branching, showing habit and receptacles, × 0.45. 191

FIG. 4. *Fucus evanescens:* old receptacles, × 0.45 193

For additional figures of Fucus see Plates 24–26.

1
4
2
3

PLATE 24

PAGE

Fig. 1. *Fucus spiralis:* individual plant with part of lower branching removed, showing habit and acutely margined receptacles, × 0.45. 191

Fig. 2. *Fucus evanescens:* individual plant with part of branching removed, showing habit and flattened receptacles, × 0.45....... 193

For additional figures of Fucus see Plates 23, 25–26.

1
2

PLATE 25

PAGE

FIGS. 1–3. *Fucus vesiculosus:* Fig. 1, portion of fertile plant, showing holdfast with incipient proliferation, denuded lower axis, upper branching with very turgid vesicles and young receptacles, × 0.45. Fig. 2, portion of sterile plant, showing somewhat wider branches and less swollen vesicles, × 0.45. Fig. 3, individual plant approaching v. *spiralis* in somewhat advanced fertile condition, with inflated receptacles and moderate twisting in the upper segments, × 0.45.. 192

FIGS. 4–5. *Fucus vesiculosus* v. *spiralis:* Fig. 4, portion of plant in sterile condition; a slender form with numerous air bladders and moderate spiraling, from a rocky shore, × 0.45. Fig. 5, portion of sterile plant from a salt-marsh tidal ditch, showing reduction of bladders and fully developed spiral form, × 0.45............... 193

FIG. 6. *Fucus vesiculosus* v. *sphaerocarpus:* small proliferating portion of a plant, showing form of branching and receptacles, × 0.45.... 193

For additional figures of Fucus see Plates 23–24, 26.

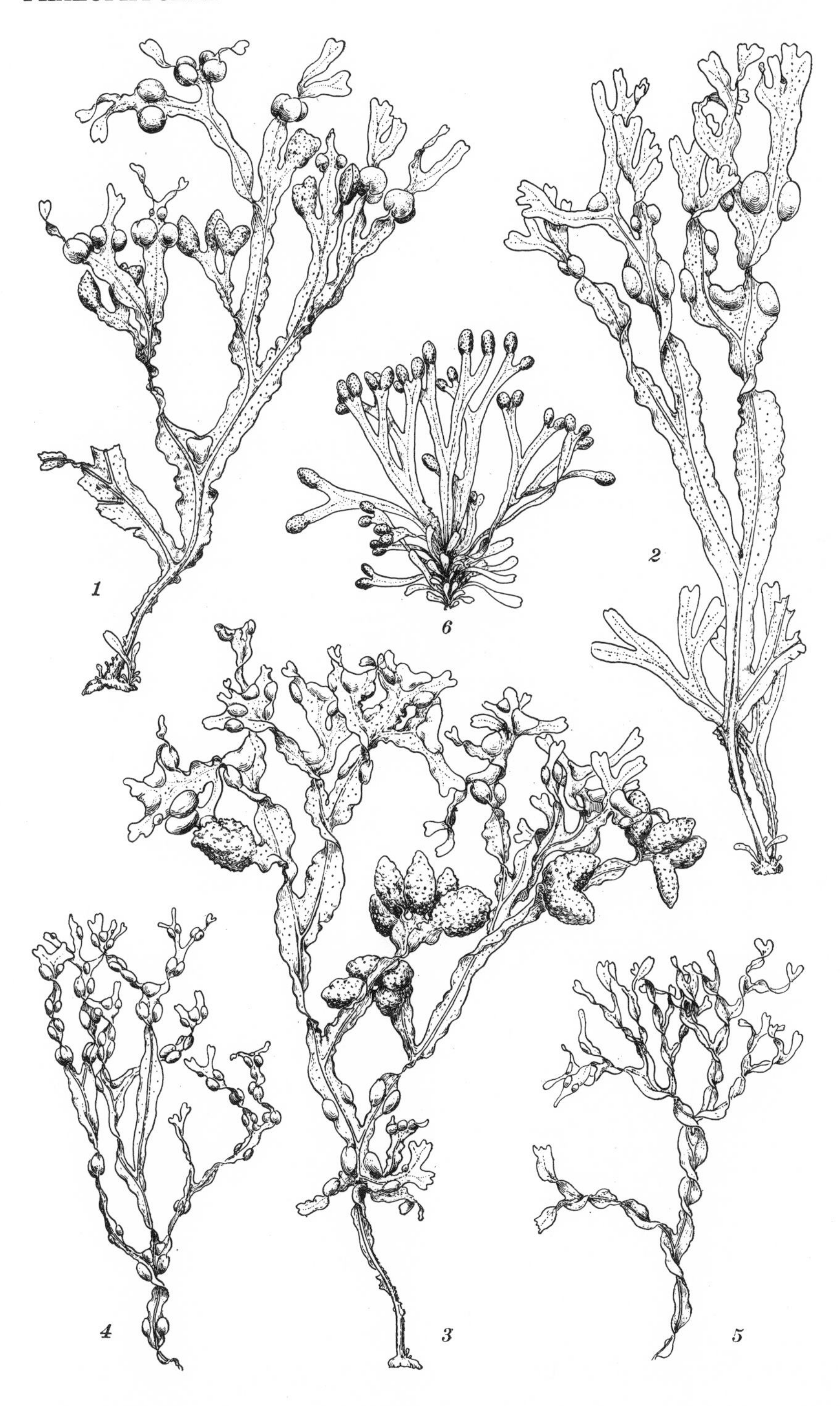
1
2
6
4
3
5

PLATE 26

PAGE

FIG. 1. *Fucus serratus:* habit of portion of a small fruiting plant, × 0.45 194

FIG. 2. *Ascophyllum mackaii:* habit of portion of a fertile plant, × 0.45 196

For additional figures of Ascophyllum see Plate 27; Fucus, Plates 23–25.
Drawings by the author.

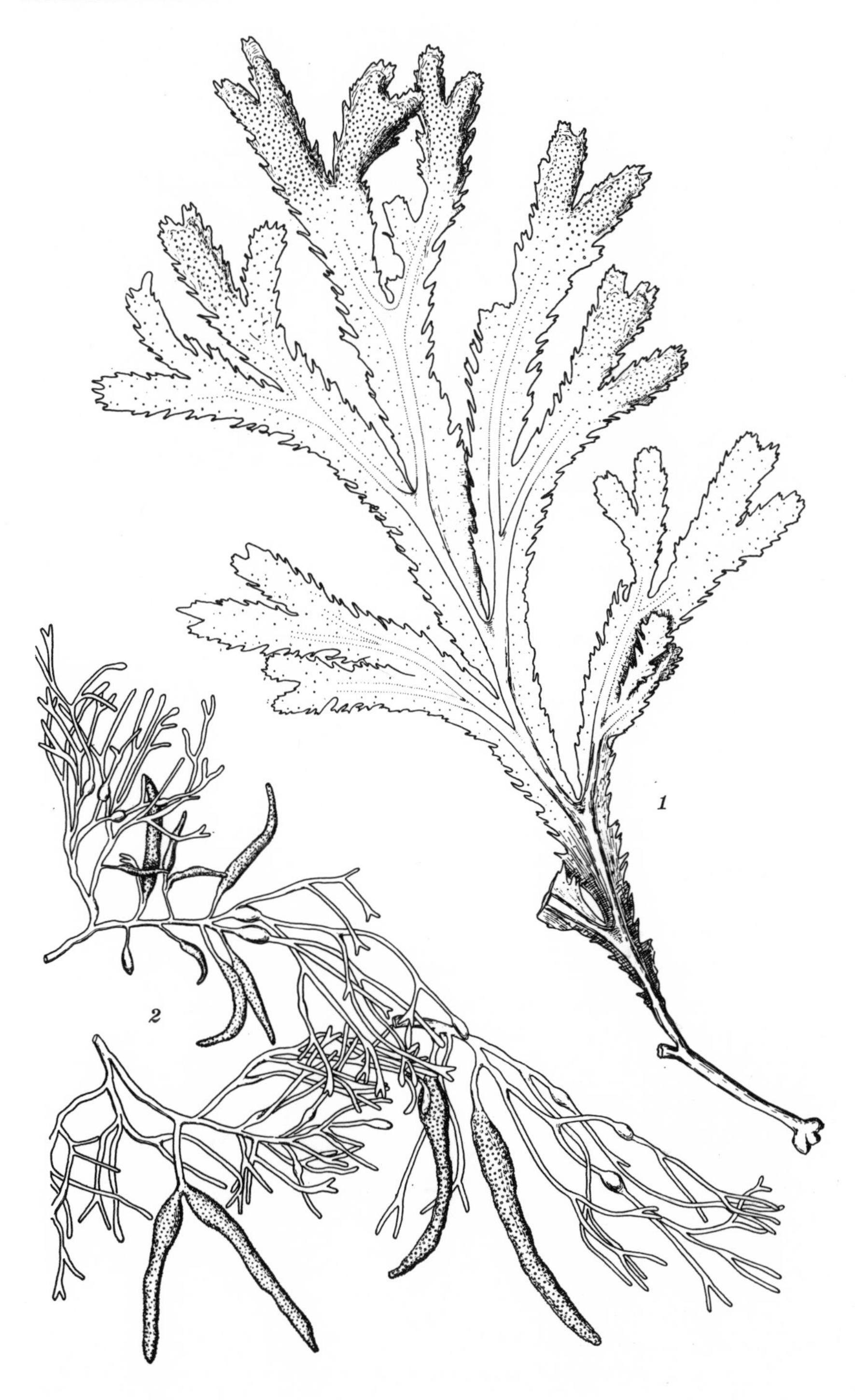
1
2

PLATE 27

PAGE

FIGS. 1–2. *Ascophyllum nodosum:* Fig. 1, small portion of the distal branching of a plant in a late summer condition, at the nodes showing groups of short branchlets and young receptacles, × 0.45. Fig. 2, branch of a slender form in summer condition, × 0.45.... 195

FIG. 3. *Ascophyllum nodosum* f. *scorpioides:* part of plant, showing characteristic subcylindrical form, close-spreading branching, and almost complete absence of bladders, × 0.45.................... 196

FIGS. 4–6. *Sargassum filipendula:* Fig. 4, portion of the upper branch system, showing branching, leaves, stalked vesicles, and receptacles, × 0.9. Fig. 5, single large leaf, showing midrib and cryptostomata × 14. Fig. 6, holdfast with dense lower branching, × 0.45...... 197

For additional figures of Ascophyllum see Plate 26.

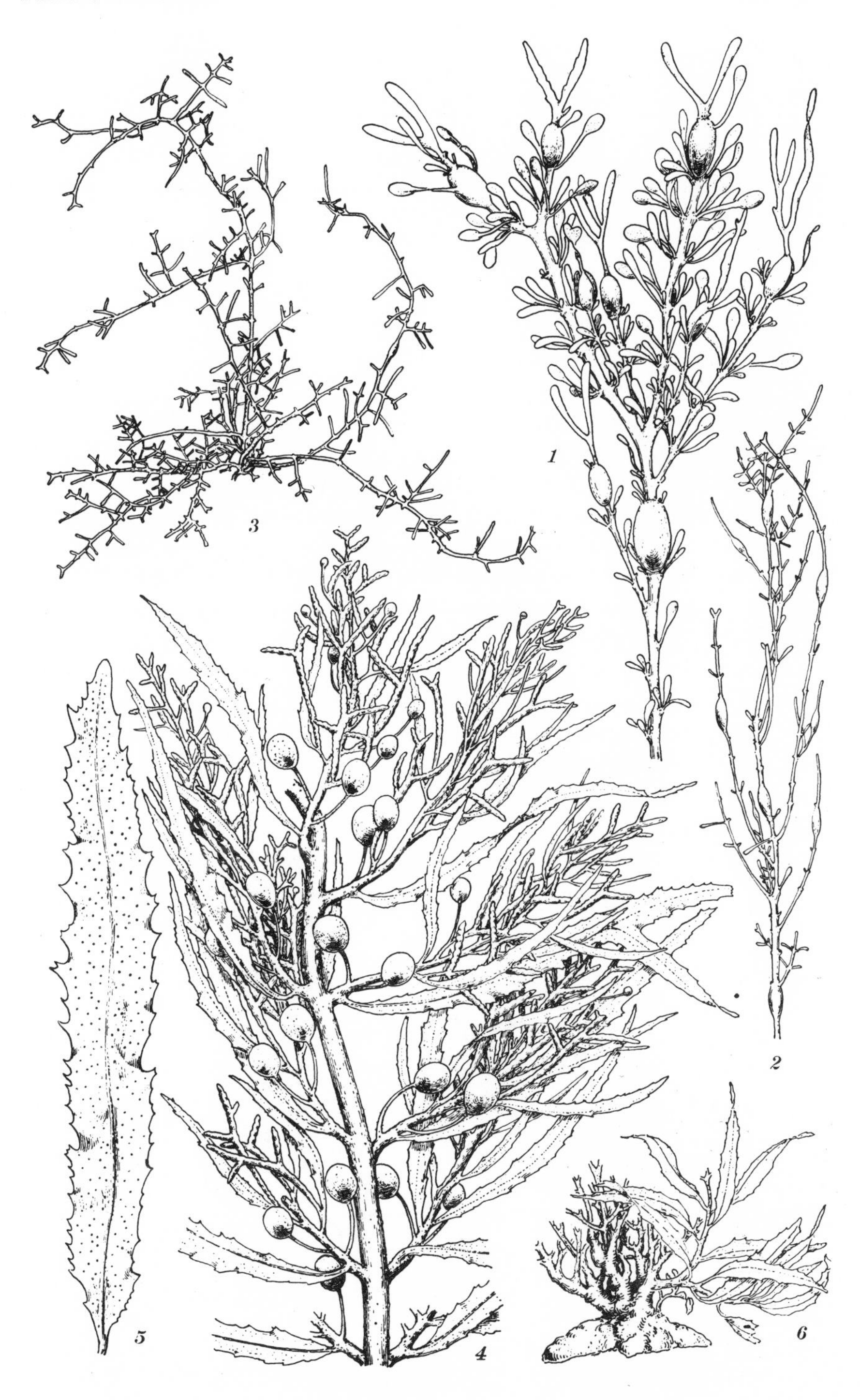
1
2
3
4
5
6

PLATE 28

PAGE

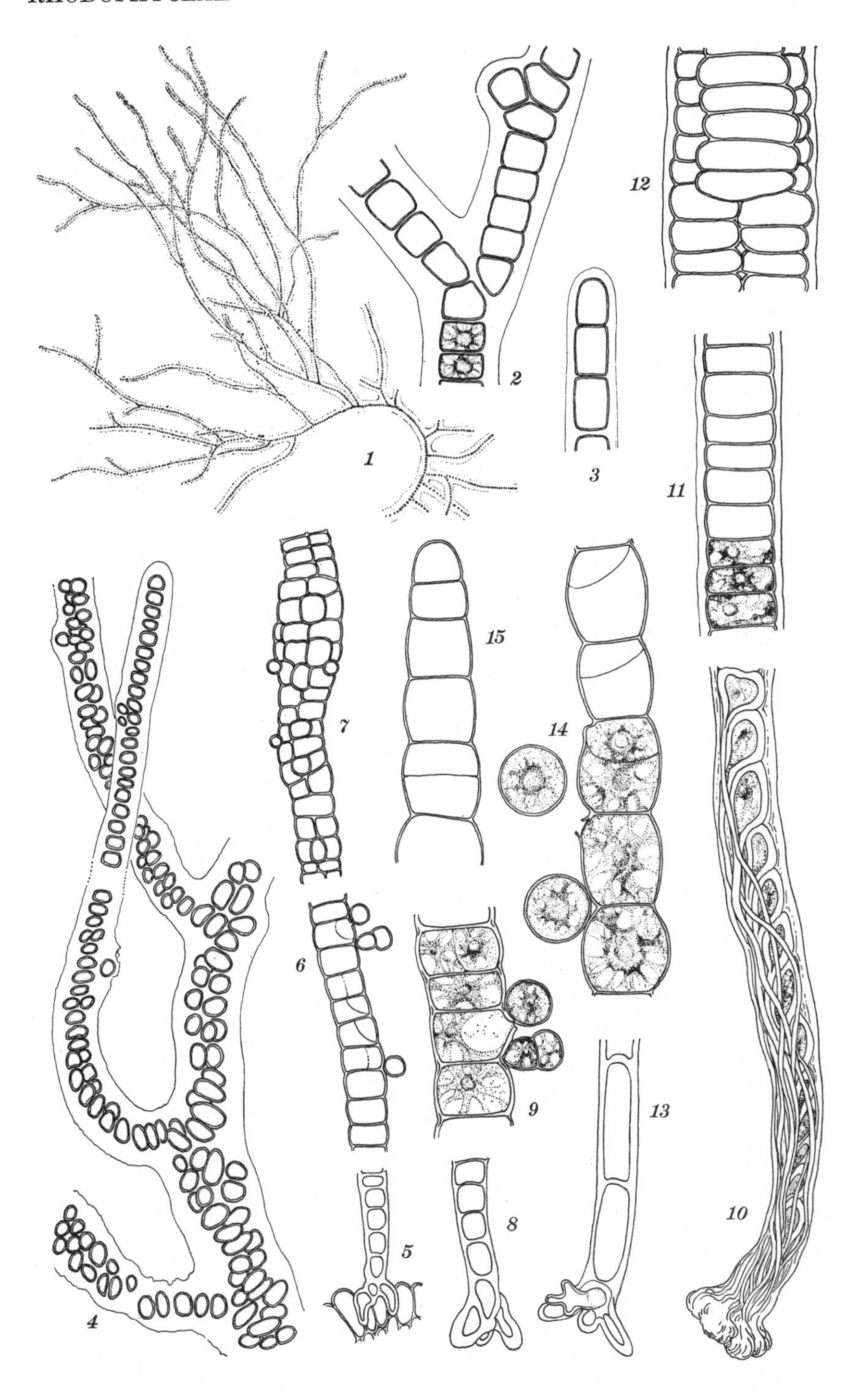
1
2
3
4
5
6
7
8
9
10
11
12
13
14
15

PLATE 29

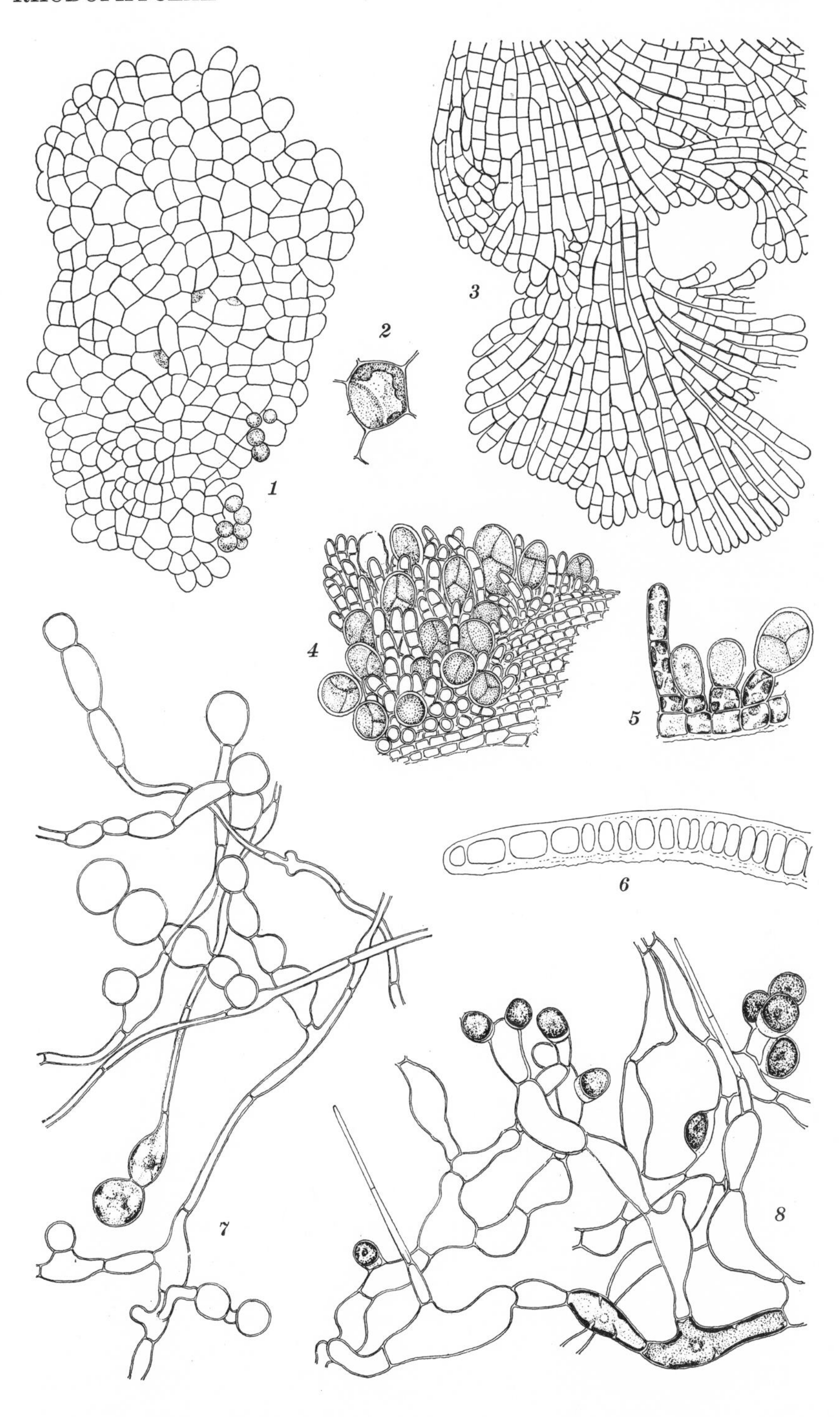
1
2
3
4
5
6
7
8

PLATE 30

PAGE

Figs. 1–3. *Porphyra umbilicalis:* Fig. 1, habit of a plant, × 0.5. Fig. 2, portion of a blade, showing cell arrangement, chromatophores, and carposporangium formation, × 320. Fig. 3, portion of blade, showing spermatangium formation at the margin, × 320... 206

Figs. 4–7. *Grinnellia americana:* Fig. 4, portion of a blade, showing cell form and chromatophores, × 320. Fig. 5, surface view of small sori of spermatangia, × 78. Fig. 6, surface view of pericarp, × 78. Fig. 7, oblique view of sori of tetrasporangia in the later protruding condition, × 78... 324

Fig. 8. *Phycodrys rubens:* small portion of a margin of fertile blade, showing serrations and sori of tetrasporangia, × 11............. 323

Figs. 9–10. *Plumaria elegans:* Fig. 9, branchlet with tetrasporangium, × 150. Fig. 10, portion of axis with branchlets and paraspores, × 150 305

For other figures of Grinnellia see Plates 42, 59; for Phycodrys, Plates 42, 46; for Plumaria, Plate 52.

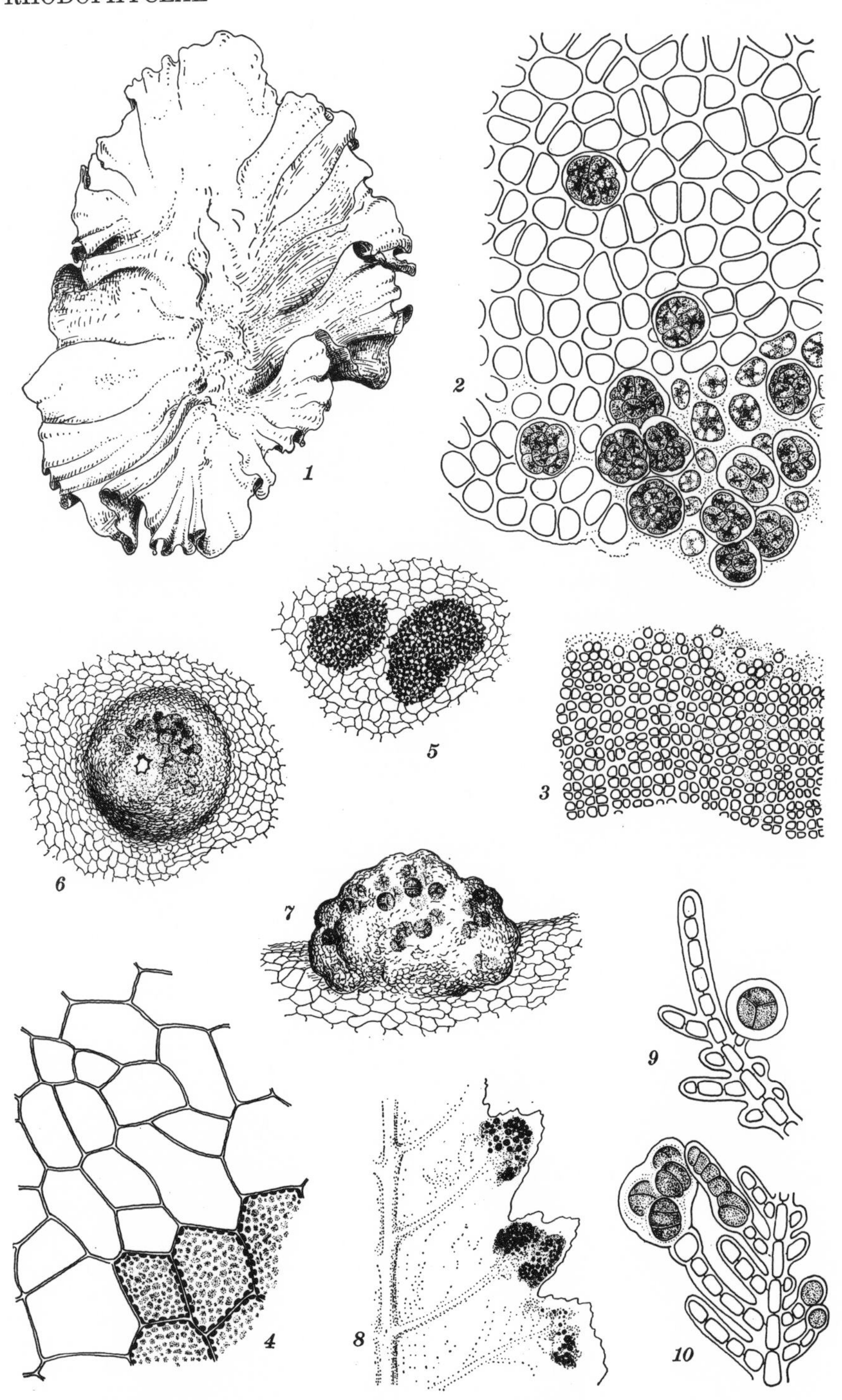
1
2
5
3
6
7
9
4
8
10

PLATE 31

PAGE

FIGS. 1–3. *Kylinia secundata:* Fig. 1, habit of a fruiting plant, × 320. Fig. 2, detail of a branchlet bearing a monosporangium and showing chromatophores, × 540. Fig. 3, detail of holdfast, × 320.... 214

FIGS. 4–7. *Kylinia virgatula* f. *luxurians:* Fig. 4, habit of a fruiting plant, × 76. Fig. 5, portion of branchlet, showing monosporangia, colorless hair, and chromatophores, × 320. Figs. 6–7, basal holdfasts, × 320.. 214

FIGS. 8–10. *Acrochaetium daviesii:* Fig. 8, habit of a fruiting plant, × 38. Fig. 9, branchlets, showing axillary groups of monosporangia, and chromatophores, × 320. Fig. 10, holdfast, × 320.... 221

FIGS. 11–12. *Audouinella membranacea:* Fig. 11, habit of the plant on an animal host, × 76. Fig. 12, detail of branching filaments, showing tetrasporangia and chromatophores, × 540............. 224

For additional figures of Acrochaetium see Plates 32–34; for Kylinia, Plate 32.

1
2
3
4
5
6
7
8
9
10
11
12

PLATE 32

PAGE

FIGS. 1–5. *Acrochaetium microfilum:* Fig. 1, two mature plants with sporangia, × 270. Figs. 2–5, stages in germination of spores, × 270 219

FIGS. 6–14. *Kylinia compacta:* Figs. 6–8, three mature plants with sporangia, × 270. Figs. 9–14, stages in germination of spores, × 270 212

FIGS. 15–17. *Kylinia moniliformis* v. *mesogloiae:* Fig. 15, a young filament, × 270. Figs. 16–17, mature plants with sporangia, × 270. Figs. 15–16 especially show relation of the basal cell wall to the wall of the supporting host cell.............................. 212

FIGS. 18–25. *Acrochaetium radiatum:* Figs. 18–23, stages in development of the plant from the germinating spore, × 270. Figs. 24–25, mature plants with sporangia, × 270......................... 223

FIGS. 26–32. *Kylinia unifila:* Figs. 26–27, early stages in germination of the spore, × 270. Figs. 28–29, 31, mature plants of usual size, × 270. Figs. 30, 32, plants which have matured at a very early stage in growth; that in Fig. 30 has only one basal cell, one hair, and one monosporangium, which has already discharged its monospore, × 270.. 212

For additional figures of Acrochaetium see Plates 31, 33–34; for Kylinia, Plate 31.

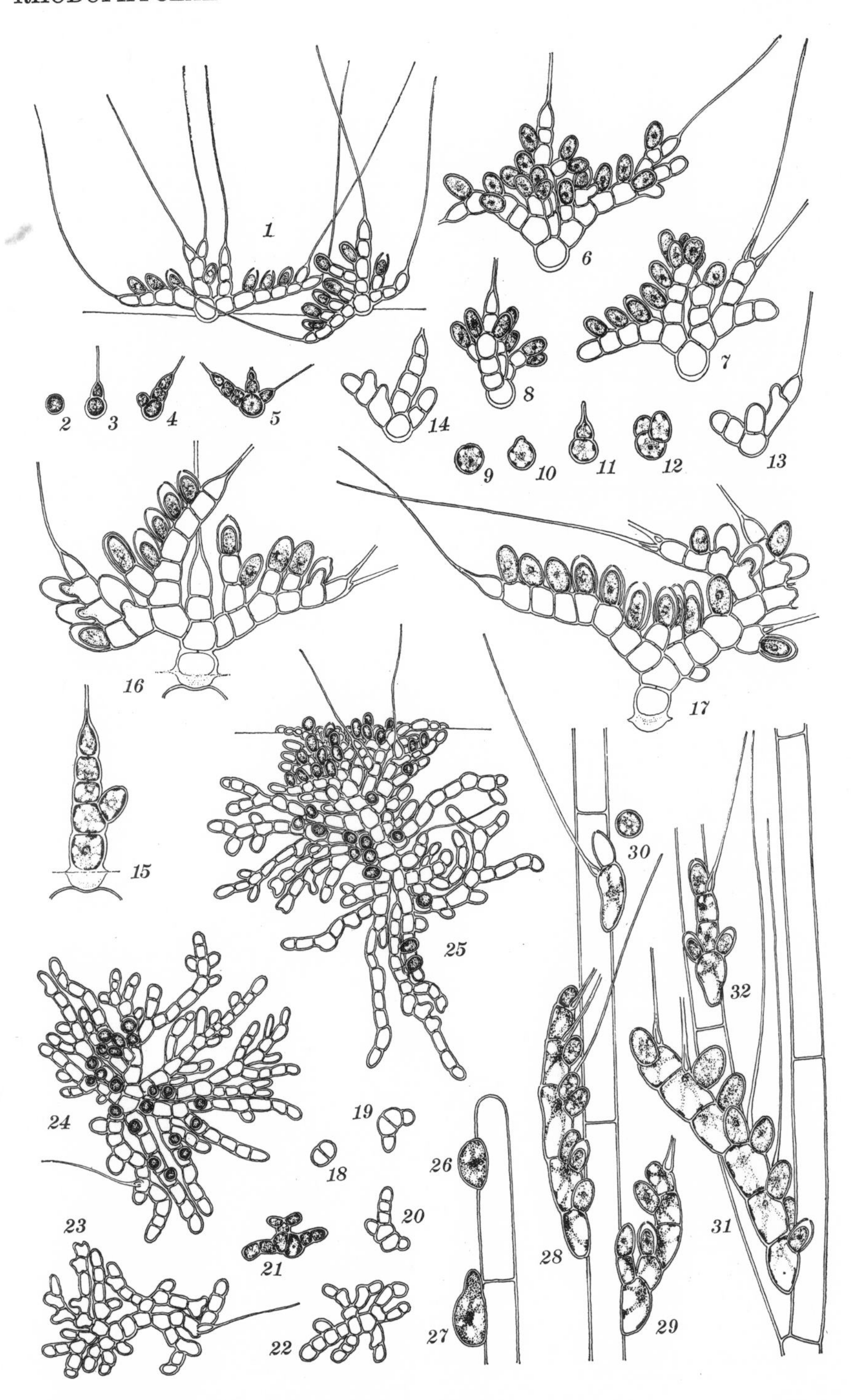
1
2
3
4
5
6
7
8
9
10
11
12
13
14
15
16
17
18
19
20
21
22
23
24
25
26
27
28
29
30
31
32

PLATE 33

PAGE

For additional figures of Acrochaetium see Plates 31–32, 34.

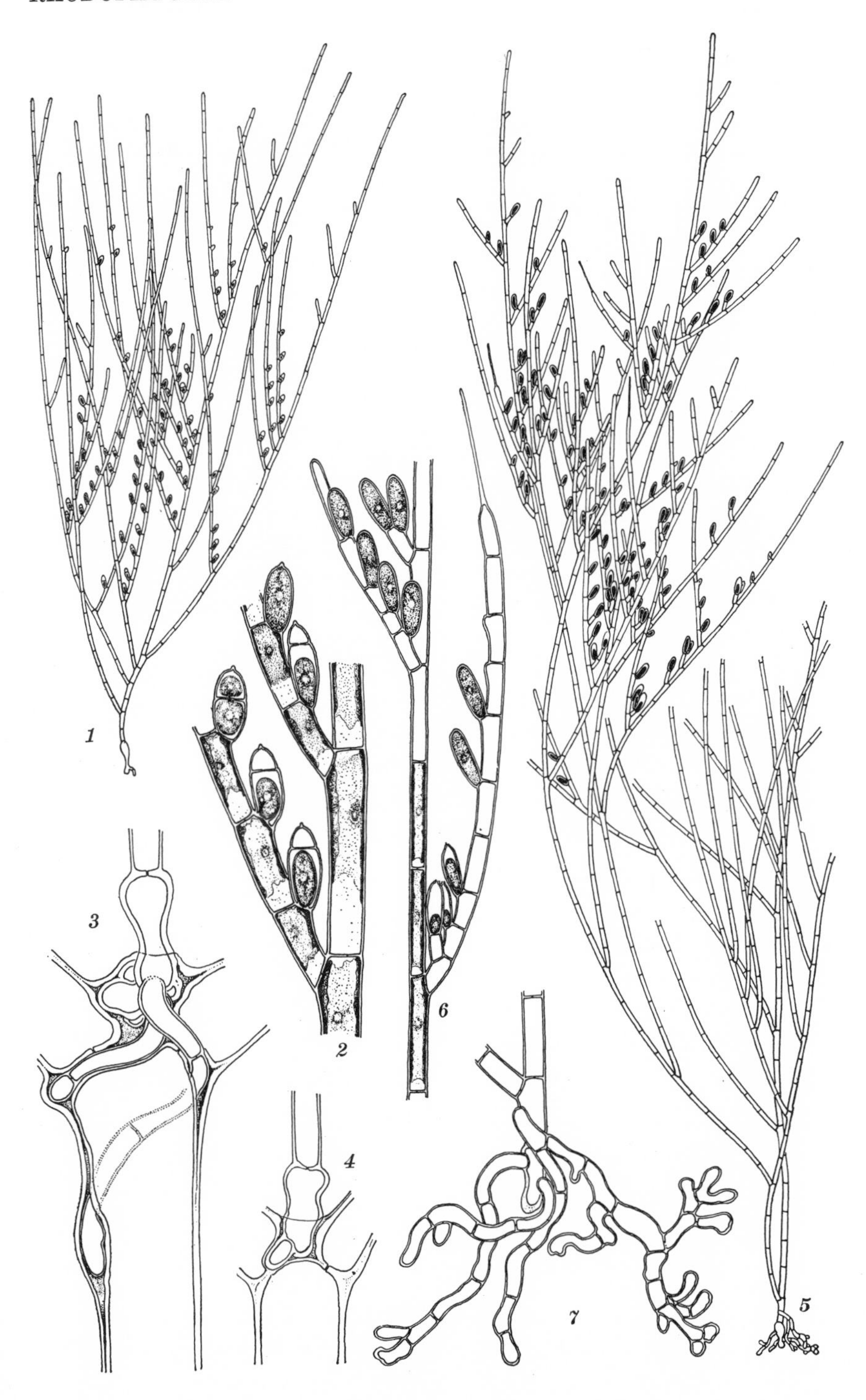
1
2
3
4
5
6
7

PLATE 34

PAGE

For additional figures of Acrochaetium see Plates 31–33.

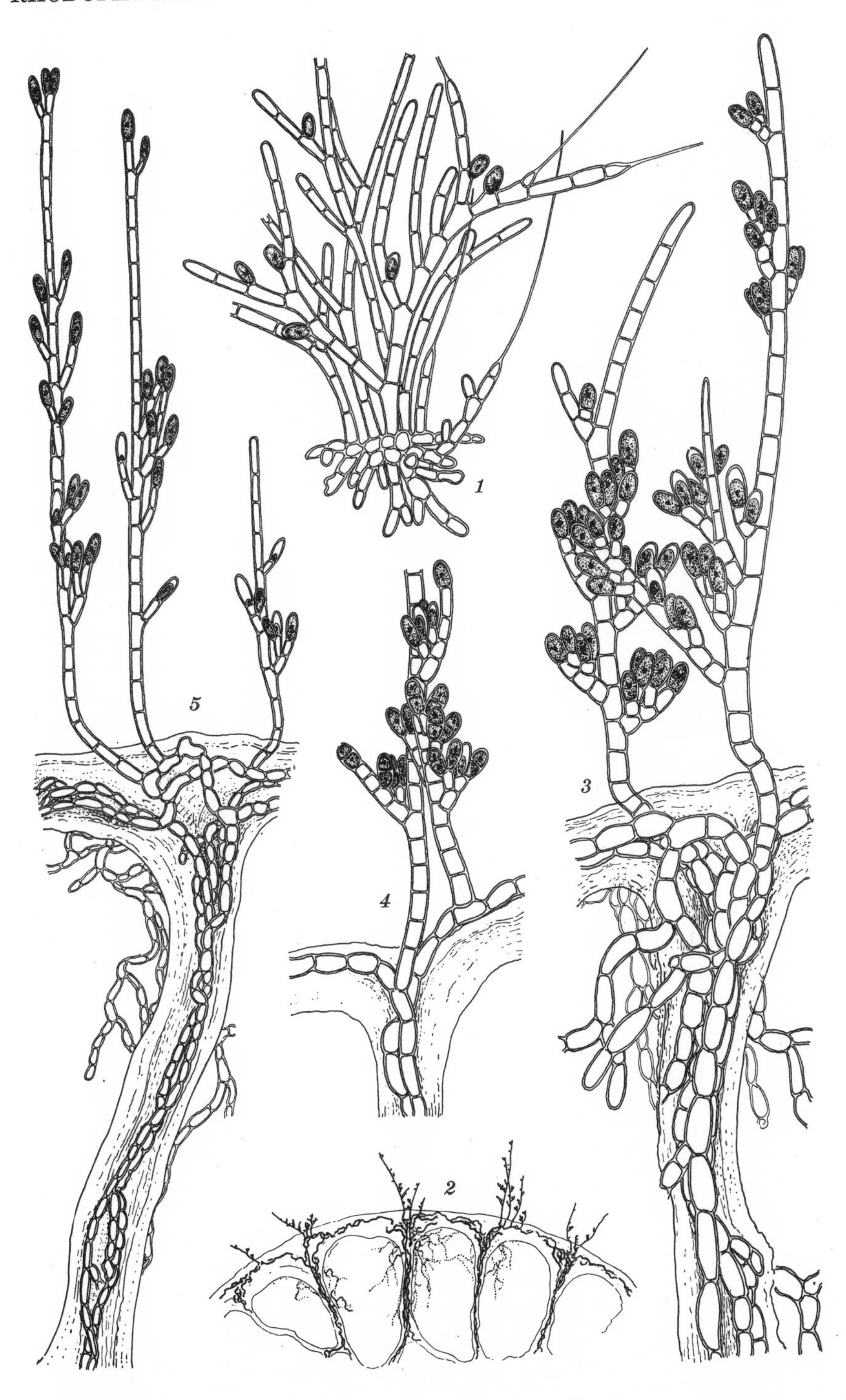
1
2
3
4
5

PLATE 35

PAGE

For additional figures of Asparagopsis see Plates 40, 44, 52; for Dumontia, Plate 41; for Gelidium, Plates 40–41; for Lomentaria, Plates 40–41, 43; for Polyides, Plate 40. Figures 3–4, 9 drawn by the author.

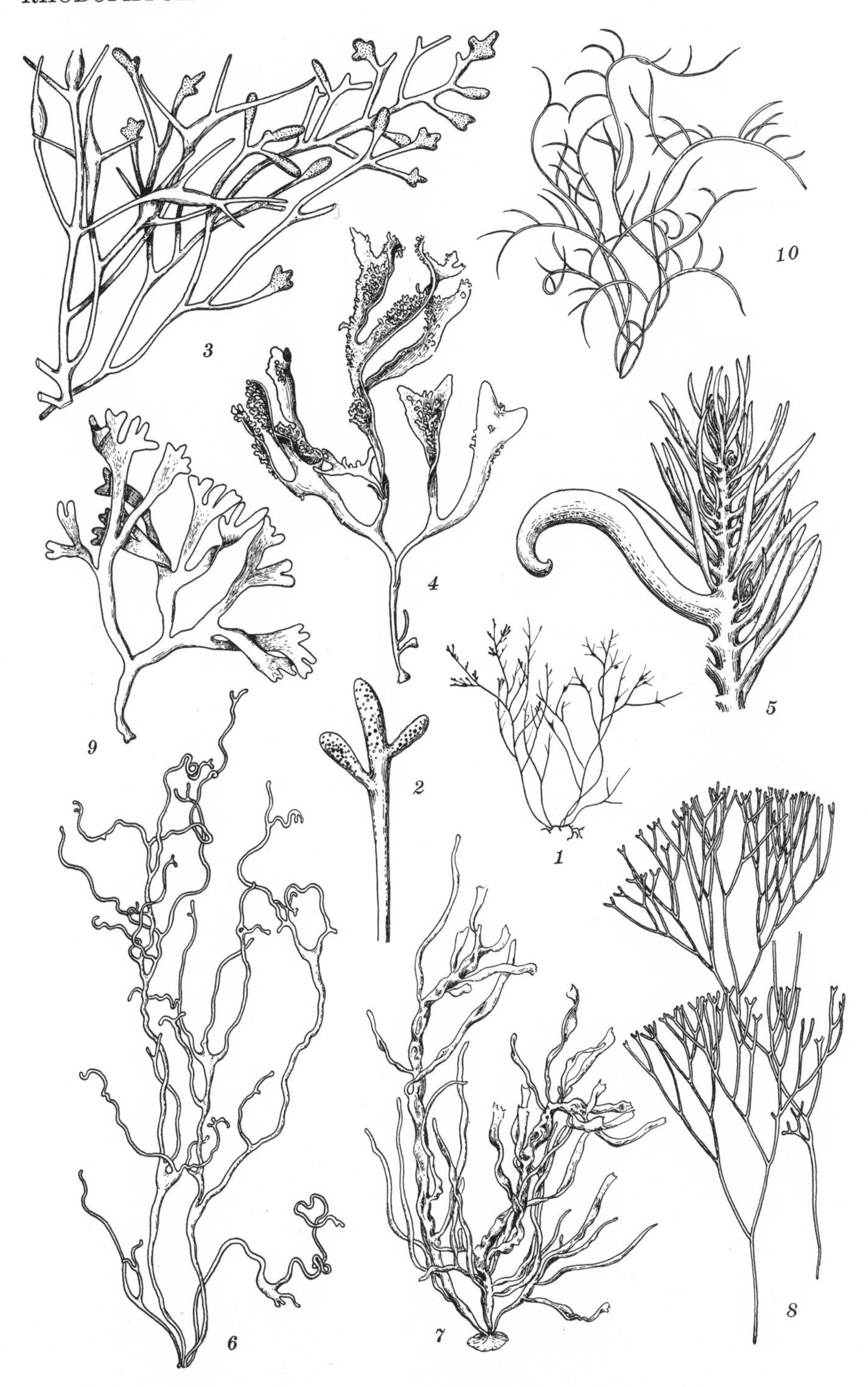
3
10
4
5
9
2
1
6
7
8

PLATE 36

PAGE

FIGS. 1–5. *Corallina officinalis:* Fig. 1, habit of small portion of a plant, × 0.9. Fig. 2, habit of upper branching of a broad form, × 4. Fig. 3, habit of branching of ordinary type of plant, with young conceptacular branchlets, × 4. Figs. 4–5, habit of branching of plants from deep water with attenuate branchlets, × 4.... 254

FIGS. 6–8. *Fosliella lejolisii:* Fig. 6, habit of several plants on a leaf of Zostera, × 0.9. Fig. 7, habit of part of a plant, × 150. Fig. 8, surface view of a tetrasporangial conceptacle, × 320............ 253

FIGS. 9–10. *Hildenbrandia prototypus:* Fig. 9, vertical section of a plant through a tetrasporangial conceptacle, × 320. Fig. 10, vertical section of plant, showing rapidly growing surface region, × 320 241

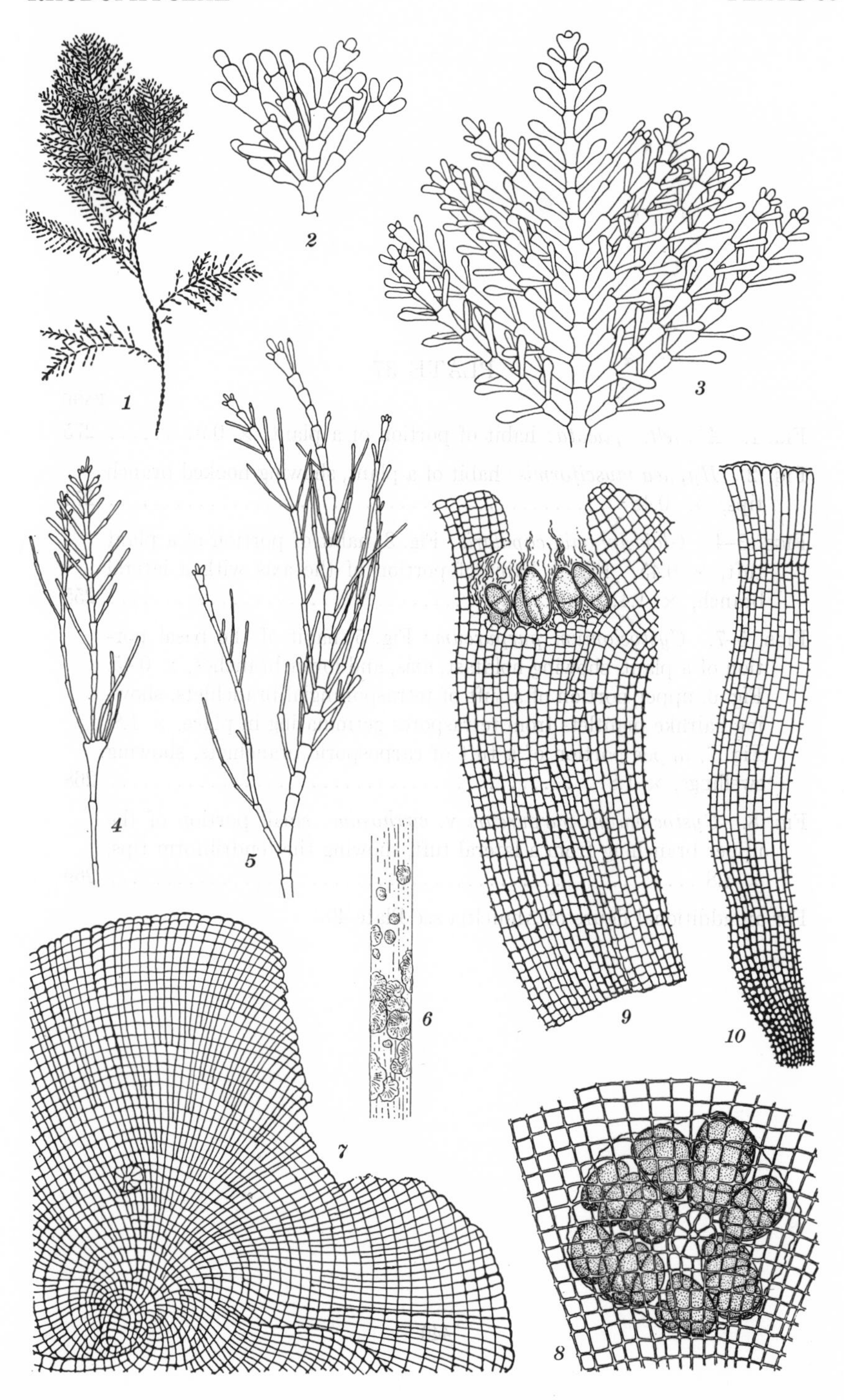
1
2
3
4
5
6
7
8
9
10

PLATE 37

PAGE

FIG. 1. *Ahnfeltia plicata:* habit of portion of a plant, × 0.9........ 275

FIG. 2. *Hypnea musciformis:* habit of a plant, showing hooked branch tips, × 0.45.. 271

FIGS. 3–4. *Gloiosiphonia capillaris:* Fig. 3, habit of portion of a plant tuft, × 0.45. Fig. 4, habit of portion of the axis with a lateral branch, × 1.4... 255

FIGS. 5–7. *Cystoclonium purpureum:* Fig. 5, habit of the basal portion of a plant, showing holdfast, axis, and lower branches, × 0.45. Fig. 6, upper portion of a tuft of tetrasporangial branchlets, showing hairlike plantlets from tetraspores germinating in place, × 1.8. Fig. 7, upper portion of a tuft of carposporic branchlets, showing swellings, × 1.8... 268

FIG. 8. *Cystoclonium purpureum* v. *cirrhosum:* small portion of the upper branching from a lateral tuft, showing the tendriliform tips, × 1.8 269

For an additional figure of Ahnfeltia see Plate 40.

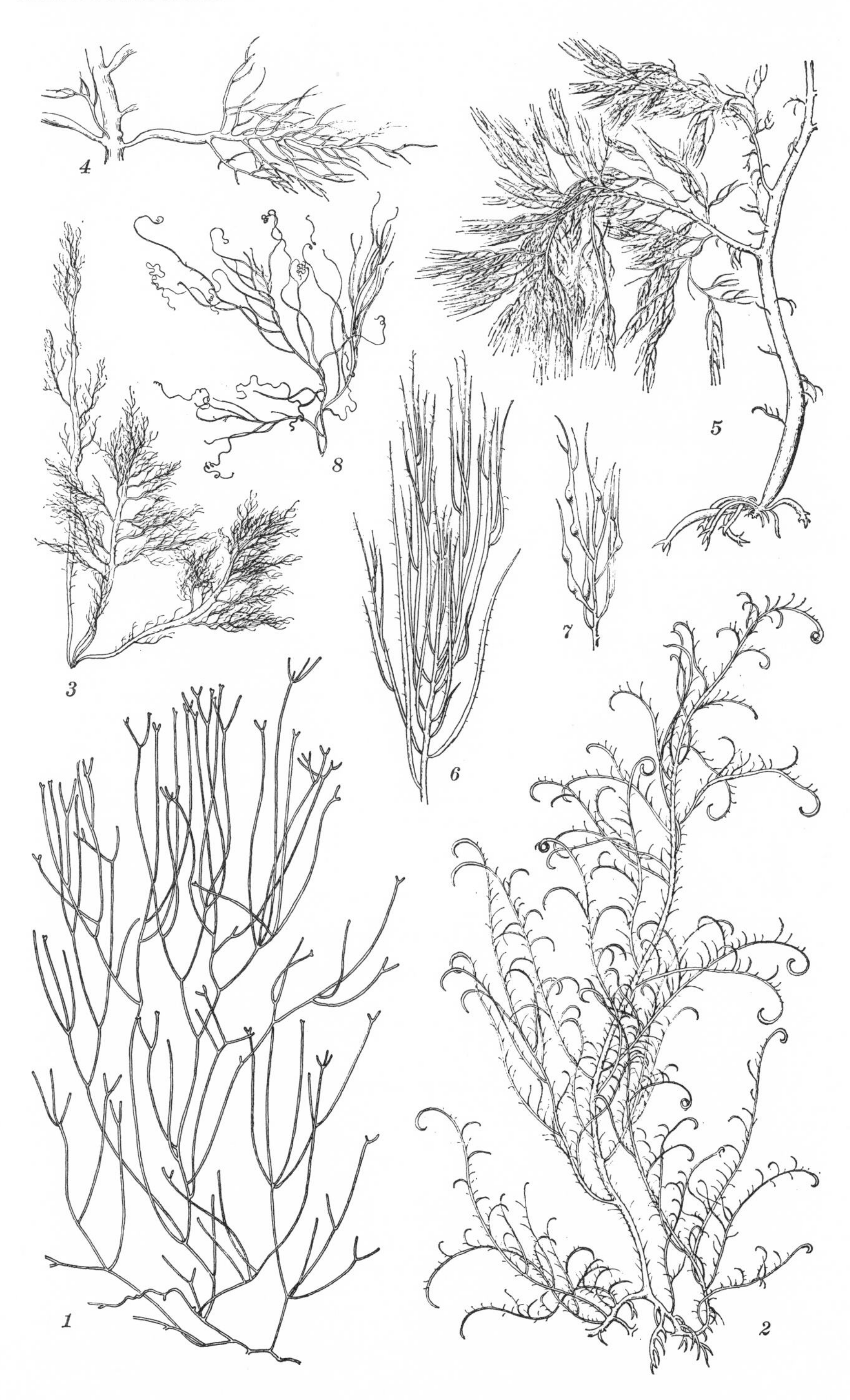
4
8
3
6
7
5
1
2

PLATE 38

For additional figures of Agardhiella see Plates 40–41, 59; for Gracilaria, Plates 41, 59.

1
2
3
4
5

PLATE 39

PAGE

FIG. 1. *Phyllophora membranifolia:* habit of portion of a plant, showing rather narrow lobes in the ultimate branching, × 0.45........ 278

FIG. 2. *Phyllophora brodiaei:* habit of a plant, showing tetrasporiferous nemathecia, × 0.45.................................... 278

FIGS. 3–6. *Chondrus crispus:* Figs. 3–5, habits of plants, showing variation in width of branches and closeness of forking, × 0.45. Fig. 6, a plant with nemathecia, × 0.45........................ 281

For additional figures of Chondrus see Plate 40.

5
6
3
4
1
2

PLATE 40

Drawings of cross-sectional views of axes of several species of marine Rhodophyceae

For additional figures of Agardhiella see Plates 38, 41, 59; for Ahnfeltia, Plate 37; for Asparagopsis, Plates 35, 44, 52; for Chondria, Plates 54, 55; for Chondrus, Plate 39; for Gelidium, Plates 35, 41; for Lomentaria, Plates 35, 41, 43; for Polyides, Plate 35. Figure 3 drawn by the author.

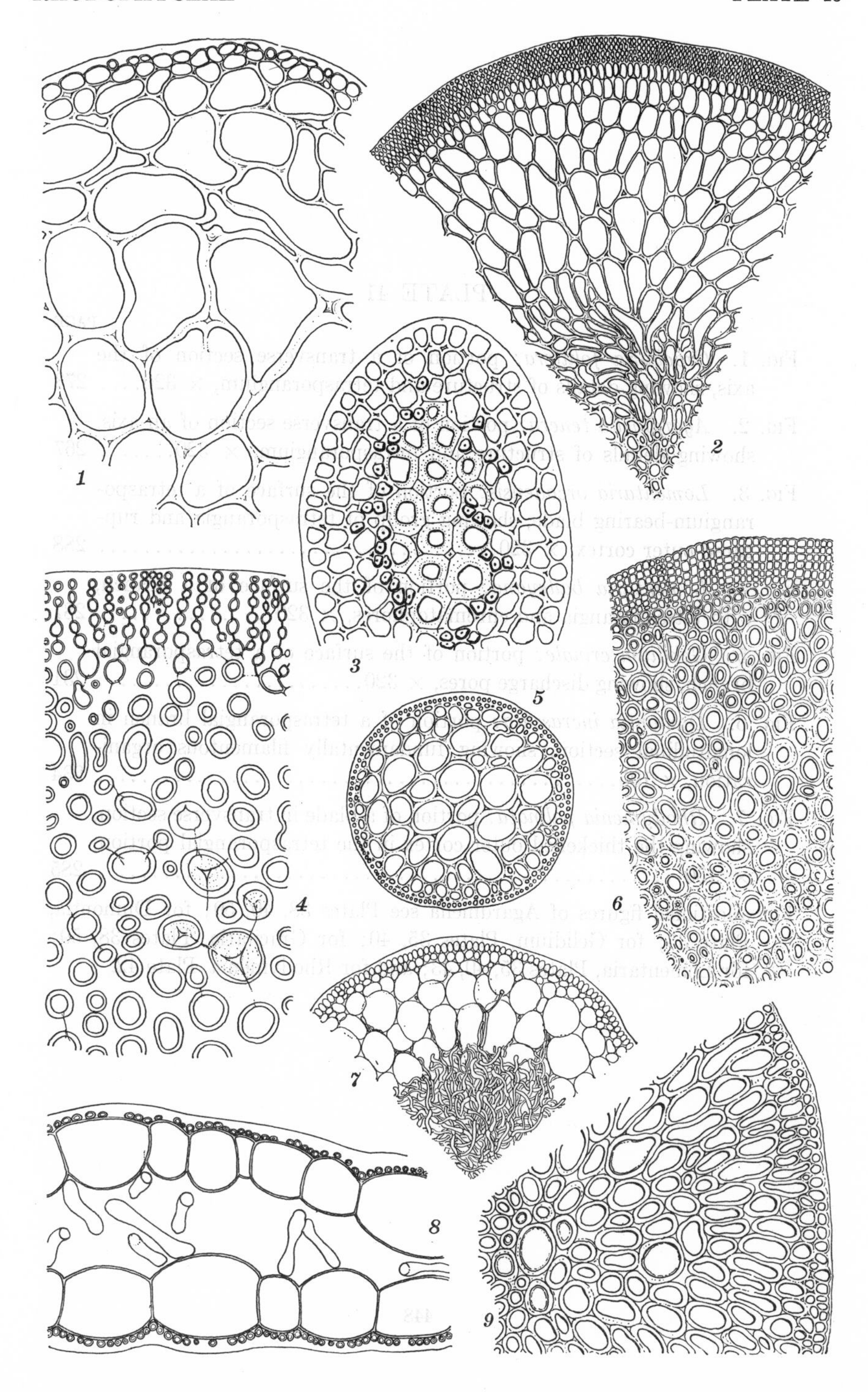
1
2
3
4
5
6
7
8
9

PLATE 41

PAGE

FIG. 1. *Gracilaria foliifera:* portion of a transverse section of the axis, showing details of structure and tetrasporangium, × 320.... 273

FIG. 2. *Agardhiella tenera:* portion of a transverse section of an axis, showing details of structure and tetrasporangium, × 320....... 267

FIG. 3. *Lomentaria orcadensis:* portion of the surface of a tetrasporangium-bearing blade, showing group of tetrasporangia and ruptured outer cortex, × 320.................................. 288

FIG. 4. *Lomentaria baileyana:* portion of the surface of a branch, with tetrasporangia and chromatophores, × 320................ 287

FIG. 5. *Gelidium crinale:* portion of the surface of a tetrasporangial branch, showing discharge pores, × 320........................ 231

FIG. 6. *Dumontia incrassata:* portion of a tetrasporangial branch in longitudinal section, showing fundamentally filamentous organization, × 195.. 234

FIG. 7. *Rhodymenia palmata:* portion of a blade in transverse section, showing the thickened outer cortex in the tetrasporangial portion, × 150 285

For additional figures of Agardhiella see Plates 38, 40, 59; for Dumontia, Plate 35; for Gelidium, Plates 35, 40; for Gracilaria, Plates 38, 59; for Lomentaria, Plates 35, 40, 43; and for Rhodymenia, Plate 42.

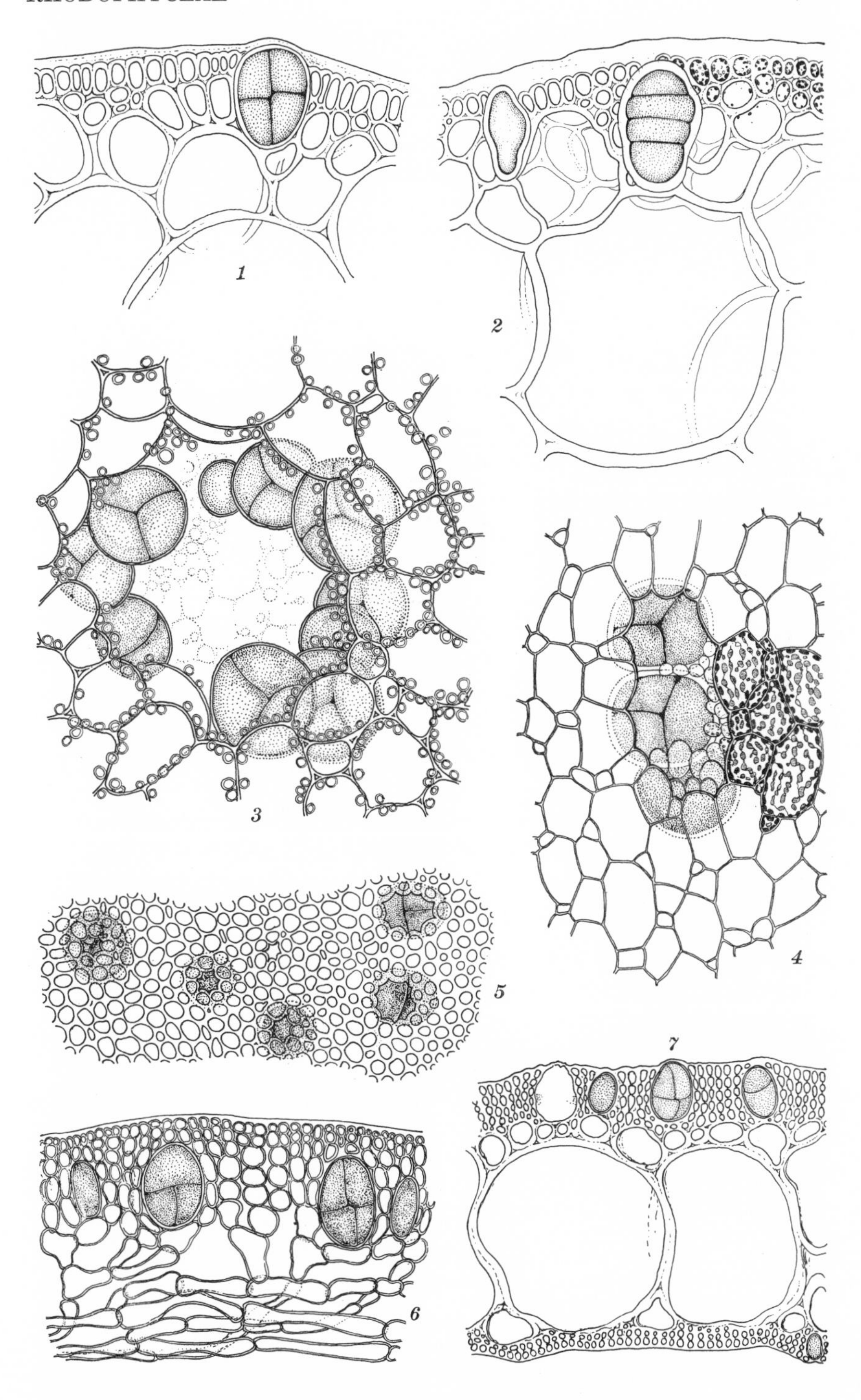
1
2
3
4
5
6
7

PLATE 42

PAGE

FIG. 1. *Phycodrys rubens:* habit of a small plant, showing a moderately branched blade, × 0.9 323

FIG. 2. *Grinnellia americana:* habit of part of a tuft of plants, showing (from left) tetrasporangial, cystocarpic, and sterile blades, × 0.45 324

FIG. 3. *Rhodymenia palmata:* habit of a plant, showing well-matured blade with forking tip and proliferating base, × 0.34 285

For other figures of Grinnellia see Plates 30, 59; for Phycodrys, Plates 30, 46; for Rhodymenia, Plate 41.

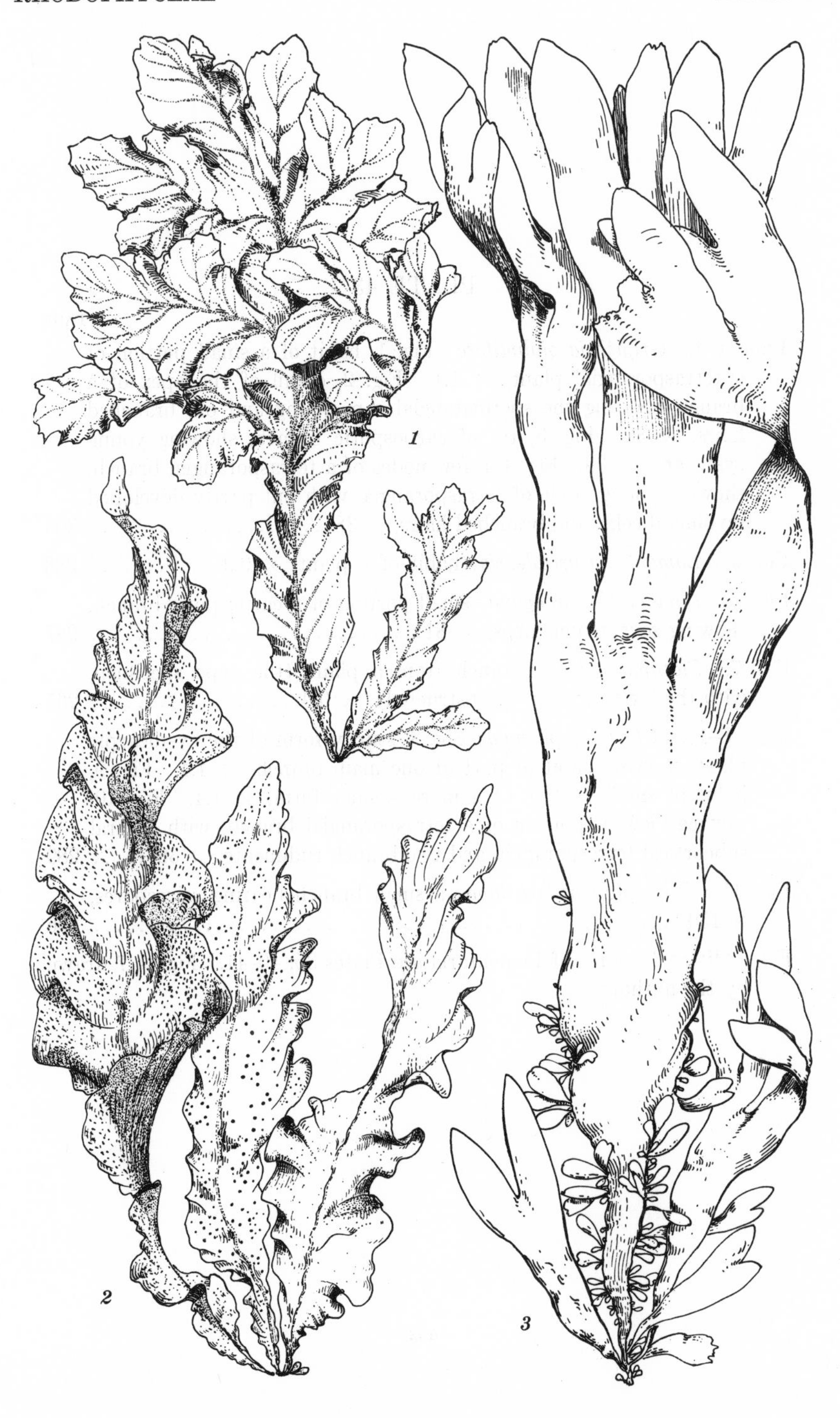
1
2
3

PLATE 43

PAGE

Figs. 1–4. *Griffithsia globulifera:* Fig. 1, habit of a small portion of a tetrasporangial plant, × 1.1. Fig. 2, detail of the apex of a branch, showing the spermatangial cap, young and old branched hairs, × 38. Fig. 3, tip of carposporic branch, showing young cystocarp, × 38. Fig. 4, a few nodes of a tetrasporangial branch, showing young circle of tetrasporangia, with but partly developed involucral cells, and branched hairs, × 38...................... 303

Fig. 5. *Lomentaria orcadensis:* habit of a plant, × 0.9............ 288

Fig. 6. *Lomentaria baileyana:* small portion of a carposporic branch, showing several pericarps, × 81.............................. 287

Fig. 7. *Platoma bairdii:* branch from a plant (the type specimen, with the lower portion reconstructed), × 0.9.................. 265

Figs. 8–10. *Champia parvula:* Fig. 8, sturdy form of the cystocarpic plant, showing habit of part of one main branch, × 1.4. Fig. 9, habit of small portion of a more slender form, × 1.4. Fig. 10, surface view of portion of a tetrasporangial branch, with several subcortical tetrasporangia and two branch rudiments, × 38...... 289

Fig. 11. *Seirospora griffithsiana:* a small branch bearing seirospores, × 150 296

For additional figures of Lomentaria see Plates 35, 40–41. Figure 7 drawn by the author.

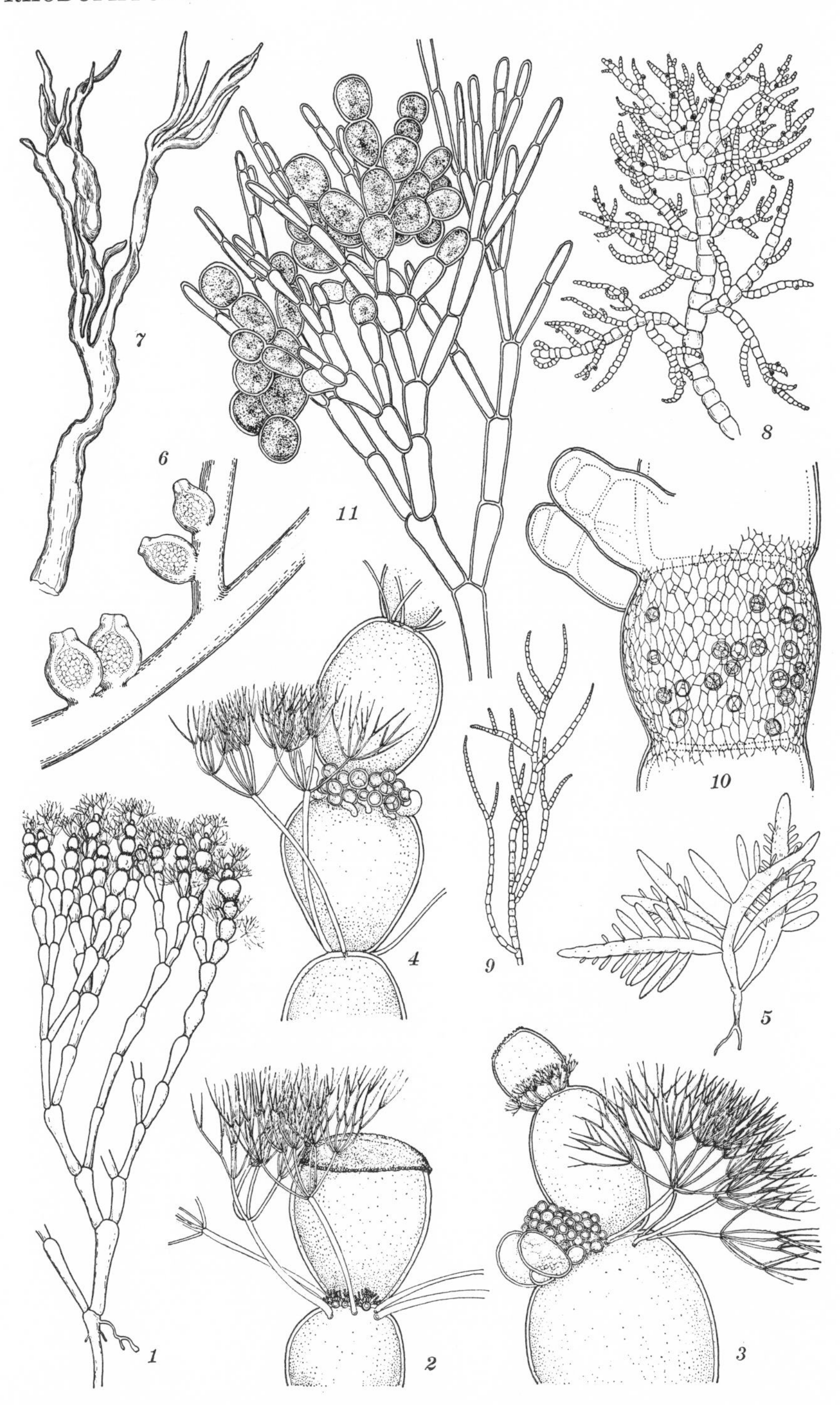
7
8
6
11
10
4
9
5
1
2
3

PLATE 44

PAGE

Fig. 1. *Spermothamnion turneri:* habit of a tuft on Fucus, × 0.45.. 302

Fig. 2. *Spyridia filamentosa:* habit of part of a plant, × 0.45...... 317

Fig. 3. *Antithamnion cruciatum:* habit of one main axis and its divisions, × 2.7.. 294

Fig. 4. *Asparagopsis hamifera:* habit of part of a plant, × 0.9..... 230

Fig. 5. *Callithamnion baileyi:* habit of part of a slender plant, × 0.9. 299

Fig. 6. *Pantoneura baerii:* habit of part of a plant, × 0.9.......... 320

For additional figures of Antithamnion and Spermothamnion see Plate 45; for Asparagopsis, Plates 35, 40, 52; for Callithamnion and Spyridia, Plate 46. Figure 6 drawn by the author.

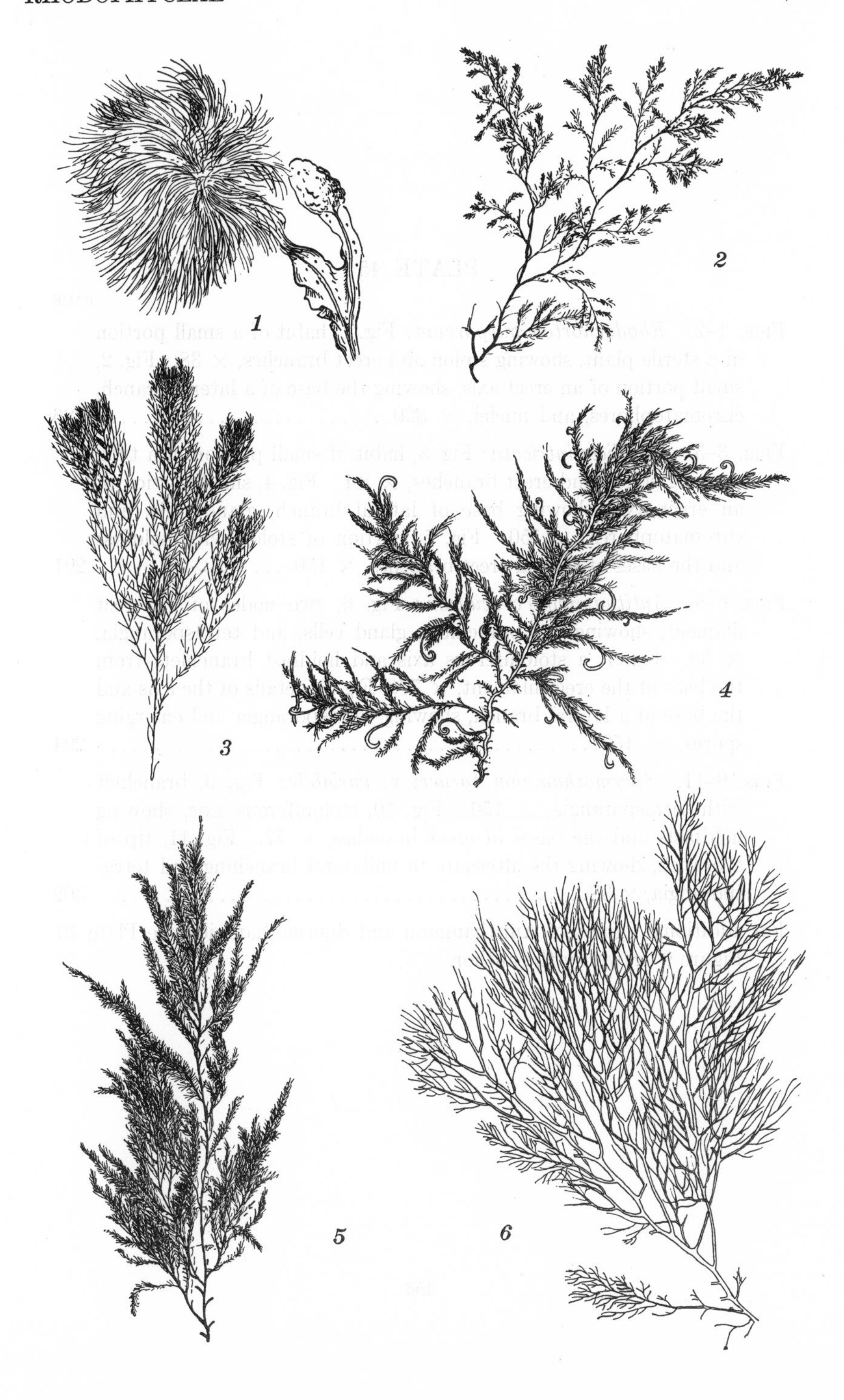
1
2
3
4
5
6

PLATE 45

PAGE

FIGS. 1–2. *Rhodochorton purpureum:* Fig. 1, habit of a small portion of a sterile plant, showing stolon and erect branches, × 38. Fig. 2, small portion of an erect axis, showing the base of a lateral branch, chromatophores, and nuclei, × 320 226

FIGS. 3–5. *Trailliela intricata:* Fig. 3, habit of small portion of a tuft, showing stolon and erect branches, × 8.1. Fig. 4, small portion of an erect axis, showing base of lateral branch, gland cells, and chromatophores, × 150. Fig. 5, portion of stolon with holdfast and the bases of several erect branches, × 150 291

FIGS. 6–8. *Antithamnion cruciatum:* Fig. 6, two nodes of an erect filament, showing the branchlets, gland cells, and tetrasporangia, × 38. Fig. 7, a stoloniferous axis and holdfast branchlets from the base of the erect filament, × 77. Fig. 8, details of the axis and the base of a lateral branch, showing tetrasporangia and emerging spores, × 150 294

FIGS. 9–11. *Spermothamnion turneri* v. *variabile:* Fig. 9, branchlet with tetrasporangia, × 150. Fig. 10, stoloniferous axis, showing holdfasts and the bases of erect branches, × 77. Fig. 11, tip of a branch, showing the alternate to unilateral branching and tetrasporangia, × 38 302

For additional figures of Antithamnion and Spermothamnion see Plate 40. Figure 9 drawn by the author.

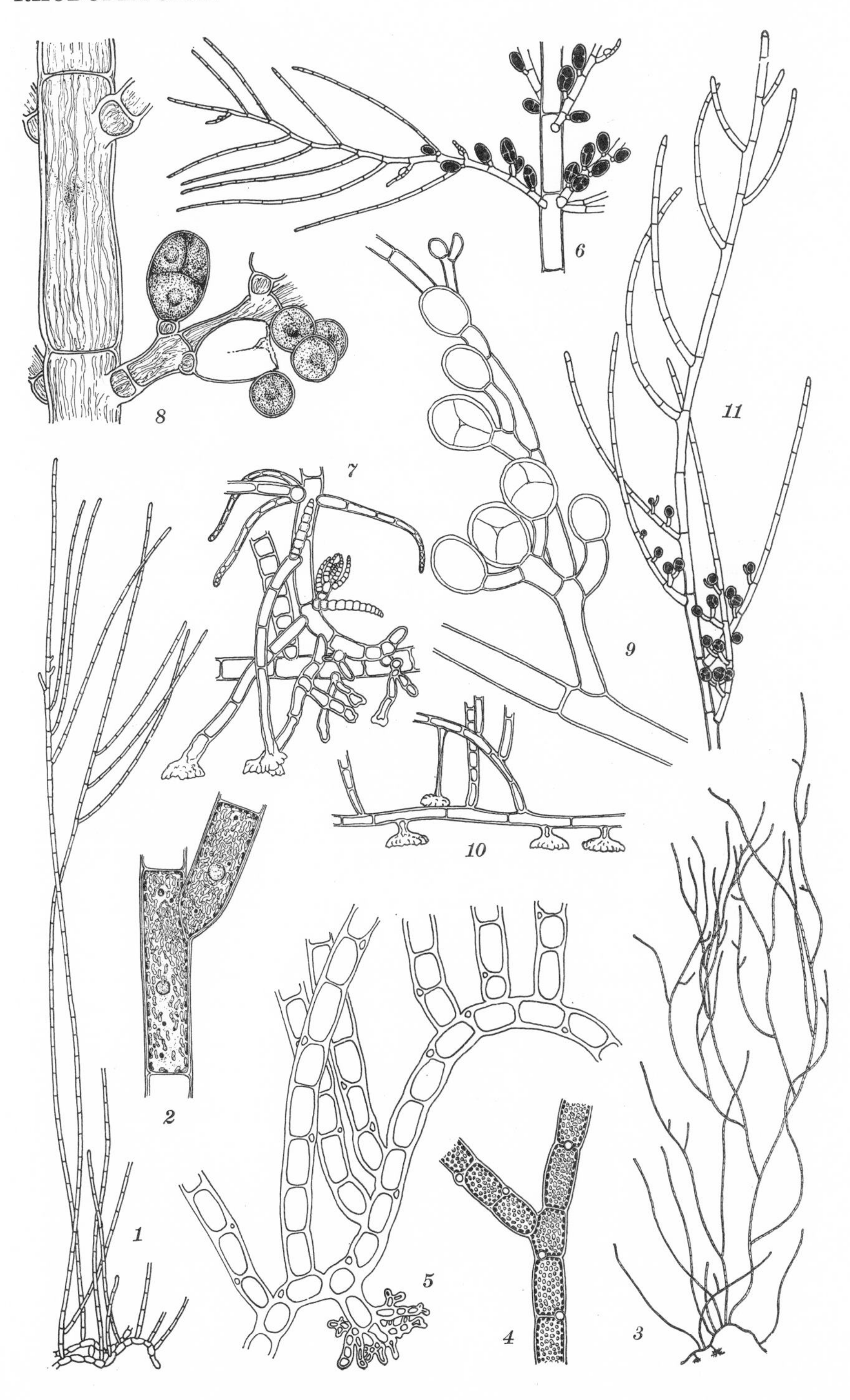
8
6
7
9
11
10
2
1
5
4
3

PLATE 46

PAGE

FIG. 1. *Phycodrys rubens:* tip of a proliferous branch, × 320...... 323

FIGS. 2–5. *Spyridia filamentosa:* Fig. 2, detail of a small portion of a branch tip, showing determinate branchlets, × 11. Fig. 3, portion of a main branch and the base of a lateral branch, with a few determinate branchlets and tetrasporangia, × 76. Fig. 4, a small portion of the surface of a main branch, showing the arrangement of the cortical cells and the chromatophores, × 320. Fig. 5, tip of a lateral branchlet, showing development of the nodes and the internodal cells, × 320.. 317

FIGS. 6–9. *Callithamnion baileyi:* Fig. 6, a small portion of a branch, showing lateral determinate branches and branchlets, with tetrasporangia, × 38. Fig. 7, small portion of a branch with branchlets and cystocarps above, spermatangia below, × 76. Fig. 8, small portion of a main axis and the base of a lateral branch, showing development of the cortication, × 38. Fig. 9, base of plant, showing extreme rhizoidal cortication continuing as the holdfast over the support, × 38.. 299

For additional figures of Callithamnion see Plate 44; for Phycodrys, Plates 30, 42; for Spyridia, Plate 44.

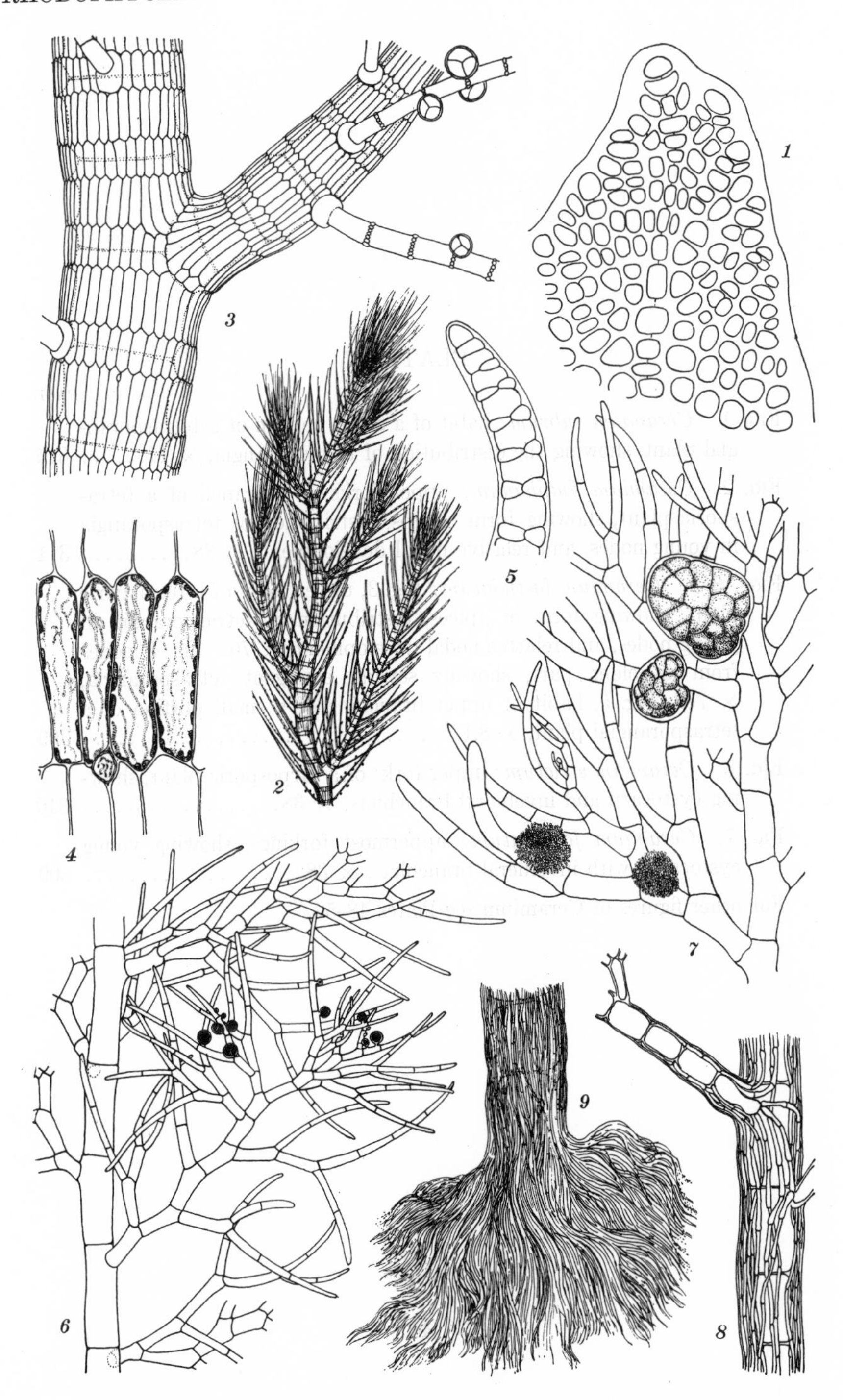
1
2
3
4
5
6
7
8
9

PLATE 47

PAGE

FIG. 1. *Ceramium rubrum:* habit of a small portion of a tetrasporangial plant, showing the distribution of tetrasporangia, × 11...... 315

FIG. 2. *Ceramium diaphanum:* upper forks of a branch of a tetrasporic plant, showing form of tips, distribution of tetrasporangia in young nodes, and relative nodal development, × 38.......... 311

FIGS. 3–5. *Ceramium fastigiatum:* Fig. 3, upper forks of a tetrasporic plant, showing form of apices, distribution of tetrasporangia in young nodes, and relative nodal development, × 76. Fig. 4, nodes from an older part, showing several emergent tetrasporangia, × 76. Fig. 5, habit of upper branching of a small portion of a tetrasporangial plant, × 8.1................................ 309

FIG. 6. *Ceramium strictum:* upper forks of a carposporic plant, showing cystocarp and involucral branchlets, × 38.................. 310

FIG. 7. *Ceramium fastigiatum:* uppermost forkings, showing young cystocarps with involucral branchlets, × 38.................... 309

For other figures of Ceramium see Plates 48–52.

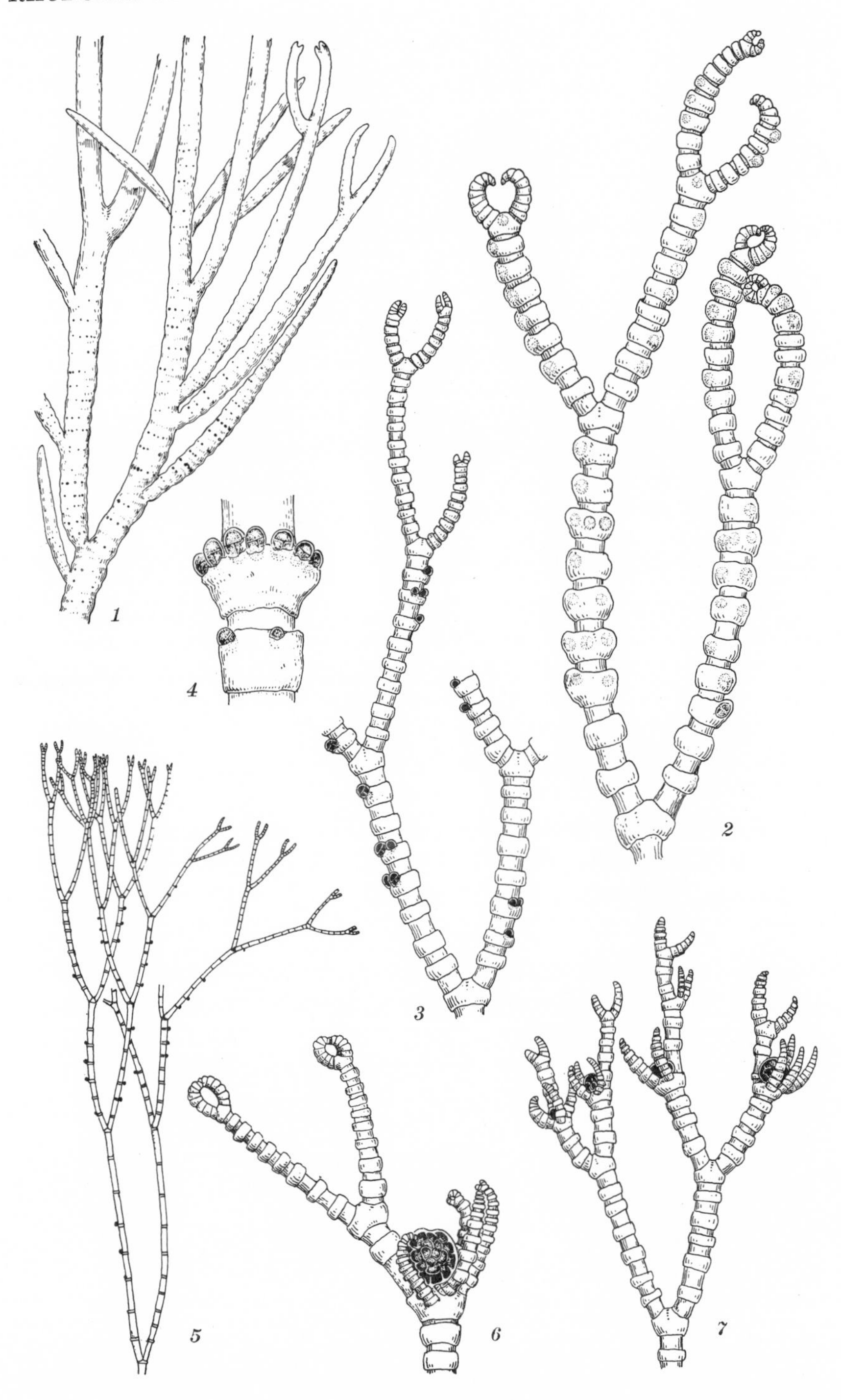
1
2
3
4
5
6
7

PLATE 48

PAGE

FIG. 1. *Ceramium deslongchampii:* forked tip of a branch, showing the erect apices and development of nodal cortication, × 260..... 312

FIGS. 2–4. *Ceramium fastigiatum:* Fig. 2, curved branch tip of the forcipate type of this species, the apical cell but recently divided, × 240. Fig. 3, forked tip of the type with divergent apices, × 240. Fig. 4, two nodes with cortication, from an old portion of the axis, × 240 309

FIGS. 5–6. *Ceramium strictum:* Fig. 5, node of intermediate age from below second fork, × 240. Fig. 6, node from the lower part of the plant, × 160... 310

FIGS. 7–9. *Ceramium diaphanum:* Fig. 7, curved branch tip, showing roughened contour, developing cortication, and hairs, × 240. Fig. 8, node of intermediate age from below the second fork, showing the cortication, × 240. Fig. 9, node from the lower part of an old plant, with broad exposure of the deeply placed cortical cells, × 86 311

FIG. 10. *Ceramium rubriforme* (*prox.*): curved branch tip, showing the greatly incurved limbs of the incipient fork and the rough outer margin, × 250... 314

FIG. 11. *Ceramium areschougii* (*prox.*): branch tip, × 240.......... 313

FIG. 12. *Ceramium elegans:* tip of an axis, showing asymmetrical development of the fork, × 155.............................. 311

For other figures of Ceramium see Plates 47, 49–52. Figures 1, 4, 10–12 drawn by the author.

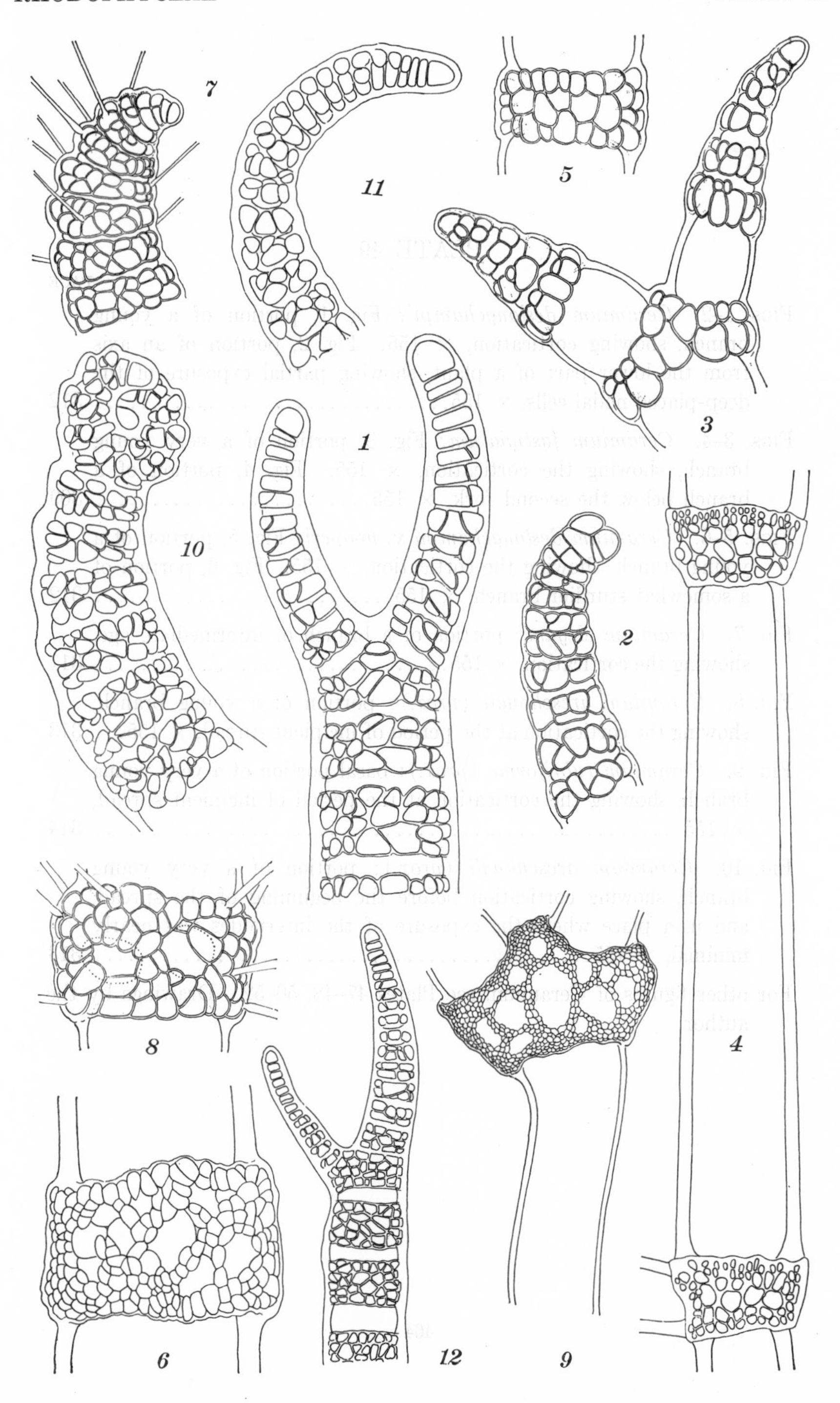
7
11
5
3
1
10
2
4
8
12
9
6

PLATE 49

PAGE

FIGS. 1–2. *Ceramium deslongchampii:* Fig. 1, portion of a young branch, showing cortication, × 155. Fig. 2, portion of an axis from the lower part of a plant, showing partial exposure of the deep-placed nodal cells, × 155 312

FIGS. 3–4. *Ceramium fastigiatum:* Fig. 3, portion of a very young branch, showing the cortication, × 155. Fig. 4, portion of a branch below the second fork, × 155 309

FIGS. 5–6. *Ceramium deslongchampii* v. *hooperi:* Fig. 5, portion of a young branch, showing the cortication, × 155. Fig. 6, portion of a somewhat sturdier branch, × 155 312

FIG. 7. *Ceramium elegans:* portion of a branch of intermediate age, showing the cortication, × 155 311

FIG. 8. *Ceramium areschougii* (*prox.*): portion of a young branch, showing the cortication at the period of incipient spread, × 155... 313

FIG. 9. *Ceramium rubriforme* (*prox.*): basal portion of a very young branch, showing the cortication at the period of incipient spread, × 155 314

FIG. 10. *Ceramium areschougii* (*prox.*): portion of a very young branch, showing cortication before the beginning of the spread, and in a place where the exposure of the internodes was nearly minimal, × 155 313

For other figures of Ceramium see Plates 47–48, 50–52. Drawings by the author.

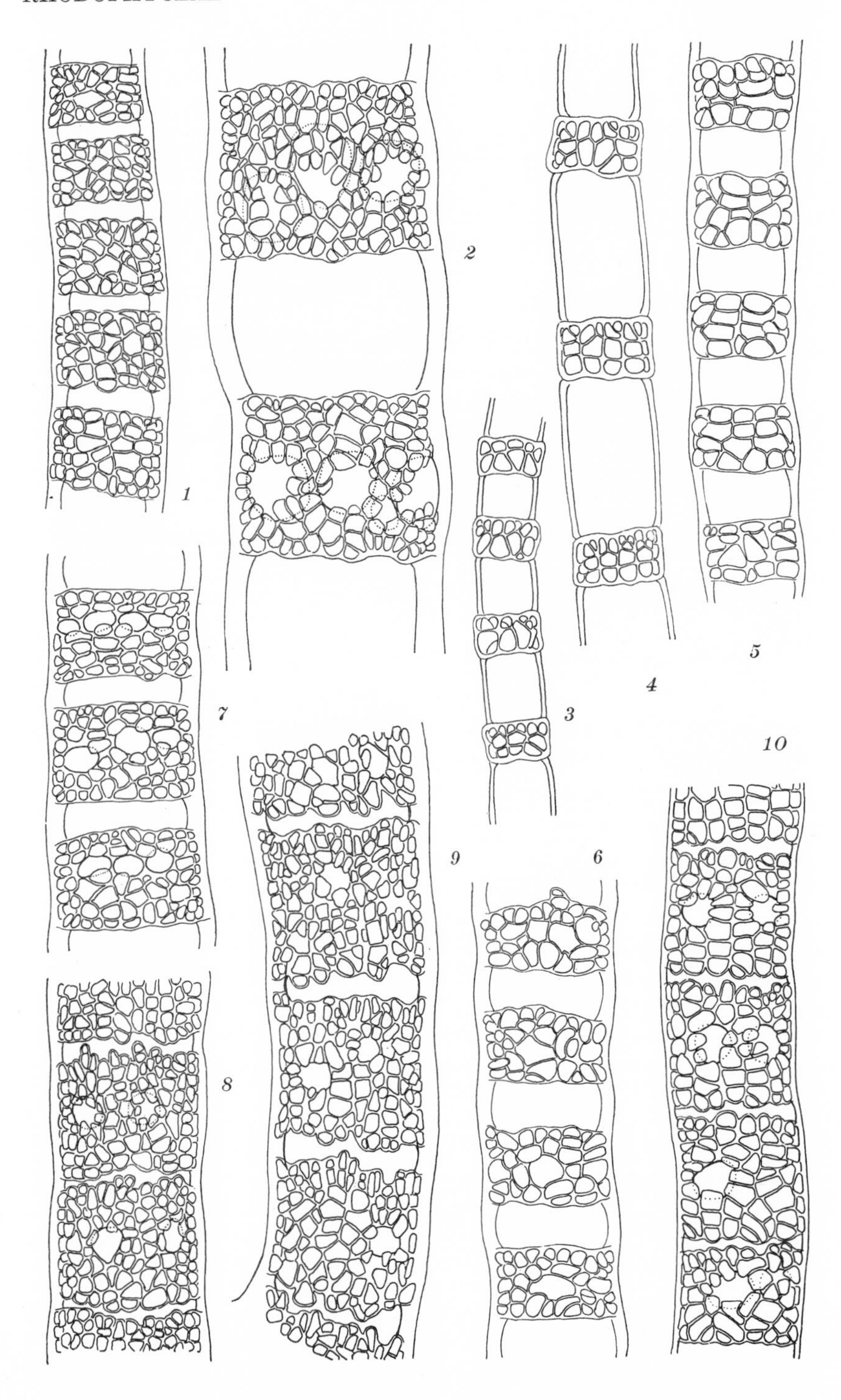
1
2
3
4
5
6
7
8
9
10

PLATE 50

PAGE

FIG. 1. *Ceramium areschougii* (*prox.*): small portion of a branch of intermediate age, showing the region of three nodes with the nodal cortication (marginally omitted) well advanced into the upward spreading phase, × 120 313

FIG. 2. *Ceramium deslongchampii* v. *hooperi:* small portion of an old axis, showing mature cortication, the bases of an adventitious branch and of a rhizoid, × 120 312

FIG. 3. *Ceramium elegans:* two nodes from a portion of an old axis, showing marked exposure of the deep-lying cortical cells, × 120 .. 311

FIG. 4. *Ceramium fastigiatum:* detail of a node from a portion of an old axis, showing subdivision of the cells along the upper margin, × 250 309

FIGS. 5–6. *Ceramium areschougii* (*prox.*): Fig. 5, two nodes in a younger part of the plant at the period of incipient spread, × 120. Fig. 6, details of one node and a portion of another with the nodal cortication well advanced in the general spreading phase, × 120 .. 313

FIG. 7. *Ceramium rubriforme* (*prox.*): details of cortication from near the base of a branch in a well-advanced spreading stage, × 120 314

For other figures of Ceramium see Plates 47–49, 51–52. Drawings by the author.

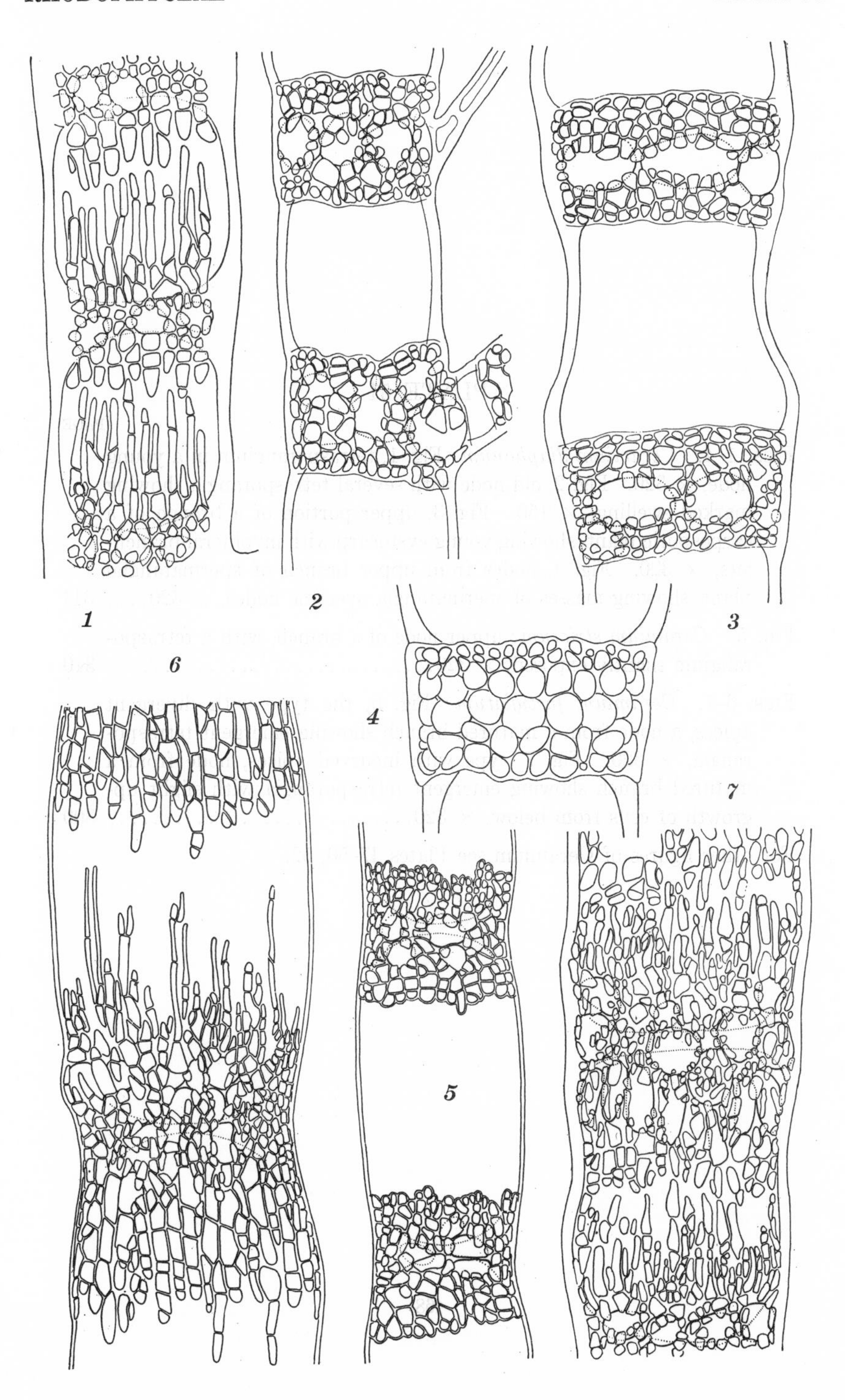
1
2
3
4
5
6
7

PLATE 51

PAGE

FIGS. 1–4. *Ceramium diaphanum:* Fig. 1, tetrasporangium in a young node, × 320. Fig. 2, old node with several tetrasporangia showing marked swelling, × 150. Fig. 3, upper portion of a branch of a carposporic plant, showing young cystocarp with involucral branchlets, × 320. Fig. 4, nodes from upper branch of spermatangial plant, showing masses of spermatangia over the nodes, × 320.... 311

FIG. 5. *Ceramium strictum:* upper node of a branch, with a tetrasporangium slightly exposed, × 320.............................. 310

FIGS. 6–7. *Ceramium fastigiatum:* Fig. 6, the type with divergent apices, a node from a matured branch showing emergent tetrasporangia, × 320. Fig. 7, type with incurved apices, node from a matured branch showing emergent tetrasporangia with slight upgrowth of cells from below, × 320........................... 309

For other figures of Ceramium see Plates 47–50, 52.

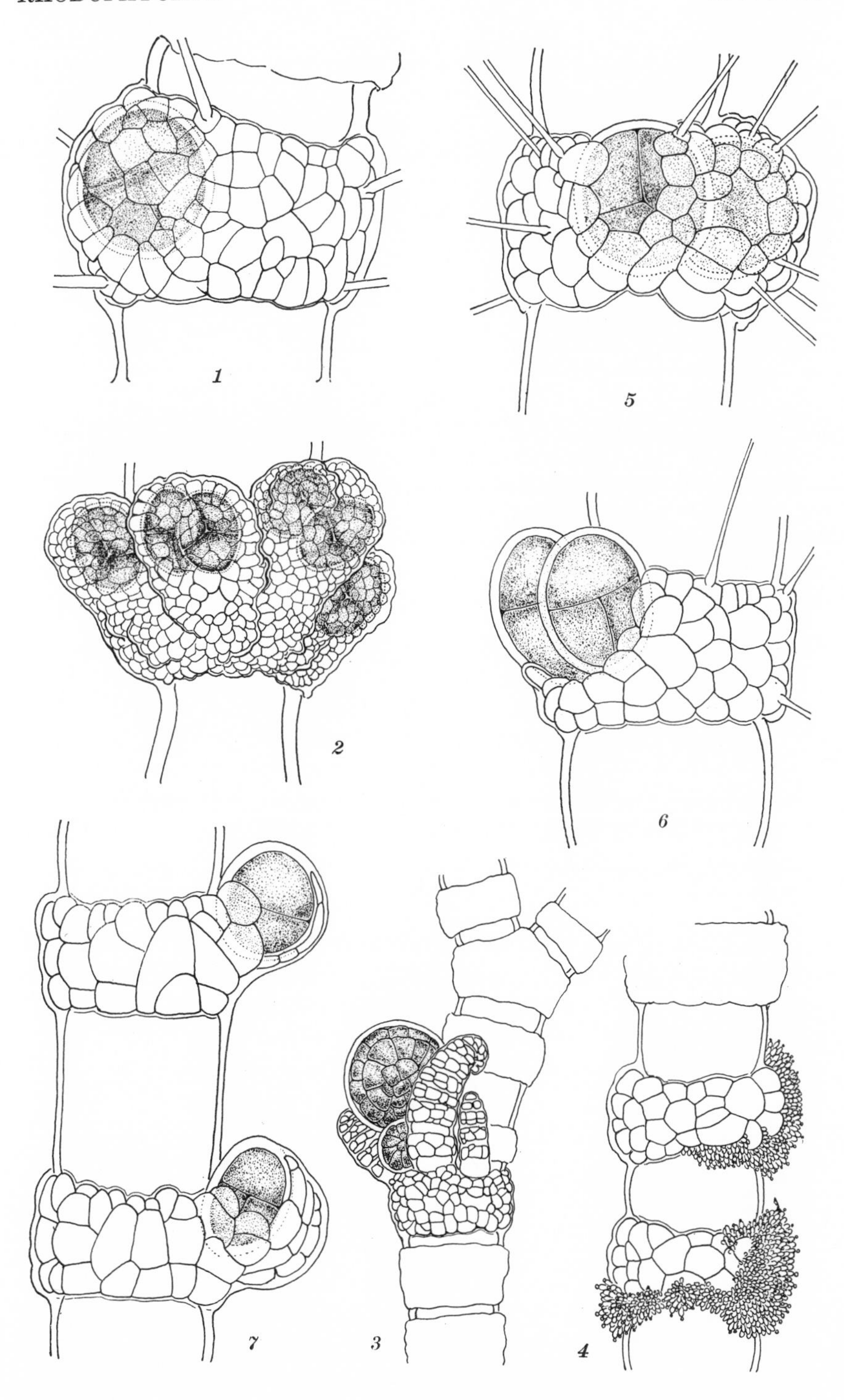
1
5
2
6
7
3
4

PLATE 52

For other figures of Asparagopsis see Plates 35, 40, 44; for Ceramium, Plates 47–51; for Plumaria, Plate 30.

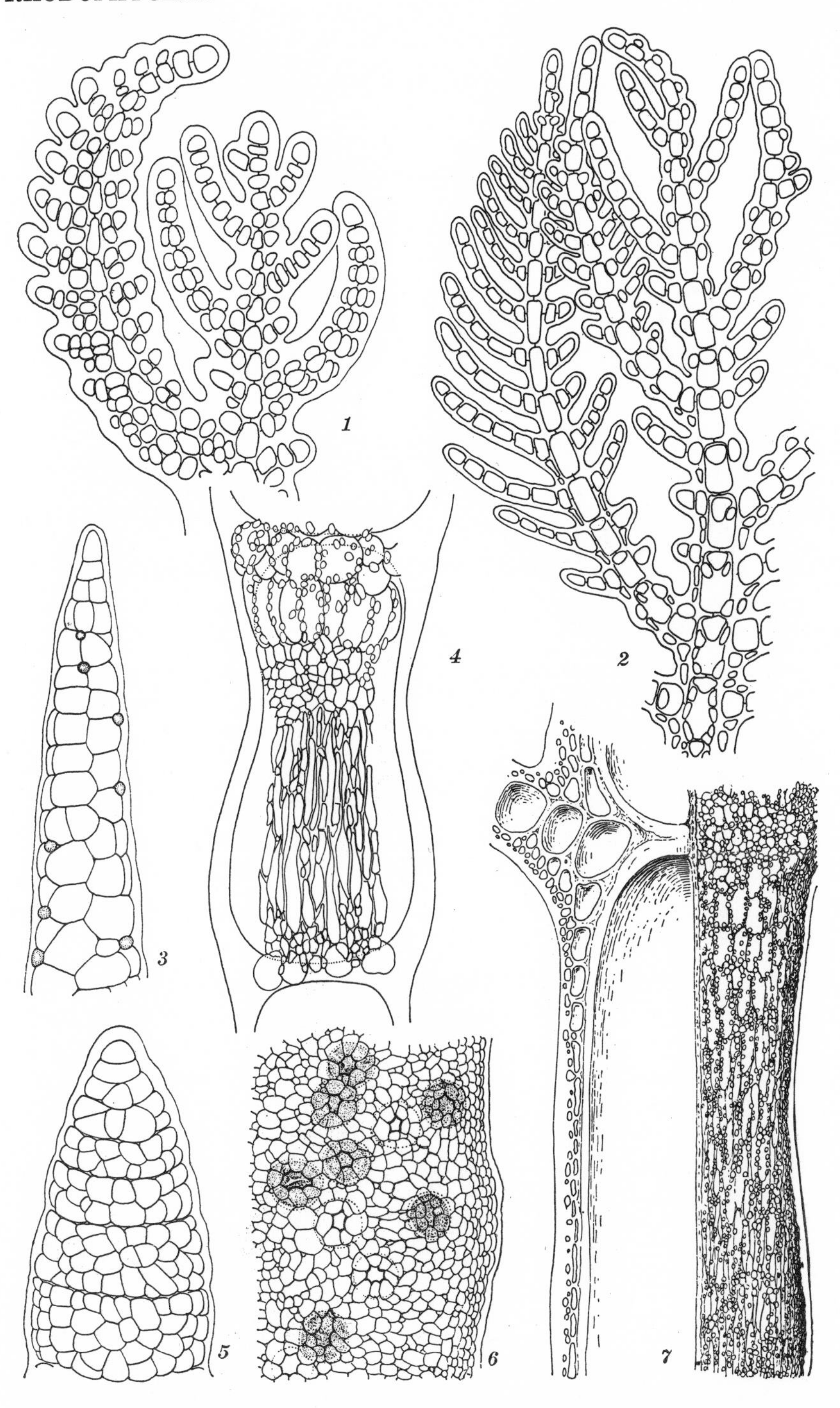
1
2
3
4
5
6
7

PLATE 53

PAGE

FIG. 1. *Scinaia furcellata:* habit of a portion of a carposporic plant, × 0.9 229

FIGS. 2–3. *Caloglossa leprieurii:* Fig. 2, habit of a portion of a slender carposporic plant, × 2.5. Fig. 3, habit of a portion of a broad tetrasporic plant, × 2.5 (both from tropical America).......... 319

FIGS. 4–6. *Membranoptera alata:* Fig. 4, habit of a portion of a large broad-branched plant, × 1.6. Fig. 5, portion of a slender-branched plant, × 1.8. Fig. 6, tip of a slender branch, × 14..... 321

FIGS. 7–8. *Membranoptera denticulata:* Fig. 7, small portion of a plant, × 3. Fig. 8, denticulate tip of a branchlet, × 9........... 322

Drawings by the author.

1
4
2
7
8
6
5
3

PLATE 54

PAGE

FIGS. 1–4. *Dasya pedicellata:* Fig. 1, habit of a small carposporic plant, × 0.45. Fig. 2, portion of a branching filament, with stichidium and tetrasporangia, × 76. Fig. 3, portion of a branching filament, with spermatangia, × 76. Fig. 4, short segment of a principal branch, showing branched filaments and young pericarps, × 8.1 326

FIGS. 5–6. *Chondria dasyphylla:* Fig. 5, small portion of lesser branch with branchlets, showing young pericarps and many tetrasporangia on the same plant, × 8.1. Fig. 6, small portion of plant, showing details of habit, × 1.8.................................... 329

For other figures of Chondria see Plates 40, 55.

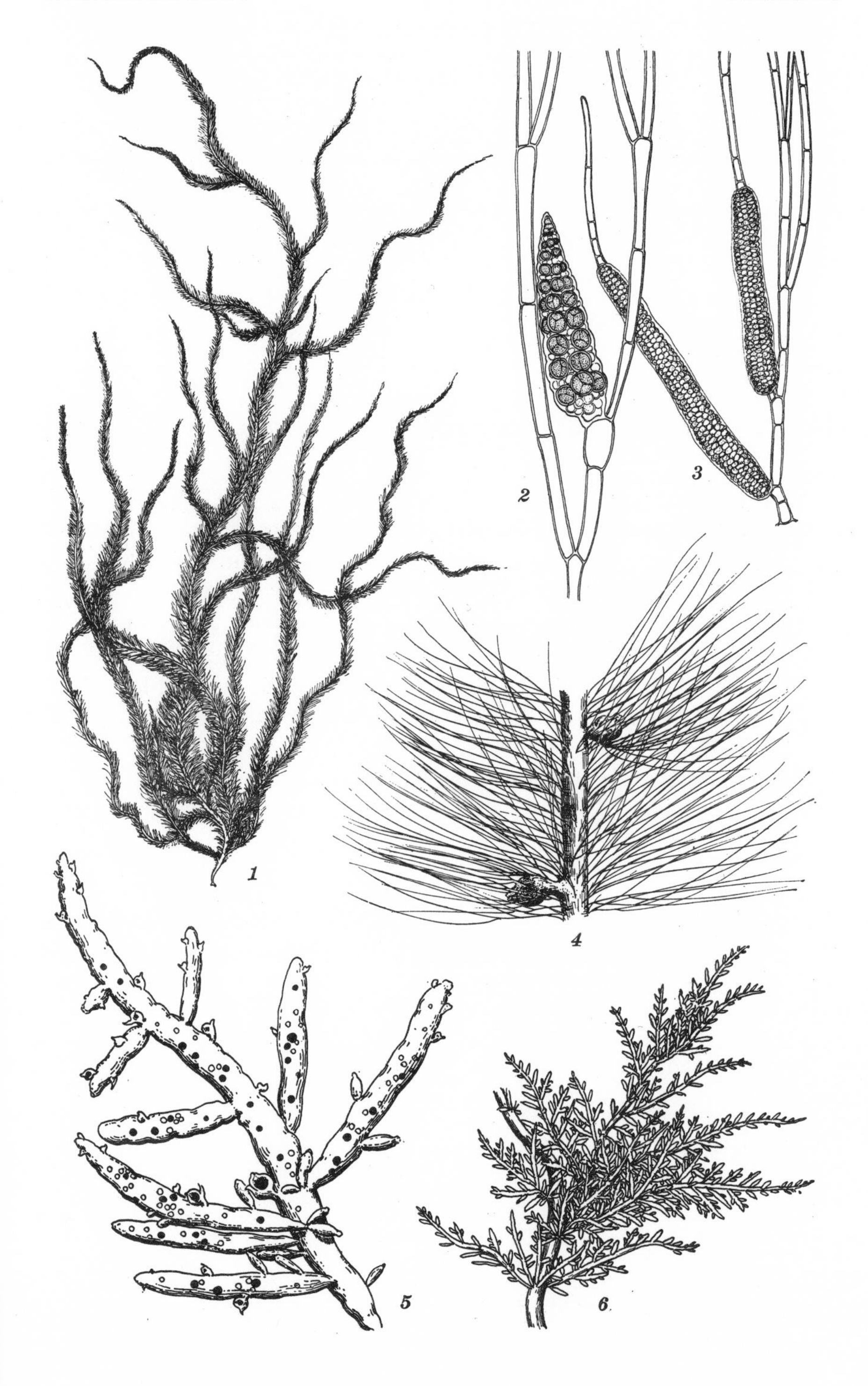
1
2
3
4
5
6

PLATE 55

PAGE

For additional figures of Chondria see Plates 40, 54.

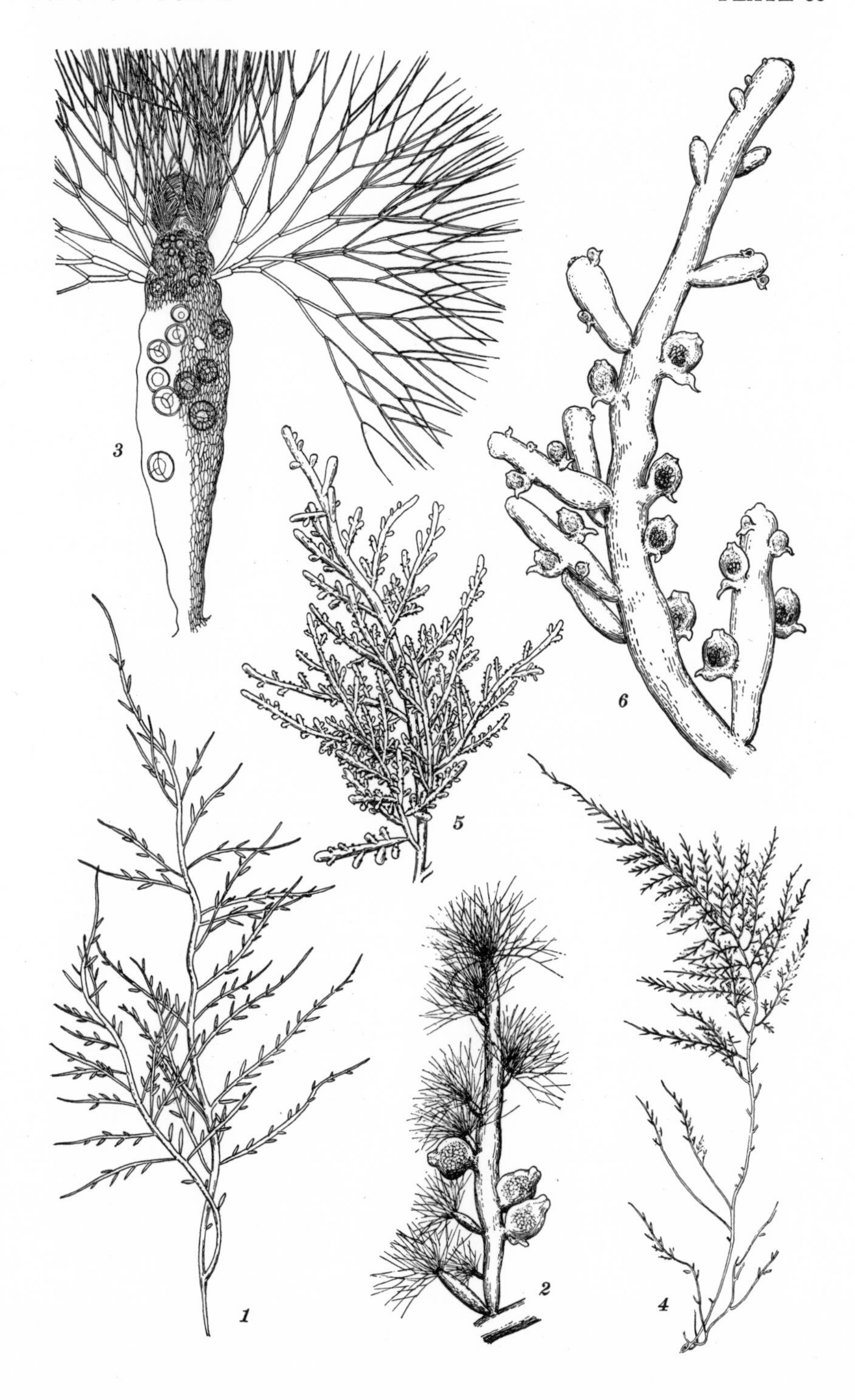
3
5
6
1
2
4

PLATE 56

For additional figures of Polysiphonia see Plates 57–59.

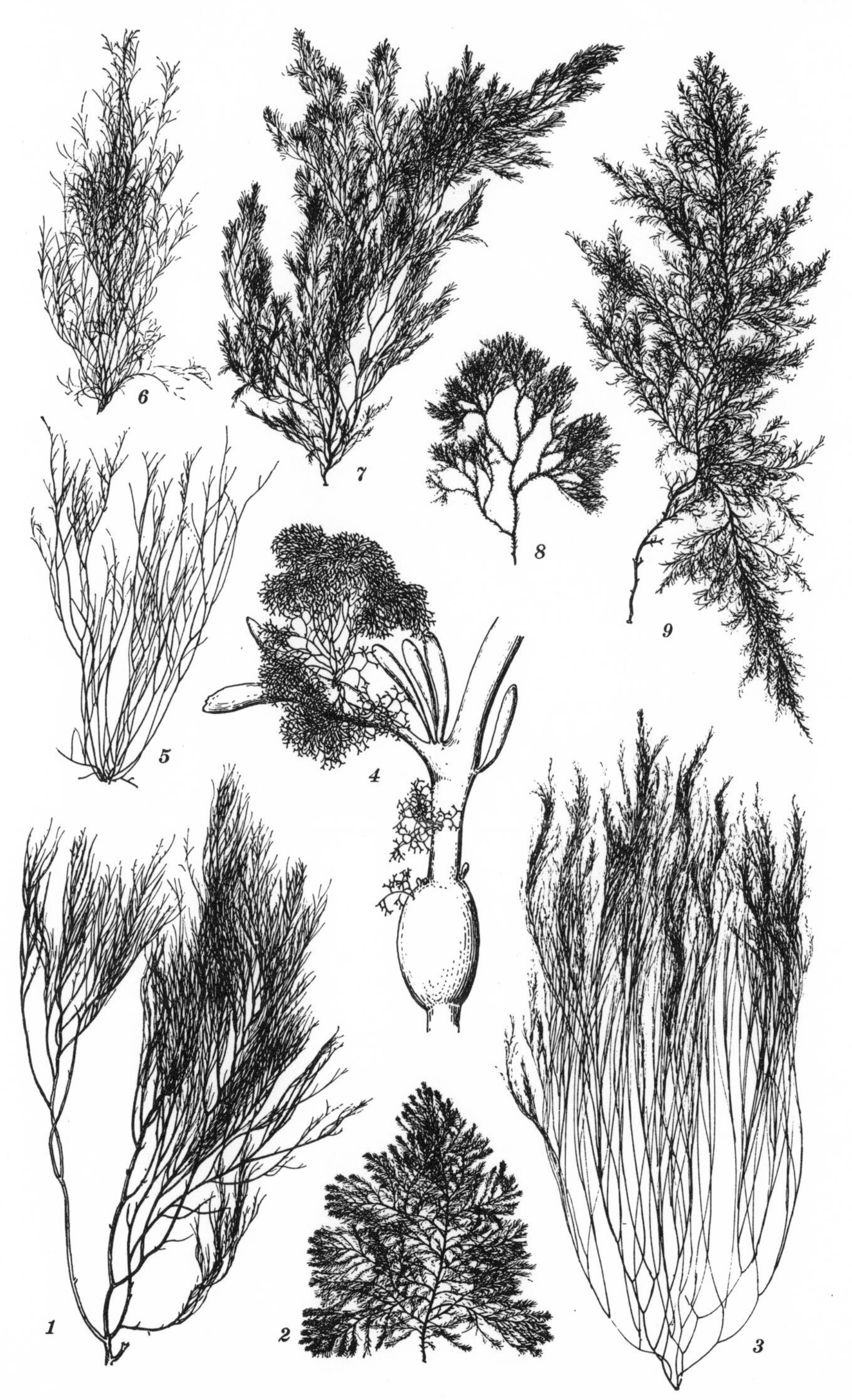
6
7
8
9
5
4
1
2
3

PLATE 57

PAGE

FIGS. 1–5. *Polysiphonia fibrillosa:* Fig. 1, portion of branch, showing cortication, × 78. Fig. 2, portion of small branch, showing branchlets and partial cortication, × 78. Fig. 3, small branch, showing origin of lateral branchlet and branched hairs, × 78. Fig. 4, tip of branchlet, showing young spermatangial clusters, × 78. Fig. 5, small portion of branchlet, showing young pericarp, × 78...... 334

FIGS. 6–10. *Polysiphonia denudata:* Fig. 6, portion of branch, showing base of small branch, × 78. Fig. 7, small portion of tetrasporangial branchlet, showing tetrasporangia, × 78. Fig. 8, detail of tip of a branch, showing origin of branched hairs and of branchlet, × 150. Fig. 9, small portion of branchlet, showing spermatangial clusters and branched hair, × 150. Fig. 10, small portion of branch with young pericarp, × 78.......................... 339

FIGS. 11–13. *Polysiphonia elongata:* Fig. 11, small section of main branch, showing complete cortication, × 38. Fig. 12, small branch with branchlet, showing beginning cortication, × 38. Fig. 13, small portion of branch, showing young pericarp, × 38.............. 336

FIGS. 14–15. *Polysiphonia lanosa:* Fig. 14, small portion of upper branching, showing detail of habit, × 8.1. Fig. 15, small portion of a branch with base of a branchlet, × 78..................... 341

For additional figures of Polysiphonia see Plates 56, 58–59.

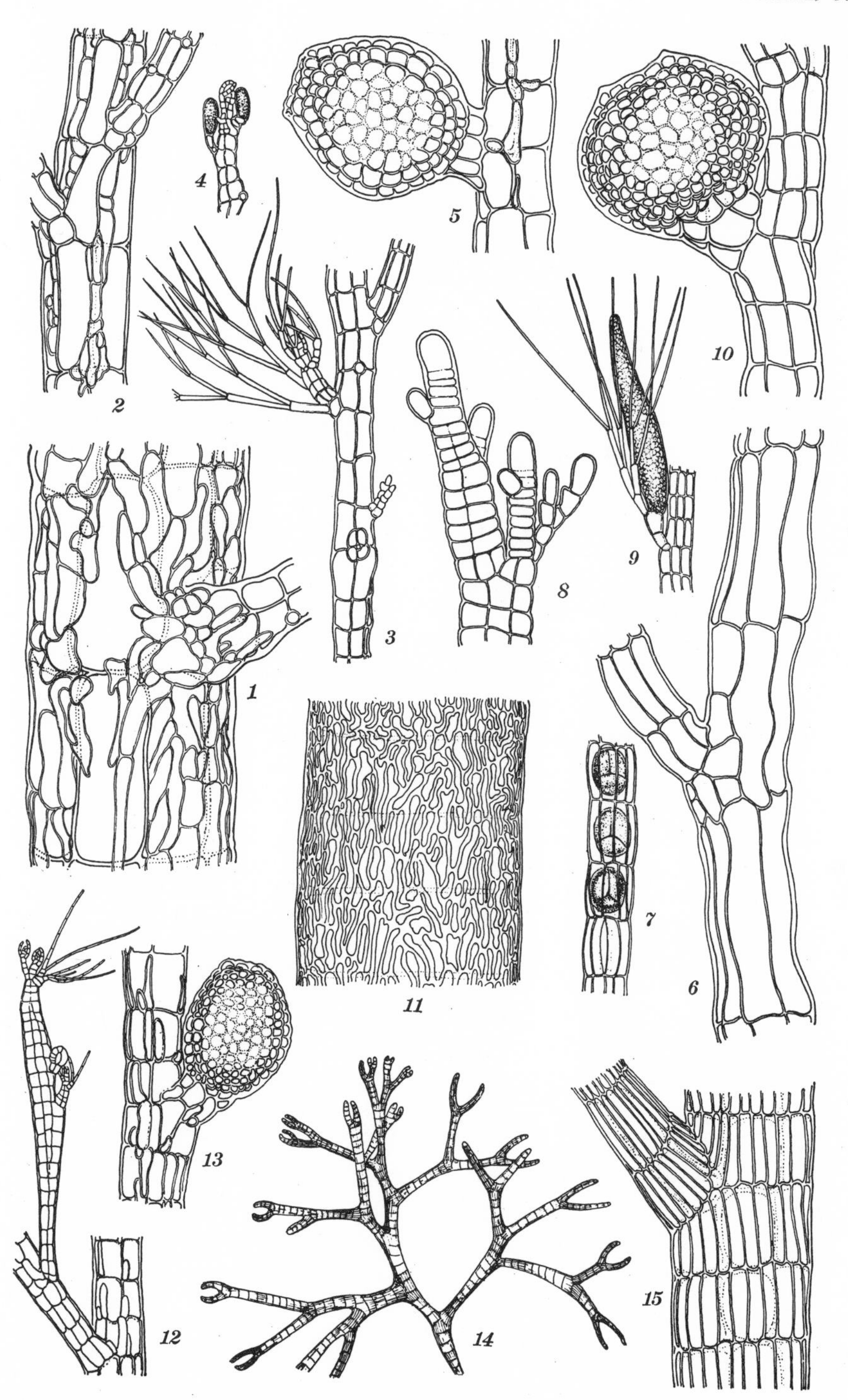
1
2
3
4
5
6
7
8
9
10
11
12
13
14
15

PLATE 58

For additional figures of Polysiphonia see Plates 56–57, 59.

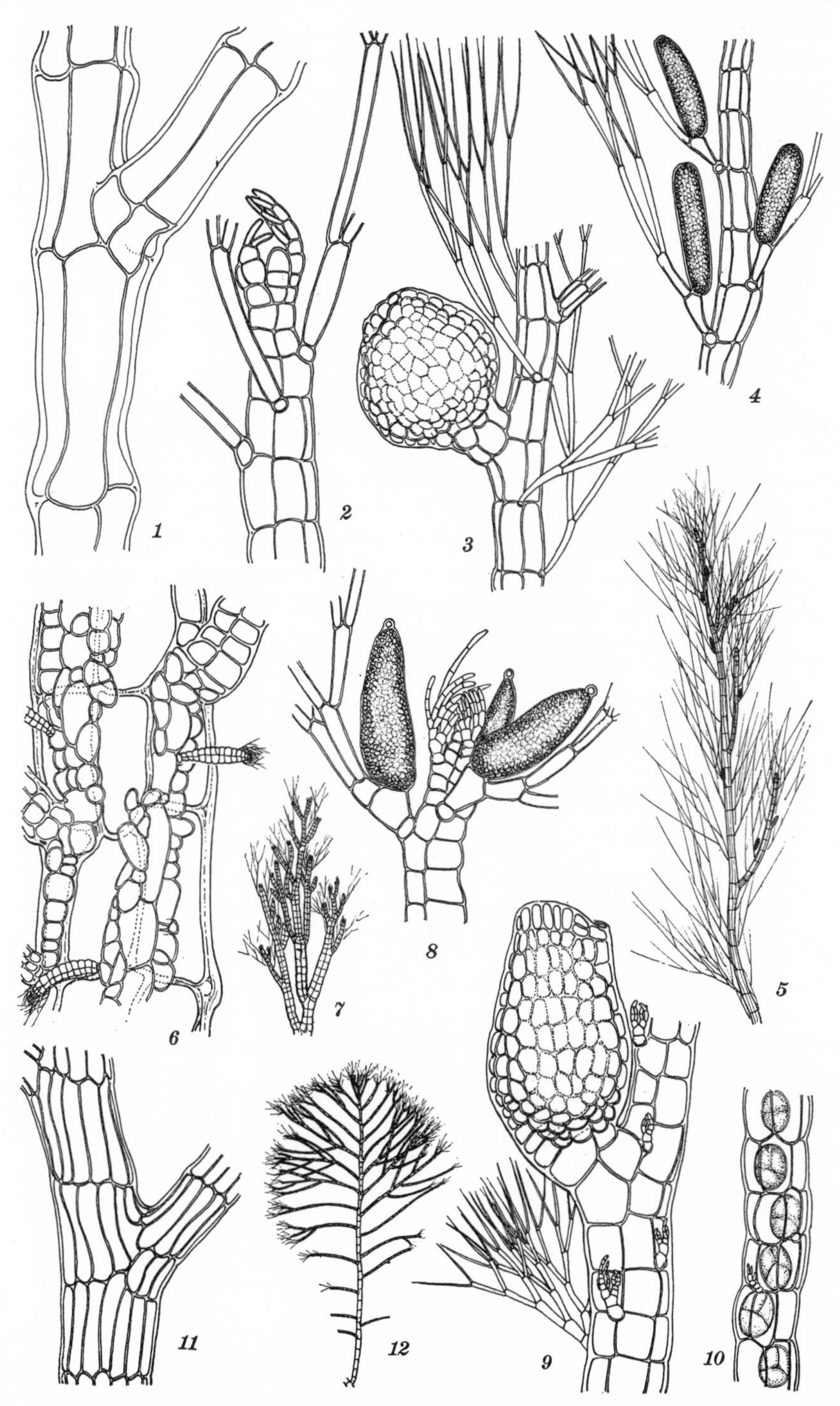
1
2
3
4
5
6
7
8
9
10
11
12

PLATE 59

PAGE

FIG. 1. *Polysiphonia denudata:* tranverse section of an axis, × 150.. 339

FIGS. 2–3. *Polysiphonia nigrescens:* Fig. 2, transverse section of young axis, × 150. Fig. 3, transverse section of old axis, showing sparse development of cortical cells, × 150................. 340

FIG. 4. *Polysiphonia lanosa:* transverse section of a mature axis, × 150 341

FIGS. 5–6. *Polysiphonia novae-angliae:* Fig. 5, young branch in transverse section, × 78. Fig. 6, old axis in transverse section, showing corticating cells, × 78.. 336

FIG. 7. *Gracilaria foliifera:* transverse section of part of an axis, showing general structure and a young cystocarp, × 38.............. 273

FIG. 8. *Grinnellia americana:* transverse section of a blade through a small spermatangial sorus, showing basal layer and spermatangia, × 320 324

FIG. 9. *Agardhiella tenera:* transverse section of portion of an axis through a cystocarp, showing general structure, × 76........... 267

For additional figures of Agardhiella see Plates 38, 40–41; for Gracilaria, Plates 38, 41; for Grinnellia, Plates 39, 42; for Polysiphonia, Plates 56–58.

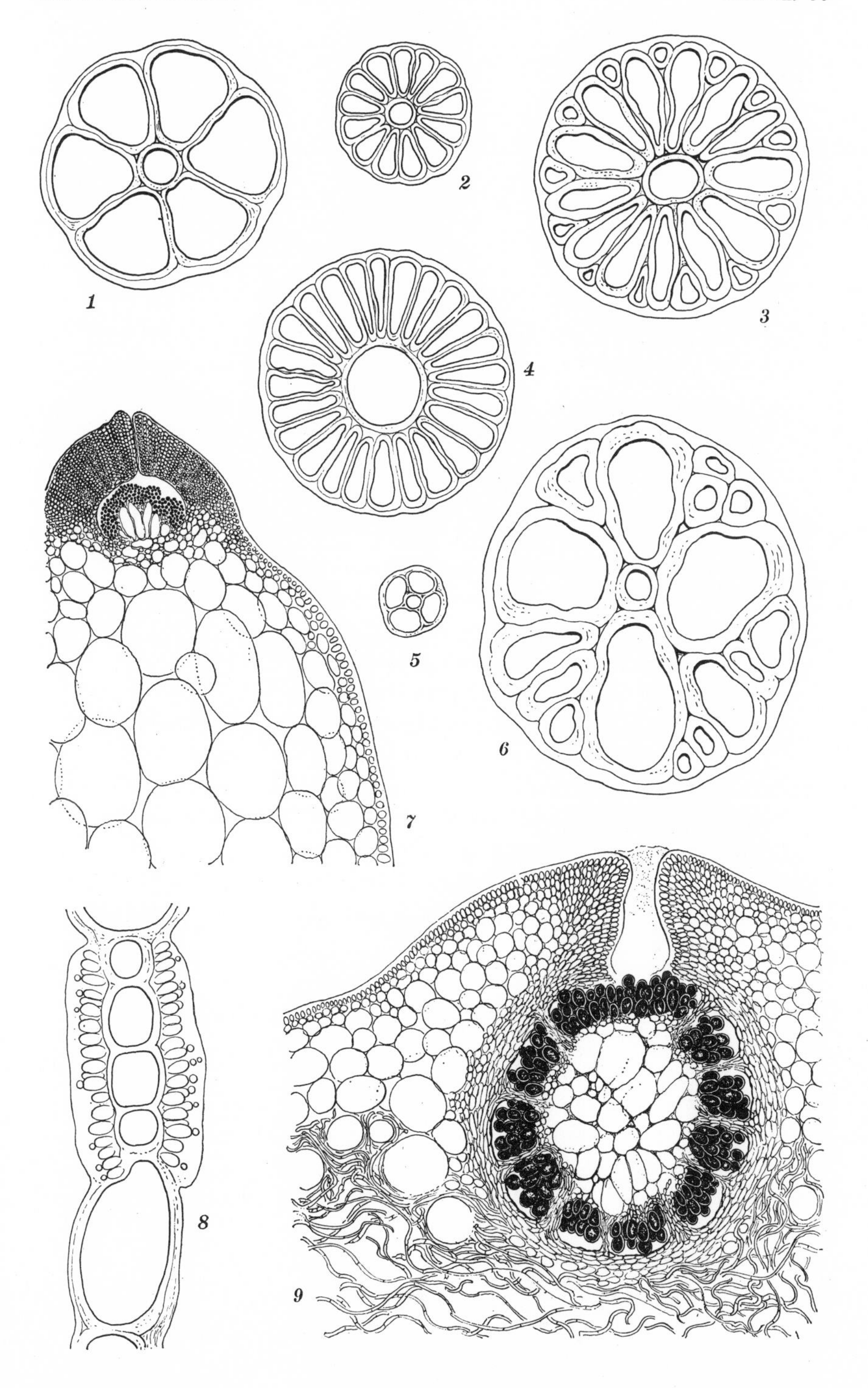
1
2
3
4
5
6
7
8
9

PLATE 60

PAGE

Drawings by the author.

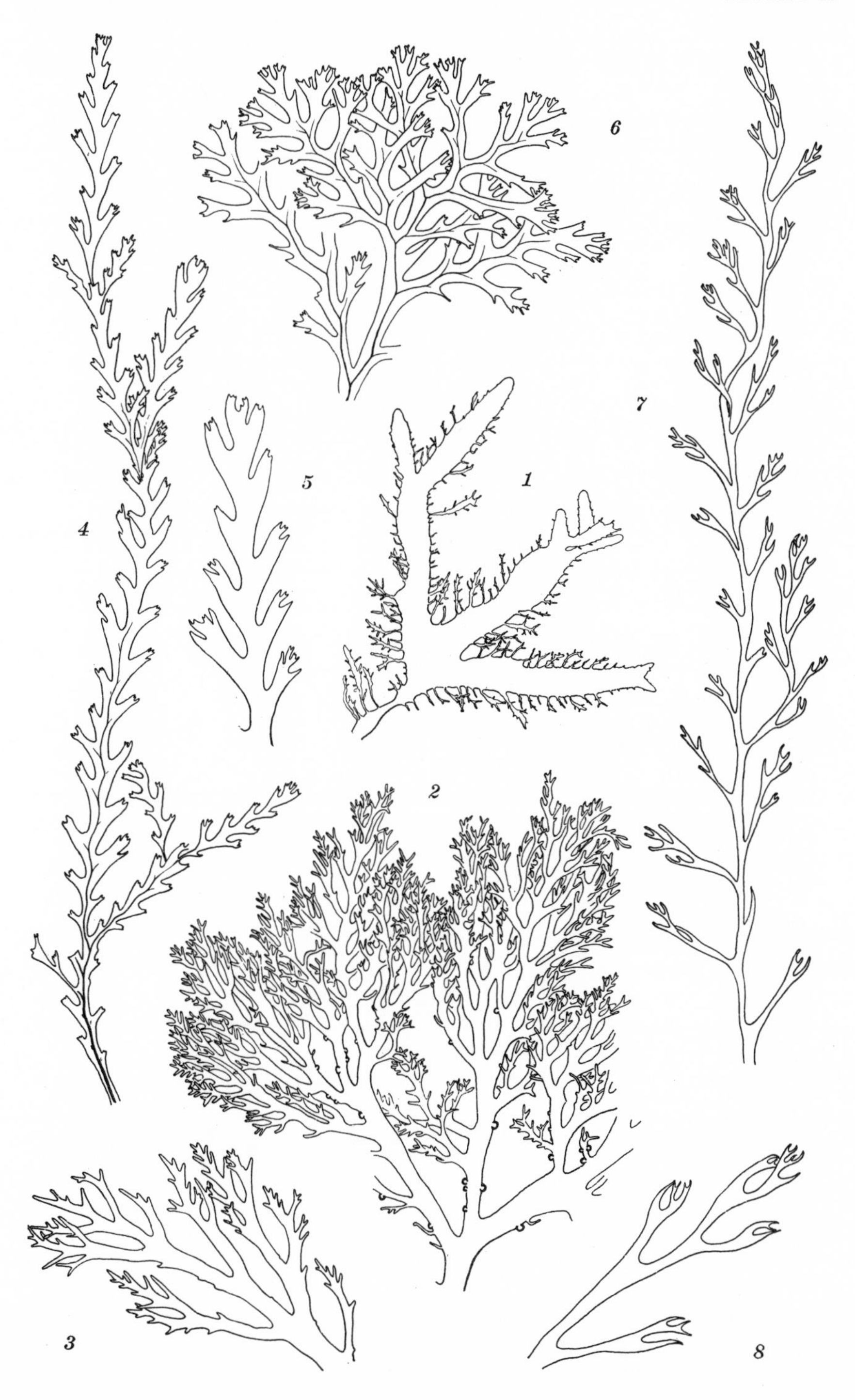
6
7
5
1
4
2
3
8

ADDENDA

The method of reproduction selected for reprinting this book has prevented any revision involving dislocation of the text. Numerous changes of dates and measurements have been made, but unpublished data, such as changes in range, are necessarily omitted. There follow some published records and a few urgent corrections.

Bernatowicz, A. J. 1958. A Bermudian Marine Vaucheria from Cape Cod. Biol. Bull., 115(2) : 344. Reports *V. nasuta* Taylor et Bernat. from Massachusetts.

Blum, J. L. 1960. A New Vaucheria from New England. Trans. Amer. Micros. Soc., 79(3) : 298–301. 10 figs. Describes *V. vipera* Blum from Massachusetts.

—— and R. T. Wilce. 1958. Description, Distribution and Ecology of Three Species of Vaucheria Previously Unknown from North America. Rhodora, 60 : 283–288. 17 figs. Reports *V. sphaerophora* Nordst. and *V. submarina* Berk., and describes *V. compacta* (Coll.) Coll., v. *koksakensis* Blum et Wilce from the Ungava Bay area of northern Quebec.

Bouck, G. B., and E. Morgan. 1957. The Occurrence of Codium in Long Island Waters. Bull. Torrey Bot. Club, 84(5) : 384. 2 figs. Reports *Codium fragile* (Sur.) Hariot ssp. *tomentosoides* (van Goor) Silva as well established on Long Island, N.Y., a particularly striking addition to the northern flora. In the winter of 1961–62 it appeared in Massachusetts. The bushy, dark green plants grow in rather shallow water to heights of 3.5 dm., being fairly regularly dichotomously branched with the terete divisions spongy but tough in texture, about 5–8 mm. diam.

Erskine, D. 1955. Two Red Algae New to Nova Scotia. Canad. Field Nat. 69(4) : 150–151. Reports *Kylinia alariae* (Jónss.) Kylin (as *Acrochaetium*) as new to Nova Scotia and *Trailiella intricata* (J. Ag.) Batt. as new to Prince Edward Island.

Wilce, R. T. 1959. The Marine Algae of the Labrador Peninsula and Northwest Newfoundland (Ecology and Distribution). Nat. Mus. of Canada, Bull. 158, Biol. Ser. 56. iv + 103 pp., 11 pl., 1 fig.

Page 254. The report of *Corallina officinalis* from Bermuda should not be credited.

Pages 261, 274. The genus Ahnfeltia will not key out through the present sequences. Until the whole key can be rewritten the following addition may prove helpful, with an appropriate addition to the family description:

5. Slender, wiry plants; monosporangia the reproductive organs; medulla filamentous.................. Phyllophoraceae, p. 274

Page 292. In the key, from the second entry 4 omit the words 'or otherwise,' retaining the comma which follows.

INDEX